CAUGHT UNAWARES

CAUGHT UNAWARES
The Energy Decade in Retrospect

MARTIN GREENBERGER

in collaboration with
Garry D. Brewer, William W. Hogan, and Milton Russell

Ballinger Publishing Company
Cambridge, Massachusetts
A Subsidiary of Harper & Row, Publishers, Inc.

333.79
G79c

International Standard Book Number: 0-88410-916-X

Library of Congress Catalog Card Number: 82-16374

Printed in the United States of America

Library of Congress Cataloging in Publication Data

Greenberger, Martin, 1931–
    Caught unawares.

    Includes index
    1. Energy policy—Research—United States.
I. Title.
HD9502.U52G73   1982   333.79′072073   82-16374
ISBN 0-88410-916-X

To the many people who made the writing of this book such a remarkable learning experience and to the special people who made it possible.

CONTENTS

LIST OF FIGURES

LIST OF TABLES

FOREWORD

The 1970s were the Decade of Energy and of big energy studies. Martin Greenberger and his colleagues have written a fascinating book evaluating what fourteen of the studies accomplished, singly and collectively. I commend it particularly to the attention of the hundreds who participated in the studies and the thousands who observed from the sidelines, cheering or hurling brickbats as their reactions to results moved them. I commend the book also to the much larger group concerned with the processes of public decisionmaking generally and with the nation's ways of dealing with its weighty policy problems. Energy provided a mine of rich ore for students of the policy process. The authors have illuminated this subterranean source with care and discernment.

What was accomplished by all the frenzied activity in energy studies in the 1970s? Substantively, I would say, a good deal. The participants, whom the authors call the energy elite, learned from each other and ended the decade in far greater agreement than when the oil embargo was imposed in 1973. The chorus of dissidence which greeted the first of the fourteen studies, the Ford Foundation project directed by David Freeman, had largely subsided by the latter years of the decade. I do not want to exaggerate. The authors learned that the experts, when arrayed from traditionalist right to reformist left, tended to cluster around two poles. But at the end of the decade, while still differing on many policy issues, the traditionalists and reformists exhibited a considerable degree of mutual respect and tolerance. They were in many respects closer to each other in their views than either was to the general informed or uninformed public.

More important in the long run than the reasoned conclusions about energy policy (the best studies were often the least influential) was the contribution the studies made to the development of policy analysis and its constituent disciplines in science, engineering, operations research, and economics. This was unhappily less true in the other social sciences.

A special and most important case of the participants learning from each other involved the technologists and the economists, whom I have been trying to get working in harness for thirty years—at RAND, the Department of Defense, and Resources for the Future. The energy studies of the 1970s, by demonstrating that both kinds of talent were needed, succeeded where earlier attempts had achieved only modest and mainly temporary results. Economists learned that their models and their advice were pretty empty without a lot of technological content. Engineers and scientists learned about the behavior, uses, and failures of markets; about elasticities of demand and supply and how potentially massive their effects can be; and about cost/benefit and cost/effectiveness concepts which can be useful operational tools in helping to choose among policies.

I expect that this desirable cross-fertilization of disciplines will, this time, have permanent effects. Economists and engineers have not only talked and worked together; each has started to think and talk a little like the other. Economists who have acquired a respectable knowledge of energy technologies are not likely to return to games played with empty boxes. And some scientists and engineers have actually become more economist in their perspective than many economists. The aphorism is confirmed that the best interdisciplinary research is research by interdisciplinarians.

If the component disciplines are profiting from this experience, what of the interdisciplinary exercises themselves? What about the future of the activity commonly called policy analysis? It should be clear by now that I agree with Martin Greenberger and his colleagues that the quality of the analysis being performed, as exemplified by these fourteen energy studies, has improved, is improving, and will continue to improve. But the authors argue that major progress is possible in highly politicized areas like energy only if policy analysts give more explicit treatment in their studies to political factors and more conscious recognition to the political context in which their work is immersed. This insight creates a role for the often neglected social sciences. It suggests a further widening of the mutual understanding and productive interaction achieved by economists and engineers to include political scientists, historians, sociologists, and psychologists.

The problem for the future as now and in the past is how to use the studies constructively, how to make them more influential without making them too influential, how to reverse the apparently negative correlation between quality and influence. This nest of questions is terribly difficult to deal with at both operational and deeply philosophical levels; it raises again all the issues involving the role of the expert in a democratic society. When rival camps of experts, as here, show greater agreement among themselves than either with the public, what is the proper direction for policy and how is it to be achieved?

As the authors point out forthwith, there are no simple answers to these questions on the role and use of policy analysis. But the authors reach some thought provoking conclusions. You will find, running through their history and in their reasoning, many insights to stimulate your thinking on a subject that sorely needs examination and discussion.

Charles J. Hitch
Berkeley, California

PREFACE

Economist Craufurd D. Goodwin, in reflecting on the reasons for the fitful response of the United States to national energy problems during the 1970s, pointed to the "failure to develop in America a sizable and capable body of disinterested and broad-gauged specialists competent to deal with complex issues of energy policy and lodged in powerful administrative units."[1] That was indeed the situation prior to the Arab oil embargo of 1973/74. The relatively few policy analysts well versed in energy matters at the time were not generally in positions of influence. Nor were they as a group expecting energy to produce the kind of economic and political trauma that developed in the wake of the actions from the Mideast.

After the shock of the sharp rise in oil price, there ensued a period of intense activity in the development of energy expertise. Numerous energy policy studies were launched, some ambitiously comprehensive and detailed in their treatment. The mobilization of analysts and analysis achieved by these studies forms a saga in U.S. political history. It was a momentous occasion in the evolution of policy analysis, and in the continuing attempt to apply analysis to affairs of state. This book contains an account of fourteen of these energy policy studies.

Many of the persons involved in the studies came to occupy key government positions. They may not have been "disinterested," at least not after their second day in office, but such is the nature of political life. Such, it must be recognized, is the nature of much of policy analysis as well. The role of policy analysis in policymaking and partisan politics is the other subject of this book.

Policy analysis, unlike engineering and theoretical science (with which it has some similarities), does not have to its credit a string of universally celebrated successes. It is by nature too problematic and controversial to inspire widespread acclaim. Nonetheless, policy analysis does have some very astute practitioners, and it has come to be an indispensable part of the modern policymaking apparatus. Its high visibility in the energy debates of the seventies underscored just how distinctive it has become as a feature of contemporary political life.

Yet, policy analysis is not an unqualified success. The difficulty in making it work well is not due to a lack of expert analysts and analysis. Rather, the difficulty relates to the way analysis does and does not get applied. The problem is not with the nation's ability to develop expertise as much as it is with how this expertise is employed. The central question is not how to stimulate analysis, but how to put it to greatest service. What can it in fact provide, and how is it that it so often gets directed to ends and purposes for which it was not intended?

Examination of the energy studies helps us address these questions. We look both at the substance of the studies and the influence they had on attitudes and policy discussions. Analysis was a much more constructive response to the energy problems of the seventies than was its alternative—the most common outlet for frustration and discontent—the search for scapegoats. Blaming and battling became almost a national pastime as people felt recurring distress from energy problems. Chapter 1 describes the situation at the time.

Chapter 2 looks back to the period before the seventies to review the economic and political background of the energy debates, making clear that the issues were not new but had deep and intricate roots. Chapter 3 briefly outlines the fourteen studies in our review, summarizes their objectives and outcomes, and notes their significance. It then presents the results of a questionnaire administered to people associated with or affected by the studies—people whom, in the unabashedly direct style of social science research, we call the "energy elite." They are the subjects of an attitude analysis described in Appendix A. We characterize the elites broadly, then describe how they scored the fourteen studies by quality, attention, and influence. The interesting results foreshadow discussion of the studies in subsequent chapters.

Chapter 4 begins the story of the energy studies, describing the first two to appear after the oil price rises, at a time when energy policy in the Nixon White House was at best chaotic. Chapter 5 continues the review with several analyses conducted during the Nixon-Ford era in search of a national policy for energy R&D. Chapter 6 considers the different policy approaches adopted during the Carter years and discusses some of the analyses fastened upon to promote that administration's energy program.

One of the studies was an effort conducted by the Committee on Nuclear and Alternative Energy Systems (CONAES) of the National Academy of Sciences. Taking almost half the decade to complete, it ended up in a virtual reenactment of much of the energy debate. Embracing over 300 experts from many fields, it exemplifies the unprecedented amount of analysis devoted to energy problems during the seventies, the wide variety of people involved, and the strongly held views that complicated political accommodation. We single out CONAES for special attention in Chapter 7.

In the third quarter of 1979, at the end of a decade of vigorous energy debate, three substantial energy policy studies were released in quick succession, just months before appearance of the final CONAES report at year-end. All three were ambitious in scope and statesmanlike in tone. Chapter 8 discusses the situation in the country at the time, and briefly describes the three studies.

The next portion of the book shifts emphasis from descriptive to deliberative, from a recounting to learning from it. Chapter 9 cuts across studies to compare methodological approaches in dealing with the central analytical issues of the day. Chapter 10 proceeds to dissect the analytical process, exploring the differences between the use of analysis in policy and its more familiar use in science. The use of analysis in policy, as characterized, is largely a potential use. It is a role that analysis might conceivably play—in theory. We then describe the role it has played—in fact. Given political realities, this is the role it will continue to play if nothing changes.

Coming out of our investigations, a perplexing question remains. Is there not something a nation can do to realize greater payoff from analysis in policymaking? Chapter 11 discusses the problem in the context of the special needs and characteristics of a democratic society. We have some recommendations to put forward.

In Appendix A, we present attitude profiles of 150 energy elite based on interviews and a specially structured opinion questionnaire that asked subjects to express their levels of agreement and disagreement with predesigned statements. We call the two main bodies of belief that emerged from the results of this questionnaire "traditionalist" and "reformist."

Although over half the book is concerned specifically with the energy studies, the scope of inquiry is broader than the place of analysis in energy policy. The energy studies of the seventies provide a well-stocked laboratory for examining the role of analysis in policy generally. The fundamental question is what contribution can analysis realistically be expected to make in the creation of policy and the guiding of public-sector actions, whatever the priority policy issues may be during a given period in the nation's history. This is a large subject—one that a free society intent on preserving its privileged status cannot afford to put aside.

Major funding for the collaborative research effort that led to this book was by the Alfred P. Sloan Foundation; auxiliary support came from the Rockefeller Foundation. The help and good fellowship of Stephen White at the Sloan Foundation was especially appreciated. Others very helpful included Anders Richter at the Johns Hopkins Press, Carol Franco and Steven Cramer at Ballinger, Harvey Brooks, Emery Castle, Edgar Dunn, Tom Glennan, and Charles Jones.

The project, referred to as the RAPP study (Role of Analysis in Public Policy), was administered by Resources for the Future (RFF) in Washington, D.C. Directing the project was Martin Greenberger of the UCLA Graduate School of Management, formerly at the Johns Hopkins University. Greenberger conceived the idea for the project in 1977, and in subsequent months enlisted the collaboration of Garry D. Brewer of the Yale School of Organization and Management, William W. Hogan of the Energy and Environmental Policy Center at Harvard's Kennedy School of Government, Alan S. Manne of the Stanford Department of Operations Research, and Milton Russell of RFF's Center for Energy Policy Research. Manne was a valued participant during the project's early phases, but was unable to play an active role during the later stages. Serving as advisers to RAPP were Kenneth J. Arrow, Charles J. Hitch, Lawrence R. Klein, Hans H. Landsberg, and Thomas C. Schelling. Debbie Groberg was a vivacious administrative assistant. Sandra Glatt and Philip Sisson helped painstakingly with research, and Angela Blake and Ann Fowler assisted masterfully at the computer. Liz Greenberger gave priceless aid during the final period of writing editing, and revision.

Martin Greenberger wrote chapters 1, 3, 4, 5, 6, 7, 8, and 11. Chapter 2 was written by Russell, Chapter 9 by Manne, and Appendix A by Brewer. Greenberger prepared Chapter 10 from drafts by Brewer and Hogan and Chapter 11 from the ideas of several collaborators and testimony he presented to Congress.[2]

Scores of people associated with the fourteen studies under review generously granted interviews, some several times. Brewer and Greenberger did the interviewing and supervised administration of the attitude questionnaire. In this task, they had the very capable help of Debbie Groberg and the expert counsel and processing skills of Ronald D. Brunner of the University of Colorado, then associated with the Institute of Public Policy Studies at the University of Michigan. Greenberger was responsible for the questionnaire on world oil and the one rating the fourteen studies.

In a quest for readability and consistency, Greenberger edited all parts of the book through more drafts than he would care to count. Adding substantially to the number of revisions required were the conscientious, earnest comments received from over 100 readers and parties-at-interest sent preliminary copy as a check on historical accuracy and fairness of interpretation. Each of the collaborators contributed to the book in important ways, as did many readers and advisers. Nevertheless, because of the practical necessity to grant proxy in a project of this nature, only Greenberger should be held accountable for views, content, and style.

1 REACTION TO THE CRISIS

The swift increase in oil prices and the ascending market power of oil-exporting countries early in the seventies caught the United States unawares. From the ensuing confusion came two antithetic responses: a stream of energy studies and persistent naming of scapegoats. To paraphrase an aphorism of the blind, in the world of the confused, an easy answer makes a king. A scapegoat is an "easy answer," a simple means for explaining a problem away. It provides temporary relief, a catharsis of sorts. It is something on which to fix the blame yet nothing with which to solve the problem. This fact is waved aside with the same sweep of the arm that dismisses a place for analysis. OPEC, charged with responsibility for soaring oil prices, became a diversion from the real reasons for the energy problem. So did the oil companies, the federal government, and numerous individuals. Our review of the constant search for scapegoats during the seventies is prelude to considering the attempt to reason and understand that is the subject of subsequent chapters.

Policy analysis—systematic examination of the complex problems facing a nation and its decisionmakers—was an active and repeat performer in the United States during the 1970s, a decade of much confusion and suspicion over energy problems. Analysis was no newcomer to Washington. It appeared in the attempts of the New Deal to steer the country out of depression during the time of FDR; it flourished in the evaluation of weapon systems during the Kennedy years; it was a conspicuous part of the efforts of Lyndon Johnson to achieve a Great Society; and it has been used without fanfare in facing a myriad of policy issues at all levels of government. With the onset of the energy crisis, analysis rose once again to national prominence. This was to be its most intense, pervasive, and sustained exercise yet.

What distinguished policy analysis in the seventies more than anything else were the sharp political realities it encountered—realities that were sharper, more impassioned, and more diverse than any experienced before. The energy crisis fragmented society, unleashing forces pulling in many directions at once, alienating large segments of the population, pitting region against region, interest group against interest group, institution against institution, and business against business. The technological and economic problems associated with abruptly raised energy prices and the fear of energy scarcities were perplexing enough by themselves, but they were vastly complicated by the political divisions they produced and intensified. The situation stimulated the nation's appetite for analysis at the same time that it rendered the public frame of mind inimical to the acceptance of analytical results.

The energy studies began in earnest in 1972 and continued for eight years in a political atmosphere of recurrent urgency punctuated by spells of complacency. Impassioned calls for action and analysis were mixed with disbelief or dimissal of what the analysts had to say. It was a time of testing for policy analysis. How one grades its performance depends on how one views its mission.

This book contains an account of fourteen of the energy studies, including those we regard as the most significant and influential. It is, in effect, a study of studies. In telling the story, we have two purposes. The first is to provide an intellectual history of what was certainly an important episode in the application of analysis to the problems of a democratic society, one significant to future development of the still young policy-research activity. The second objective is to learn from this episode about the nature of policy analysis generally. We ask how analysis must improve in order to reach its fuller potential for service to a nation and world beset with problems of ever increasing complexity and urgency.

The attempt to comprehend policy analysis poses two difficulties. First is the simple fact that the field is still developing. Were it to mature along lines suggested in our concluding chapter, its role in the political process could change in ways that cannot fully be foreseen. Second, policy analysts do not yet have a satisfactory paradigm for what they do or even what they would like to do (which is often quite different). Without such a model and the standards it provides, it is not easy for people knowledgeable and experienced in the field to be self-critical and introspective. Fair and discerning external evaluation is harder still.

Policy analysts might offer any of several models to describe their craft. Empirical science is one. Systems analysis is another. Engineering is a third.[1] Policy analysis exhibits similarities with each. It uses the observations and results of science, inquires into the intricacies and interactions of complex systems, and is creative, design-oriented, and utilitarian. But it goes beyond these possible descriptions in working within a process that it itself seeks to improve or redesign. There is a dualism between prescription and participation in policy analysis, like that between mind and body in philosophy. It is a source both of stress and of strength.[2]

The inadequacy of the various potential paradigms for policy analysis testifies to the lack of a satisfactory standard for evaluation. We believe any attempt to force policy analysis within one or another of the existing molds could be destructive and misleading at the present time. Such a goal is neither necessary nor desirable. Still maturing, the craft evolves as it adapts to its mistakes and limitations.

Under the circumstances, policy analysis can justify or reward only so much analysis of itself. Thus, our account will tend to be more descriptive than dissecting or dialectic. We point this out in advance—without apology—for the reader who might be expecting to find herein a tightly knit, scholarly statement about the practice of policy analysis. The field is hard enough just to plow. Others may wish to cultivate formal plantings in our furrows. If so, we are pleased and wish them well.

THE SETTING

The energy crisis came at a time in the political history of the United States when the nation was suffering grave doubts about the legitimacy of its institutions and the scope of its international influence and physical resources. The Vietnam War had split the country asunder and weakened U.S. hegemony in world affairs. The healing that began after the war left scars of alienation and festering suspicion. Watergate intensified the mistrust.

In extricating itself from Vietnam and in its preoccupation with Watergate, the United States paid insufficient attention in the early seventies to the energy problems starting to form. Domestic oil production had reached a plateau. Oil imports from the Mideast were increasing precipitously. Oil-exporting countries were nationalizing foreign oil operations and reclaiming rights to the petroleum beneath their lands. Some of these countries were openly militant. Even friendly ones had political interests and allegiances in conflict with those of the United States.

Simultaneously, an antinuclear constituency was growing at home and abroad in opposition to ambitious plans for the expanded use of nuclear energy. Concerns about radiation, the environment, safety, and weapons proliferation were causing many Americans to wonder whether nuclear technology was worth the risks it carried with it.

In thermodynamics, entropy is a measure of disorder in nature. It must always rise as energy is utilized or converted. In the universe as a whole, natural chaos constantly increases. Put so, this inviolable law sounds like a statement about the divisiveness and confused course of U.S. energy policymaking in the seventies. Energy as a political issue proved stubbornly deranging in effect. Through most of the decade, it confounded moves toward accommodation, agreement, and order.

In the years since World War II, cheap and abundant energy had come to be assumed by Americans almost as an inalienable right. Energy fueled the economy and was the means for providing amenities. Now there was the prospect of costlier, scarcer energy. Concern grew that this was to be the beginning of a decline in the quality of life. Some wondered if Atoms for Peace was a Trojan horse carrying perils for mankind instead of progress. Were big industry and large-scale operations a mark of modern achievement or only tyranny tuned to the twentieth century? Small was beautiful according to this perspective, and large centralized sources of energy were the dinosaurs of industry. Business, government, science, and technology were not to be trusted. As the energy problem surfaced, such suspicions and anxieties came with it.

The era of cheap energy was at an end; a fundamental transition to uncertain, highly contested, alternative sources was needed. Coal, fission, fusion, synthetics, solar, and biomass all had their advocates and their detractors. To every argument—whether technical, economic, or political—there was a counterargument; to every point of view, an opposite point of view. There was no one expert, no one authority, and no one energy study that everyone—or even a majority of people—believed.

Politics has been called the art and practice of moving from one emotionally satisfying and enabling consensus to the next. The energy crisis jolted comfortable consensuses and challenged accepted truths—the ballasts of political stability. Both domestically and internationally, the status quo was seriously shaken.

Especially divisive were the sharp rises in energy prices that occurred after the Arab oil embargo in 1973/74 and then again after the Iranian revolution of 1978/79. The energy price hikes caused massive transfers of wealth and income, not only between oil importing and exporting countries, but within countries as well. Some gained, some lost—a condition bound to awaken age-old hostilities and produce fierce tugs of war. OPEC became a target of resentment and envy. The wrath of energy consumers was now vented on it as it had been upon the oil companies ever since the trust-busting days earlier in the century. Instinctively, OPEC and the oil companies blamed one another.

The anger and alienation felt by the public showed up plainly in the evening news, letters to the editor, tirades in state assemblies and Congress, and in social science research on public attitudes and opinions. There was mistrust of government and industry and outrage at the alleged conspiracy between them. Media coverage fanned the flames and gave the public "only a murky understanding of what may have caused the crises."[3]

The angry mood was not reasoned or constructive. It was not based on facts or analysis, nor did it come about for want of facts or analysis. Energy studies *were* taking place in the seventies in great number. But those studies did not always pay a great deal of attention to the differential impact at the heart of citizen unrest. Their technically based messages could not provide solace to the alienated and confused.

One might wonder if the energy studies recognized the full meaning of what was taking place. A "crisis" is a change of context where simple extrapolations no longer apply and former truisms must be rethought. Expectations were in a state of rapid flux, and the previously accepted basis for analysis was no longer acceptable. Trusted experience was becoming obsolete, and the experts of the past were not necessarily going to be the experts of the future.

The analysts who participated in the energy studies did not generally regard their subject the same way as did the public and the press. Their attitudes were not the anxieties and hostilities of the average person. This was clear from the questionnaire we gave to explore the viewpoints of energy analysts and decisionmakers—a group we called the "energy elite." In Appendix A, we compare the results of this questionnaire to those from a similar one given to members of the general public. As one might expect, the outlooks of the elite were on the face of it more rationally based than were those of the public. The elite tended to focus on substantive rather than emotional issues, although they were not immune from strong partisan feelings. They could cast blame too.[4]

The crisis of the seventies, we believe, was more one of policy than of energy. The energy problems that filled the news for much of the period were due largely to a mismatch of government actions with physical and economic circumstances. There was an incongruity between regulations and conditions, a disparity between policy and fact. These inconsistencies were more important than any actual energy shortage.

If, in a democratic system, public atttitudes and private interests are mediators between policy and fact, they are at best inefficient mediators. Controls on domestic oil prices were dysfunctional in the face of soaring world oil prices and increasing oil imports. The controls made matters worse by stimulating demand and thereby further increasing imports and raising prices. Yet, public opinion favored continuing controls.

As the facts became better understood with the help of the energy studies, suspicions and hostilities gave way to realities and reason, and attitudes began to shift. Eventually it became politically feasible for President Carter to move to eliminate the controls, an action completed by President Reagan in one of his first acts after taking office.[5] But it was not a total victory for advocates of the free market. Passage of windfall profit taxes was part of the political bargain, and public sentiment against natural gas deregulation remained strong.

Early in the crisis, there was a prevalent view promoted by many in industry that reduction in the rate of increase of energy use would impair the U.S. economy. Part of the slowdown in the economy during the 1970s was attributed to more expensive energy. Yet, the notion gradually took hold (again with the help of analysis) that reduced energy growth need not stall economic progress. Energy conservation gained favor as though it were a valuable new energy source. Priorities were modified, and the changing mood was reflected in legislation.

The fall from favor of nuclear energy offers a different example of the part played by public opinion. Here opinion ran ahead of the known factual base, partly because of emotional fears about nuclear explosion and radioactivity and also because of the organized efforts of certain scientists and engineers who insistently warned of weaknesses inherent in the national program. The nuclear cause was dealt a serious blow by the costly and unnerving accident at Three Mile Island. At least as damaging was escalation of construction costs and lowered projections of electricity demand.[6] Yet the stated policy of the Reagan Administration in the early eighties was to promote continued nuclear development. The nuclear issue looked like it might be as much a political focal point in the 1980s as oil security had been in the 1970s.

One of the major functions of the energy studies—a function we believe they served both well and not well enough—was to bring questioning and logic to the discussion. The studies exposed mismatches of policy and fact and helped to correct misunderstanding of the relationships governing energy use and energy markets. Attitudes on the issues gradually changed over the course of the studies, but not without dissension and loud protest. It was fascinating to observe.

The subject matter of the fourteen studies examined in detail in this book span the range of all the principal questions being debated. Directly or indirectly, the studies influenced how these questions were framed, argued, and at times resolved. The evidence is that despite the controversy and misinterpretation they engendered, the studies as a group did have an overall salutary effect in clarifying the energy problem. Analysis served as a healthful antidote to the nigh irresistible urge to look for whipping boys and chase witches.

Public Enemy Number One

Every decade makes its own special waves in history. For Americans, Watergate would have been the big swell of the seventies had the much larger energy problem not risen up fast behind it even before Nixon resigned from office. The energy crisis encompassed many of the decade's major events and symbols: war in the Mideast, the Arab oil embargo, interminable gasoline lines, the quadrupling of crude oil prices in three months, revolution in Iran, a second upsurge in oil prices, depressed economic growth, and rampant inflation. Amidst these unsettling occurrences was the persistent hint of a possible villain: OPEC—the Organization of Petroleum Exporting Countries.

White House domestic affairs adviser Stuart E. Eizenstat wanted to have OPEC declared public enemy number one. It was the summer of 1979. Eizenstat's memorandum of June 28 to President Carter ("disclosed" to the *Washington Post* after cancellation of a scheduled presidential address on energy) sought to assign responsibility for the nation's economic problems to actions from abroad. Eizenstat wrote his boss:

> I do not need to detail for you the political damage we are suffering from all of this. It is perhaps sufficient to say that nothing which has occurred in the Administration to date ... has so frustrated, confused, angered the American people—or so targeted their distress at you personally.

Eizenstat's formula for changing this "worst of times" for the Carter Administration into a "time of opportunity" was to "shift the cause for inflation and energy problems to OPEC." Eizenstat did not mince words. "With strong steps we can mobilize the nation around a real crisis and with a clear enemy—OPEC."[7]

Many people do not realize that the origins of OPEC predate the 1970s by a full ten years and trace back to Venezuela along with other founding members in the Mideast. For much of the industrialized world, OPEC became (figuratively as well as literally) a four-letter word during the seventies, the embodiment of rapacious greed. Yet OPEC was merely a convenient scapegoat—one of several we will consider.

Venezuela, Saudi Arabia, Iran, Iraq, and Kuwait created OPEC in 1960 to withstand attempts by the major international oil companies to reduce the posted (tax reference) price of crude oil. The world took little note. More than two weeks elapsed before a brief account of the event appeared in the *New York Times*, with a title line that neglected Venezuela's key role:

Mideast Oil Lands Seek Price Stability

Bagdad, Iraq, Sept. 24 - (Reuters) - A five-nation oil conference held here earlier this month voted to demand that oil companies try to restore prices to their former level and keep them steady. The meeting, held after the major companies cut Middle Eastern crude oil prices, also voted to form a permanent Organization of Petroleum Exporting Countries to unify their oil policies and promote their individual and collective interests.[8]

This was the humble beginning of a momentous shift of power to producing countries. The dominance of the corporate forebears of British Petroleum, Exxon, and Shell dated back to the late 1920s. These three giants of the industry had avoided destabilizing competitive skirmishes by accepting and maintaining fixed shares of the market. Their international cartel ruled over world oil up until World War II.[9]

After the war, the industry became progressively more competitive as large oil finds sparked an energetic campaign to stimulate demand and open new markets. Companies extended their exploratory operations widely through concession arrangements with producing countries. Firms of all sizes joined the contest. There was a decline in real oil prices, growing discontent among the oil-exporting countries, and a resolve to mount a united defense against the petroleum companies.

Efforts at cooperation among oil-exporting nations began in the late 1940s and early 1950s. The first formal agreement was signed between Iraq and Saudi Arabia on June 29, 1953 in an attempt to improve their bargaining positions and the terms offered them by the companies. This agreement came in the wake of an abortive nationalization of the Iranian oil industry two years before, triggering a boycott by the international oil companies that brought Iranian exports almost to a standstill. There was little doubt who held the power in those years.

Oil import controls adopted in the United States in 1959 contributed to a developing surplus in Mideastern production capacity and weakened foreign markets. By the terms of the concessions granted them, oil companies had no obligation to consult Middle Eastern exporters on changes in posted prices. Venezuela was very concerned about the adverse impact that reduction in the posted prices for Mideastern crude was having on its oil market. It joined Iran as an observer at the First Arab Oil Congress, held in Cairo in April 1959. That congress provided a forum for venting unhappiness and apprehension over the price reduction and spawned the idea for OPEC.[10]

The immediate impetus for creating OPEC came in August 1960 when Mideast posted prices were further cut by from 4 to 14 cents a barrel. Even if the action was basically sound it could not be acceptable to us as long as it was taken without our consent, declared the angry Shah of Iran. A meeting in Baghdad the following month gave birth to the organization of oil producers. The meeting was conducted in an atmosphere of crisis. Indignation overcame the Shah's natural reluctance to collaborate with the Arab states, especially with the revolutionary antimonarchial regime of neighboring Iraq. But with Venezuela participating, Iran would not be the only non-Arab country in the covenant.

Eventually, in addition to Saudi Arabia, Iran, Iraq, and Kuwait, OPEC came to include Qatar and the United Arab Emirates in the Middle East, as well as Algeria, Gabon, Libya, and Nigeria in Africa, Indonesia in the Pacific, and, joining Venezuela, Ecuador in South America: thirteen nations supplying the bulk of the free world's crude oil exports.

Thirteen is OPEC's magic number it would seem, for on its thirteenth birthday in 1973—at the age that Israelis celebrate as the transition between youth and manhood—OPEC soared into public prominence. It was the time of the Yom Kippur War and the Arab oil embargo. By unilaterally quadrupling oil prices in two successive price hikes within three months, OPEC announced emphatically that crude oil prices were thereafter to be the producers' province and no longer the prerogative of the multinationals. With the stroke of a pen, OPEC upped its member countries' revenue by tens of billions of dollars, thus setting off public outrage and accusations.

HISTORY PERSONIFIED

Postmortems of the period leading up to this dramatic price action are often personal in nature, with considerable finger pointing. James E. Akins, head of the Fuels and Energy Office at the State Department during the pre-embargo years, is one of the more controversial figures of the time. He has been accorded both credit and blame and has himself been accuser as well as accused.

Akins is credited with being one of the people who foresaw the impending oil shortage and the danger of relying on foreign oil. He proposed measures to avoid a crisis in testimony to Congress and in talks as far back as 1968. His article in *Foreign Affairs* titled, "The Oil Crisis: This Time the Wolf Is Here," appeared just months before the embargo.[11]

Akins is criticized for his outspokenness and insistence that world oil production could not keep up with rising demand. History is replete with early prophets of oil exhaustion whose prognostications were proven wrong by subsequent finds. Akins warned openly of the vulnerability of Japan and the West to an oil boycott. His critics claim his warnings impressed upon OPEC countries what a potentially effective weapon they had and encouraged them to use it.

OPEC had in fact declared its intentions years before. In a 1968 resolution, member countries insisted on their rights to "participation" in the ownership of oil lifted from their ground and announced a determination to gain control of posted prices to prevent the value of oil revenues from deteriorating relative to import costs. They made no secret about wanting to recoup "excess" company profits.

Many OPEC countries were experiencing a severe financial squeeze at the end of the 1960s. Although production was expanding, expenditures were rising faster than were revenues at the oil prices then prevailing. The pressure to cut back on economic development provided a strong incentive to price aggressively.

Producing countries had their first real taste of market power in September 1970 when they were able to get companies to agree to significant increases in both tax rates and posted prices. Libya's triumph on the Mediterranean was followed by that of Persian Gulf[12] producers in separate negotiations with the companies. The pathbreaking agreements opened the way to further gains in subsequent years: participation rights in the oil produced, sharply higher revenues, and eventually complete control of oil pricing.[13]

The companies did not give in without a fight. To avoid price leapfrogging by Libya and the Gulf countries, the companies united to insist on a single global negotiation with all OPEC countries simultaneously. They were careful to clear these arrangements with the Departments of Justice (to avoid antitrust charges) and State.

Akins, unsympathetic to the companies' cause, was the official in the State Department at the time in charge of petroleum affairs. The companies asked help in getting producing nations "to moderate their demands" and "engage in fair bargaining practices." They requested that representations be made to these nations by a person "with clout," and were told that Under Secretary John N. Irwin was prepared to make such a tour.

President Nixon personally authorized the Irwin mission. Irwin met first with the Shah in Tehran on January 17, 1971. According to a 1975 Senate subcommittee report, "Irwin made it clear that the U.S. Government was not in the oil business and did not intend to become involved in the details of the producing countries' negotiations with the oil companies."[14] Irwin had met resistance by the Persian Gulf countries on the concept of unified bargaining with the Libyans, and he was not himself sure that putting the Libyan "price hawks" together with what he considered to be "price moderates" was a good idea. Impressed with the Shah's assurance that the Gulf countries would not be influenced by the deal negotiated with Libya regardless of its terms, Irwin cabled the Department of State that the companies should negotiate with the Gulf countries separately. "Thus did the State Department abandon the basic strategy of seeking an overall negotiation encompassing all of the parties at the same time," says the Senate report.

Akins and Irwin have both been called to task for the roles they did or did not play at this tumultuous juncture. But the tables were already turning, as revealed by the very desire to get the petroleum exporters to negotiate jointly. Throughout the fifties and sixties, the oil companies were able to keep crude oil prices low by dealing with the exporting countries separately, playing one off against the other when necessary. The companies refused to deal with OPEC as a unit when it was formed. Ten years later, Libya was using this same "divide-and-rule" strategy against the oil companies.

One critic of Akins and Irwin was M.I.T. economics professor Morris A. Adelman, who viewed the developments of 1970/71 with consternation. Adelman charged that the Gulf countries and the Libyans "had the backing of the State Department" in its efforts to quiet down the "turbulent oil market." According to Adelman, the fact that the State Department's mission emphasized for the Shah the plight of Europe and Japan if oil supplies were cut was "incitement to drastic action."[15] Middle East economist Robert Mabro is doubtful of that analysis. According to him, "The Shah did not need anyone to tell him that Europe and Japan were dependent. That was obvious."[16]

Now "the genie is out of the bottle," said Adelman, who at the time was completing a major study of international petroleum markets. Adelman had predicted that the result of greater importation of Mideast oil could be to *lower* crude oil prices. Although he recognized the irreversibility of what was taking place, Adelman thought the United States could exert enough market power as the dominant consumer of oil to get real prices back on their postwar downward course.[17] After all, the cost of producing Persian Gulf oil was only a small fraction of its market price, and proven reserves would last for at least another half century at the then-current production rates.

Akins was on assignment to the White House in early 1973 advising on energy policy. Pushing for production of substitute fuels and more efficient use of oil, he was critical of economists who were predicting a downturn in oil prices. So far as he was concerned, they were telling politicians just what they wanted to hear, giving govenment officials an excuse for doing nothing.

For oil prices to have "tumbled," as some economists predicted, would have required a high degree of competitiveness among producing nations and a failure of OPEC countries to maintain their coalition. Neither occurred. The market became progressively less competitive, and OPEC remained intact in its own way. It is debatable whether history would have turned out differently under policy less congenial to the Shah and Libyans. There is no way to know if the "genie" could have been contained. But there is more to the story.

In June 1973, Secretary of State Henry Kissinger appointed Akins ambassador to Saudi Arabia. A colleague of Akins in the State Department at the time called him bright and articulate— "a marvelous fellow"—but also self-centered. When Akins was notified of his appointment, he was told he would have to "behave." Observed the colleague, "Akins was honest about himself. He said he knew what his problems were. He got carried away."

Despite good intentions, Akins had a less than tranquil tenure as ambassador. A large percentage of the messages he sent in— and there were a "great many"—would go unanswered. When Secretary Kissinger visited Saudi Arabia, he is said not to have invited Akins to join in the meetings on occasion. "One of Henry's failings," said a close associate. "Very poor."[18]

Finally, Akins was informed by State Department official Joseph Sisco that his reports "were annoying the Secretary." Akins told Sisco that "if they really wanted people to report only things that they wanted to hear, then they could get themselves another boy." Added Akins, "They proceeded to do that fairly smartly."[19]

Kissinger dismissed Akins summarily in August 1975. The ambassador learned of his discharge in the newspapers. Akins feels he was fired because he was not as malleable as Kissinger wanted. Whatever the reason, there was clearly poor communication and bad feelings between the two. Knowledge of this troubled relationship is significant in interpreting the next set of events and allegations (this time by Akins). Also important is an understanding of the changing conditions that shook world economic stability in the early seventies and set the stage for the quadrupling of oil prices. We pause to review these conditions briefly.

The Early Seventies

In 1971, the oil companies finally worked out five-year pacts with the producing countries known as the Tehran and Tripoli agreements. The posted price of crude—the price on which tax income to producers was based—was then substantially more than the price for which the crude was actually sold. Since OPEC had been successful in maintaining a floor on posted prices as market prices fell, its members were obtaining a growing share of oil revenues. The Tehran and Tripoli agreements were designed to give them an even bigger share by boosting both the posted price and the tax rate. An escalation clause was included to protect them against the impact of inflation. All of this assumed that posted prices would continue to exceed market prices and inflation would be modest. Neither assumption was justified as it turned out. Inflation in the industrialized countries during the early seventies, for example, went from under 6 percent to over 13 percent.

Worsening inflation in the United States was a byproduct of the Vietnam war. It was accompanied by a fall in the exchange rate of the dollar relative to other currencies, adversely affecting revenues from dollar-denominated oil on the world market. In August 1971, President Nixon rescinded the convertibility of the dollar into gold, and on December 18, the dollar was formally devalued. The following month an agreement was reached with the OPEC countries in Geneva to increase posted prices by 8.59 percent to take account of the currency changes—the first amendment to the terms of Tehran and Tripoli. Within a year, the dollar had been devalued again. By a second Geneva agreement in June of 1973, oil prices were boosted by another 12 percent. Saudi light crude was now posted at $2.90 a barrel compared to $1.80 at the start of the decade.

Meanwhile, prices of crude climbed upward on the open market, primarily because of sharply rising demand for oil in the United States and elsewhere. The pinch was felt in many sections of the country. As the clouds of Watergate began to spread, President Nixon replaced the mandatory import quotas in force since

1959 with a fee system in an April 1973 energy message to Congress. Given growing shortages and the upward pressure on price, "there was no choice," says Stanford professor Ezra Solomon, then a member of the Council of Economic Advisers.[20] The Nixon Administration was politically committed to keeping a lid on prices. There was a swell of imported oil into the United States, and the last vestiges of the 1959 world oil surplus disappeared.

Competition for foreign oil intensified. By the summer of 1973, prices received by the exporting countries on the spot market were well in excess of posted prices. As a result, although the revenues of exporting countries were improving, their share of realized prices (based on posted rather than market prices) fell from 80 percent at the time of the Tehran agreement to 64 percent by September 1973. The OPEC countries accused the oil companies of reaping excessive profits by raising their product prices more than was warranted by the increased cost of crude. They were in no mood to continue to honor the Tehran and Tripoli agreements.

Up to this point price modifications were almost always negotiated in meetings with the companies. The OPEC countries were reluctant to take unilateral action to abrogate the Tehran and Tripoli agreements and set their own prices as Venezuela urged. Their hesitancy dissolved as their financial situation improved. Says State Department official Joseph Twinam, "key producers were for the first time in a financial position to risk confrontation with the companies in price negotiations without risking severe and immediate economic consequences."[21]

In September 1973, OPEC member countries formally requested reconsideration of the 1971 agreements. Negotiations began on October 8, two days after the outbreak of the Yom Kippur War. An impasse developed and the meeting was adjourned. A few days later, the six Gulf producers unilaterally announced a 70 percent hike in the price of Arabian light crude, from $3.01 to $5.12 a barrel. The next day, the Arab countries set production cutbacks and recommended an embargo to put pressure on Israel and its supporters. International rerouting of oil muted the embargo's effect, but the production cutbacks, mishandled by Washington, had a severe impact. Americans need no reminding of the gasoline lines that winter.

The fast-rising demand for oil was taking place not only in the United States, but throughout the industrial world. A general

GNP boom had 1973 growth rates ranging from 5.4 percent in Western Europe to 10.4 percent in Japan. Fuel consumption by Japanese industry was nearly 30 percent higher in October 1973 than the year before. The world oil market was getting progressively tighter.

Domestic oil production had peaked in the United States in 1970, and oil imports from Venezuela and Canada were no longer able to satisfy the country's growing thirst. Oil was not yet available from the Alaskan North Slope and the output of price-controlled natural gas was no longer expanding. Nuclear power plants were starting to suffer delays. The influential environmental movement discouraged the use of coal for generating electricity, and there was a shift toward oil (whose real price had been dropping up until 1970). That was the situation.

Christmas 1973

The rapid price increases of late 1973 set off a barrage of charges and countercharges that ignored the economic background. The Yom Kippur War and the ensuing embargo were big media events. It was natural to assume these political events were at the root of the price hikes. But while the embargo provided drama, as George P. Shultz, then Secretary of the Treasury observed, the underlying economic conditions combined with the cut in Saudi production were the real causes of the boost in prices.

The biggest increase took place on December 24 in Tehran when the six Gulf countries raised the posted price of crude to $11.65 a barrel. The now fourfold increase in posted prices translated into a fourfold increase in take as well—from $1.77 a barrel in September to $7.00 a barrel by the end of the year.

The Shah of Iran assumed a conspicuous role at the December meeting in Tehran. He "took the unusual step" of calling a news conference afterward to disclose the higher price while expounding his concept of linking oil prices to the cost of other energy forms. A reporter wrote that the Shah appeared intent on playing the role of "responsible leader in the Gulf and a statesman to be taken seriously."[22] What the Shah did not intend was to set himself up as a target—a scapegoat by Westerners for the oil price increase and a villain by his own people for being a tool of the imperialists. That is essentially what happened.

There is disagreement on who did what to whom. Iran was producing at maximum capacity. It was Saudi Arabia that was aggressively cutting production. Yet Saudi Arabia says it came to the December meeting urging moderation. So attests former Ambassador Akins, referred to by several publications as the "Saudi Ambassador to the United States," so closely did he identify with the interests of the Saudis.[23] Akins declared that Saudi Oil Minister Zaki Yamani warned him that oil prices would go way up unless the United States intervened with the Shah. Akins cabled the warning to Washington. He says he received no reply and believes no action was taken. He sent similar cables to other OECD capitals to suggest they also make diplomatic demarches, which he acknowledges was "bad form," but "it had to be done."[24]

When only Belgium protested to Iran, Akins recalls, the unfortunate impression was created that the price rises were not disturbing to the others. Wrote Sheik Yamani to former Secretary of the Treasury William E. Simon:

> There are some among us who are of the opinion that the United States Government has no real objection to an increase in oil prices. Some even believe that the United States actually encourages a price increase for obvious reasons of foreign policy, and that the purpose of any officially assumed opposite stance is merely to conceal these facts.[25]

According to Akins, Yamani said repeatedly both publicly and privately, "the Shah told him and King Faisal that the Americans knew why oil prices had to go up."

Akins, the much accused, got the opportunity to become accuser before tens of million of viewers May 4, 1980 on CBS's top-rated TV show, *Sixty Minutes*.[26] Correspondent Dan Rather had put together a segment on "The Kissinger-Shah Connection" developing the thesis that Kissinger wanted the Shah to have the weapons needed to protect vital U.S. oil interests and keep Iran's borders safe from Iraq and the Russians. "In order to get the weapons, they had to have the money, and the way they got the money was to raise the oil prices." The implication was that if Kissinger had not actively encouraged the price rise, he did nothing to discourage it. Akins provided testimony.

Correspondent Rather got started on the story while looking into charges made after the Iranian revolution of a special relationship between Kissinger and the Shah. He did not turn up much new that could be documented, so he decided to focus on the Shah's role in the price increase and the "commonality of interests" with Kissinger. "Here's something that will interest everyone: Kissinger, Shah, oil prices," he concluded. "Kissinger used to have an excellent press—well deserved. I helped to give it to him. But was there something we should have known? Would it have changed U.S. government policy?"[27]

The subject is touched on in Kissinger's book White House Years, an autobiographical account of U.S. foreign policy during the first term of the Nixon Administration.

> Britain at the end of 1971 had just completed the historic withdrawal of its forces and military protection from the Persian Gulf at the precise moment when radical Iraq was being put into a position by Soviet arms to assert traditional hegemonic aims . . . It was imperative for our interests and those of the Western world that the regional balance of power be maintained . . . There was no possibility of assigning any American military forces to the Indian Ocean in the midst of the Vietnam war and its attendant trauma. Congress would have tolerated no such commitment; the public would not have supported it. Fortunately, Iran was willing to play this role. The vacuum left by British withdrawal, now menaced by Soviet intrusion and radical momentum, would be filled by a local power friendly to us . . . without any American resources, since the Shah was willing to pay for the equipment out of his oil revenues.[28]

On May 30, 1972, President Nixon and Kissinger, then National Security Adviser, were returning home from a summit conference in Moscow. They stopped to visit, in Kissinger's words, "one of America's closest allies, the Shah of Iran." It was at this meeting, according to a Senate report, that Nixon and Kissinger "agreed for the first time to sell Iran virtually any conventional weapons it wanted and so instructed the bureaucracy." The next year, as oil prices quadrupled, "Iranian purchases from the United States boomed."[29]

After the Shah's overthrow in 1979, it was generally acknowledged that Iranian funds had not been spent wisely. Charges ranged from gross waste to corruption and mismanagement. The program of accelerated spending and forced modernization was said to have been a shock to Iranian society and a primary factor in the Shah's downfall.

Assessing the Charges

Had the Shah become dangerously overconfident because of the assurances by Nixon and Kissinger? Were the Shah's ambitions responsible for the rise in oil prices? Not an entirely implausible thesis, perhaps, and emotionally satisfying too, especially for those who opposed arm sales to the Mideast, did not care for Kissinger, repudiated the Shah, or were glad to have still another charge against Richard Nixon. Plausible, emotionally satisfying, but nevertheless, by any of several perspectives, wrong.

Wrong, of course, according to Henry Kissinger, who complained that the 1972 visit had "retroactively been made a scapegoat."[30] Agreed *Wall Street Journal* editor Thomas J. Bray, the Shah was "a convenient villain" on whom the Saudis could attempt to "hang the rap for oil prices" in a "poorly understood and highly successful ploy."[31] Bray emphasized the importance of Saudi Arabian exports:

> The key to OPEC is Saudi Arabia's willingness to keep production down. It cut back production two million barrels daily in late 1973, making oil scarce and driving up the price. Supply and demand, not OPEC's or the Shah's blandishments, led to the quadrupling of prices in late 1973. The United States was vulnerable because price controls had held down supply and propped up demand in this country. Nothing Mr. Kissinger could have done would have changed that.

Wrong also according to former Treasury Secretary George Shultz, who joined Bray in faulting the government for its counterproductive policies in the face of rising demand and limited supply:

> By 1973, the Texas Railroad commission [the regulatory body that traditionally limited oil output from Texas oil fields] was, for the first time, allowing production at 100 percent of capacity ... OPEC country production was also at capacity, with consumption throughout the world increasing rapidly ... The crisis is not the result of villainous behavior by malicious individuals but of erroneous policies of the

United States government, policies in which Mr. Kissinger played no part.[32]

And wrong according to economist Adelman, whose cartel theory viewed OPEC countries as collaborating to restrain output and increase their joint earnings:

I have no doubt that President Nixon and Secretary Kissinger acted ... in their illusion that Iran was our instrument ... our policeman in the Persian Gulf. But that was effect not cause of the stream of wealth flowing into Iran because of the higher oil prices. Prices had been raised by the actions of the oil producing nations as a group, with Iran as a minor partner, and the U.S. government as the major catalyst.[33]

Even James Akins, who would have differed with Adelman and Bray on Saudi Arabia's role, was uncertain whether the Shah would have responded to pressure by the United States, had it been applied, and if he had responded, whether this by itself would have made any real difference.

Kissinger was in Geneva at the time of the December 24 OPEC meeting trying to settle the Mideast war and end the Arab embargo. He was unlikely to have been thinking about prices at the time. Because of restricted oil supplies and surging demand, spot crude was bringing three to four times its posted price on the open market before the OPEC meeting. Eighty million barrels of Iranian oil had gone for $17.40 a barrel at auction on December 11, and prices well over $20 a barrel had been reported by Nigeria. One would have had to look no further than that to account for the price aggressiveness of the Shah.

To be reasonable, it was as incorrect to lay the blame for the price increase at the doorstep of Kissinger's relationship with the Shah as it was to attribute OPEC's militancy earlier on to the prophecies and warnings of James Akins. It was equally erroneous to countenance the accusations made against the oil companies for welcoming the crisis or OPEC for causing it. Preoccupation with scapegoats is more than just a mistake. By giving the illusion of understanding, it distracts attention from the real problem—and the hard decisions and constructive steps necessary to correct it. One may wish that certain representatives of government or certain institutions had acted differently at critical points in a crisis, but that does not mean they created the crisis or that they had the power to avert it.

FROM BLAME TO UNDERSTANDING

Injured parties want retribution. Parties at interest want answers. Everyone seems to want explanations and reasons. The simpler, the more direct, the better. Finding a "guilty" individual is particuarly satisfying since it is personal, concrete, and easy to grasp. Almost as good is fixing responsibility on an institution— less personal, but still specific. Oil companies have been fair game. Yet the 1975 Senate report found no reason to indict the companies for the OPEC actions of the early seventies.[34]

Criticizing the government is another popular means of venting anger and frustration. This can take the form either of indiscriminate carping or considered reasoning. Free-market economists believe that price controls on oil and gas, intended to contain inflation and soften the impact on the poor, were major culprits in the nation's growing dependence on energy. The argument is that price regulation, by stimulating consumption of oil and gas at all levels of society, contributed to an increase in the price of energy worldwide, thus worsening, not lessening, inflation. Artificially low prices domestically undermined the incentive to find alternatives for oil and gas. The entitlements program, designed to equalize the costs of foreign and domestic crude among domestic refiners, further heightened the desire to buy foreign oil.

U.S. energy policy, seemingly misguided and counterproductive, upset America's allies. One Britisher charged that the imposition of U.S. import quotas in 1959 induced a surplus in the world oil market, whereas lifting the quotas in 1970 caused a shortage, "mainly felt outside the United States."[35] Domestic price controls in the United States put further pressure on world supplies and prices.

Why did the government have so much trouble straightening out its energy policy? A pluralistic democracy has difficulty handling a divisive national problem where some must lose while others gain. Despite the energy studies, society was not inclined to educate itself sufficiently to appreciate the substance of the political disagreements. Nor were the fact base and understanding offered by the studies easily accessible to the majority of people. When the studies were used, too often they were used for advocacy rather than understanding, to advance special interests rather than to help in designing a united attack on a national problem.

Rarely were the studies used to communicate important information and understanding to the public. Of course, that is not simple to do. History is by necessity selective and inconclusive. Real life is too convoluted and multifaceted—too interwoven—to allow for definitive interpretation. Although a particular interpretation may appear *sufficient* to explain an occurrence or set of events, it is unlikely to be the only possible explanation. It may be the most parsimonious or appealing one; it may seem to be the most logical one; but who can be certain that it is the right one? Who's to contest what none can test?

How then to understand the oil crisis? The easy explanations were the ones that cast blame and sought a scapegoat. There was no lack of ready targets, as we have seen: U.S. consumers for their profligate ways, the U.S. government for playing favorites and regulating perversely, government officials for acting foolishly, oil companies for unseemly profits and conspiratorial behavior, and OPEC for greed and irresponsibility. A Gallup survey conducted in December 1973 asked, "Who or what do you think is responsible for the energy crisis?" No less than 25 percent of the national sample blamed the oil companies. A similar number (23 percent) specified the federal government, while 16 percent designated U.S. consumers. Only 7 percent named the Arab nations, despite the fact that the Arab oil embargo was in full swing at the time. But, then, most of the American public only gradually came to realize how much of its oil was imported.

A later Gallup poll done in the summer of 1979 was not quite as generous to Arab and other OPEC countries. It asked the question, "Whom or what do you blame for the gasoline crisis?" OPEC and the Arabs combined were singled out by 13 percent of those polled. The U.S. government was selected again by 23 percent, President Carter by 11 percent, and the American people by 11 percent. Oil companies were still the favorite target by an even larger margin (42 percent), probably due to the focus on gasoline, plus the fact that oil company profits were much in the news at the time. OPEC deflected criticism from itself by joining the attack on the oil companies, charging that company profits were evidence of the exploitation and injustice its member countries had long suffered. Once again, accusers alternated with accused.

The Gallup polls were administered to the general public. We compared the Gallup results to the opinions of two groups of energy professionals to whom we gave a structured questionnaire on world oil.[36] One group consisted of about forty energy policy analysts working on world oil. The other group of similar size was drawn from the management and senior staff of a large petroleum firm. Since such people might be expected not to feel the same resentment and suspicion as the general public, we asked them about causal factors rather than about blame.

Both groups considered two factors most decisive: the growing demand for world oil relative to supplies, and the changing power relationship between OPEC countries and the multinational oil companies. Respondents from the petroleum firm emphasized the significance of OPEC's actions. The analysts, and especially the economists, assigned greater weight to market forces. They questioned whether heterogeneous OPEC was really a unified cartel acting to restrict production and raise prices. Both groups believed a marked change had taken place after the Iranian revolution in the importance of: resource limitations and the rate of new oil finds, changing perceptions about the value of oil in the ground versus the value of the dollar, the cost and availability of alternative energy sources, the oil production capacity of OPEC countries, and the effect of rising oil prices on demand and energy conservation.

Was the Iranian revolution an historical watershed? To answer that question requires some attention to the events occurring toward the end of the decade.

The Late Seventies

During 1974 and 1975, worldwide recession and sluggish economic growth reduced the real price of oil, especially in Germany and Japan, whose currencies had appreciated relative to the dollar. There was talk of an oil glut and confident predictions about breaking the cartel. OPEC, blamed for causing the world's economic problems, teetered but did not fall. The petrodollar surplus was less than feared, and the international banking system was able to recycle much of this surplus through loans to non-OPEC developing nations whose oil bills had expanded sharply.

The lull in oil prices ended abruptly in early 1979 with overthrow of the Shah and drastic cutback of Iranian production. In the wake of the revolution, oil prices more than doubled and prices on the spot market reached unprecedented heights. Buyers panicked. Worried about the continuing unrest and uncertainty in the Middle East, companies and governments frantically built up inventories. Trading dislocations added to the general anxiety. A series of price boosts occurred even after other oil exporters had made up for the loss of oil from Iran.

As in the earlier round of price rises (also induced by a sudden and unexpected drop in world production), the spot market overheated. As before, real oil prices (which had not recovered to 1974 values until the third quarter of 1979) were lagging the higher import costs of OPEC nations. These countries felt fully justified in selling some of their crude at the elevated spot prices and in adjusting their official prices upward. Again OPEC petrodollar surpluses soared, and again recycling became a concern.

The spurt ended in about two years. Prices on the spot market subsided and sales began to be negotiated at less than official prices. Oil stocks were high, demand fell, and as in 1975, the world oil market softened. With a deepening recession, there was again talk of an oil glut, and futurist Herman Kahn renewed his predictions of OPEC prices tumbling.[37] Still, most experts did not expect oil prices to decrease over the long term. Only 3 percent of respondents to the questionnaire on world oil thought the real price of oil would fall in the next ten years, although half believed it would display great turbulence and instability.[38]

A Role and Goal for Analysis

OPEC was not pleased with price instability. Member countries wanted an orderly market and secure projections of revenue for planning. In an effort to stabilize the market, Saudi Arabia led an effort to set price ceilings. Others fought the move. Said one Venezuelan official,

OPEC was created to build a floor. We are not good at building roofs. But the OPEC countries live in warm climates. We don't need a roof. The industrialized countries live in cold climates. They're the ones that are going to have to build the roof.[39]

The ultimate roof would be the cost and availability of alternative energy, if the world oil market were efficiently coupled to other energy markets. Uncertainty reigned in the oil market. Oil companies were no longer balancing production with demand. Needing all the crude they could get for their own refineries, they effectively stopped selling to third parties. The closely articulated structure of the international oil industry to which the world had become accustomed was at an end.[40] The Iranian revolution was in that sense indeed a watershed in economic history.

It has been argued that oil company concessionaries in the Mideast had grown progressively more aware that their oil rights were in jeopardy during the fifties and sixties. They chose, by this line of reasoning, to accelerate production and lower prices to promote sales. National governments have a different system of values. Their incentive is not to moderate prices and increase production but to moderate production and increase prices. Recognizing this difference helps us understand the fundamental change that took place in the world oil market as producing countries assumed the reins.

Oil was the dominant (and sometimes only significant) source of revenue for most of these countries. It was the depletable resource account from which they drew to finance their imports, fund their social programs, provide education to their people, and develop their economies. The oil reserves of OPEC member countries dropped in twenty years from sixty-five years to thirty-nine years of output while that of the rest of the world remained relatively steady. OPEC countries wanted to be in a position to call on new revenue-generating industries as their oil income declined. The funds to develop these replacement industries had to come from oil.

OPEC countries had a spotty record in developing new industries. The waste and mismanagement in Iran was but one example. Capital projects in the Mideast cost two to three times as much as competing facilities elsewhere.[41] High levels of spending generated internal inflation, reducing the real value of oil revenues. OPEC nations were concerned about squandering their precious resource. They told the industrialized world, "Our oil must be lifted at *our* pace of development, not yours. Its exploitation must match *our* needs."[42]

Some called upon oil importing and exporting countries to work together in their mutual self-interest. Saudi Arabian businessman Suliman S. Olayan addressed his appeal to Americans:

> We, like you, want to see U. S. dependence on Arab oil imports reduced. We, like you, want to halt the disorderly spiral of oil prices. We, like you, want to restore order to the supply side of the market. We, like you, want to see the development of alternative energy sources. We, like you, suffer from the harm done the West's major economies by disorderly and insecure oil supplies. With so many common interests, we should be able to move beyond the totally unnecessary and self-defeating language of confrontation ... to deal with our common energy problems soberly and maturely, to the benefit of all.[43]

It was as though Olayan was asking for an end to scapegoating. Yet even he, in the selfsame article, could not resist the temptation to snipe at another familiar target: "the addiction to oil the U. S. has acquired," and that "sacred cow, the consumer," who "has an unquestioned right to unlimited consumption at subsidized prices." The urge to accuse runs deep. The fact is that it was natural for Americans to use oil liberally when it was cheap and plentiful, just as it was natural for OPEC countries to raise their price when the demand outran the amount they were able and willing to provide. These are issues of economics, not morality. They are subjects for analysis and reflection, not allegation and reproach.

OPEC and the oil companies headed the list of scapegoats for the energy crisis. The list was long, as we have seen. It included an oil expert, his superior in the State Department, the secretary of state to whom they reported, a beleaguered president, a vulnerable emperor, a leading oil-producing nation, optimistic economists, and prodigal consumers.

Readers of history will want an explanation of how the events of the 1970s came to pass. What were the causes? The search for scapegoats feeds on darkness and dissolves with understanding. Illumination from careful analysis improves the chances for constructive action. Shedding light on policy issues, evaluating alternatives, putting values in perspective, and conveying insights to decisionmakers and the public are roles analysis can play. In subsequent chapters, we examine the successes and failures of energy policy analysis during the seventies against these standards.

2 ENERGY POLITICS LOOKING BACK

Government actions—regulations, taxes, price controls, rationing, and the use of public lands—direct and bind the production and use of energy and have always done so. Looking back a half century, the stimulus to federal action (or studied inaction) came from a variety of attitudes about energy ranging from the benign neglect of the Eisenhower era to the dire concern of the 1973/74 embargo period. Policy studies, however haphazard and flawed, by helping to shape, reinforce, and modify these attitudes have influenced government action. In this chapter, we concentrate on the twenty-year period before the explosion of energy studies in the 1970s. Analysis of energy issues did not begin with the seventies. But we shall see that the nature of energy policy analysis and its reception did change in important ways at that time.

For most of the era between World War II and the oil embargo, energy studies tended to be opinion followers rather than leaders. Studies were generally designed to perfect understanding within the established world view of resource optimism. A perceived problem would lead to funding a particular approach. Only if the findings were congenial with the predisposition or sentiments of the audience were they likely to be taken seriously.

The optimistic view of energy held that while some sources would ultimately grow too expensive to use, others would come along to take their place. Hence "energy" overall was not considered a problem, although particular fuel industries might face difficult transitions. Rising oil imports, the plight of Appalachian coal miners, the prospective price and availability of uranium and natural gas—these were the issues that demanded policy decisions. With the exception of the comprehensive study undertaken at the time of the Korean War by the Paley Commission,[1] analyses were typically initiated to inform (or justify) specific actions rather than to address energy in any integrated sense.

The U.S. Government had no continuing group devoted to integrated long-term energy planning before the 1970s and no central body concerned with energy policy. Yet interrelationships did not go unrecognized. In 1939 the National Resources Committee concluded that "it is time now to take a larger view, to recognize more fully than has been possible in the past that each of these energy resources (coal, oil, natural gas, and water power) affects the others, and that the diversity of problems affecting them and their interlocking relationships requires the careful weighing of conflicting interests and points of view."[2] But work in energy and minerals was slow getting started and less successful than that directed toward the problems of land and water.[3] Attempts to pursue integrated analysis had the support of three successive secretaries of the interior, but not the Congress or the president.[4]

So it was that Resources for the Future, in its 1955 Annual Report, wrote,

> Energy commodities already are widely interchangeable, and new advances in technology are offering more and more opportunity for substituting one for another. Thus it is important to consider energy resources as a whole and to understand relationships among them. Long-range answers to the nation's problems of energy supply may rest largely upon taking full advantage of the flexibility in energy use. Yet there has been no one place for taking a comprehensive view of the whole energy field. In industry, a company or trade association is naturally concerned with its own set of energy commodities, and chiefly with problems of the present or the near future. In government, also, fragmentation is the rule. Policies for the different energy commodities are adapted piecemeal in response to pressures of the

day, and even facts and figures are collected and analyzed in separate compartments.[5]

Analysis responds to the tempo of the times. Without a constituency, analysts drift to other tasks. But with an interested audience and the funding it attracts, analytical skills quickly assemble. RAND defense analysts plan mass transit systems; scientists from the Jet Propulsion Laboratory and nuclear physicists from Oak Ridge investigate the prospects for renewable energy. The key to understanding the application of analysis to energy problems—and the change it underwent in the seventies—is to focus on the problems that commanded national priority and the questions to which policymakers wanted answers.

In this chapter we examine the nature of energy questions asked in the two decades prior to the embargo of 1973/74. Energy policy in those years related largely to four areas: breeder reactor development, natural gas price control, supply restrictions on domestic and imported oil (including questions of industry competitiveness and tax treatment), and problems of the coal industry. In reviewing these policy areas we identify the enduring economic and political interest groups that were affected. The concerns of these interest groups give a sense of the political environment within which energy policy analysis has been conducted. As our examination shows, the controversy surrounding recent energy policy has deep roots, which helps explain why it has not been easy to resolve.

THE POLITICS OF ENERGY

Energy issues have always been politically sensitive. Controversy regarding electricity, and the respective roles of government and the private sector in providing it, began with the Niagara Falls power project at the very start of the electric utility industry.[6] Operations of electric power companies have long excited suspicion—early on because of the financial manipulations of Samuel Insull and the electric utility trusts of the 1920s and more recently because of environmental impacts. Ida Tarbell sensitized a generation of readers to the predatory practices of John D. Rockefeller and the Standard Oil Trust (dissolved in 1911). The oil and gas industry has been subject to frequent scrutiny in books, congressional investigations, and court proceedings for many years.[7] It did not begin with the oil embargo.

What are the reasons behind government attention to energy industries? The first and most obvious is that energy services are critical to modern life; direct energy expenditures make up about 10 percent of personal budgets. Second, individual energy production units are large in relation to energy-consuming units. Energy firms are also large: as of the beginning of the 1980s, one-half of the top twenty firms in the Fortune 500 were oil companies. Third, special conditions surround energy acquisition and use: nuclear power presents special hazards; dependence on oil imports affects national security; coal burning degrades the environment.

Still, the depth of feeling about the energy industries, as expressed for example in numerous articles in the *Nation* over almost a century,[8] requires further explanation. Its intensity may have to do with the concentration of wealth the oil business has spawned; the overarching power of the electric and gas distribution companies (outside of the government, utilities are the most invasive faceless bureaucracies with which most people come in contact); the appearance of rapacity associated with oil wealth; the implicit belief that petroleum wealth is ill-gained because the returns from individual strikes often seem to far exceed the effort expended; and the attitude that the value of oil in the ground belongs to society as a whole.

There is a widespread view that government is manipulated by the energy industries. At the local level, utilities and other energy companies are often the largest, wealthiest, and most important enterprises. At the national level, representatives of energy firms appear to have easy entree to Congress and the executive branch because of their alleged wealth, political contributions, and common interests.[9] At the international level, the market outside the United States was dominated from the 1920s to the mid-1950s by the major international oil companies who divided concessionary areas and exercised power over the price of oil and the rate at which host governments were to be compensated.[10] With price control shifting to the oil-exporting countries during the seventies, conspiratorial suspicions assumed xenophobic overtones.

Diverse interests compete *within* the energy producing and distributing sectors, complicating any explanation in terms of simple producer-consumer or industry-citizen divisions on policy issues. Energy consumption varies widely by form and intensity across the United States. Regional interests are at stake. Consumers in the southwest, where distances are vast and the automobile vital, want gasoline prices held down even if it means extra charges being loaded onto heating oil. But New England consumers are directly affected by increases in the price of heating oil. They also suffer higher electric bills when prices for residual fuel oil rise. Until recently, neither group cared much about coal since its use and production were concentrated elsewhere.

Such regional differences add to numerous other political conflicts such as those among energy sources and fuel forms, among different stages of the same industry, and among companies whose operations are mostly domestic versus those that are heavily international. To illustrate, coal producers were harmed by imports of residual fuel oil and by the use of natural gas to generate electricity. They lobbied for policies to restrict access to both fuels. Natural gas distributors seeking to expand their markets wanted regulated gas prices in the field kept low—at least until shortages developed. Then they favored higher prices to spur the exploration necessary to give them more gas to sell. Still later they opposed the acceleration of field price decontrol, which they feared would reduce the competitive position of gas with respect to electricity and fuel oil. Control of oil imports "pitted independent (domestic) producers against the mighty integrated corporations with their far-flung producing wells in virtually every corner of the globe."[11] The independents lobbied for regulation of pipelines and production that would assure them access to markets through refineries owned by the major oil companies. Indeed, some of the most bitter rhetoric against "big oil" has come from within the energy industry.

In another context, while oil quotas were in effect, domestic oil producers were not very concerned about imports of residual fuel oil from abroad. In fact, they favored decontrol of residual oil as a relatively painless (to them) sop to the northeastern consuming interests. Small refiners called for import control provisions that would subsidize them—at the expense of larger firms—and formed an association to combat positions taken by the organization of large operators.

Petroleum industry spokesmen sometimes cite the number of stockholders and employees in the industry (down to the service-station attendant and natural-gas distributor) to suggest the potential for political influence of an aroused industry. But there have been no issues on which all got aroused—at least on the same side. Political power based on the size of the petroleum industry has remained largely immobilized for the good reason that interests run off in all directions.

Yet this has not meant political impotence. Energy-producing industries are concentrated by regions, and subsectors of the industries are even more localized. Thus, legislators could take positions on energy matters in concert with the interests of important constituents without offending others on whom their political survival depended. Senators Randolph and Byrd of West Virginia could safely fight against residual fuel oil imports; Douglas of Illinois could rail against those who sought natural gas price decontrol; Kerr of Oklahoma could favor oil import restrictions; and Baker of Tennessee, home of the Clinch River project, could support breeder reactors. The executive branch, in contrast, has constantly been torn by competing pressures within the nation as a whole, and has had to deal with conflicting views in the legislature. Energy issues have been vehicles to power for members of Congress but political headaches for presidents since the time of Truman.

Given this picture of divided interests within and among the energy industries, why is it that some special interests do better than others? First, members of Congress with vitally involved constituents can be very influential on such issues as aid for small refineries, gas deregulation, and import control. They find it worth their while to use all the power of their offices in support of policies important to their states and districts.

Second, constituents with narrow interests may place a disproportionately large amount of pressure on a pending decision. An independent refinery owner will be willing and able to use more political capital on a refinery-versus-producer issue than will the producing division of an integrated oil firm. Indeed, the amount of active political support or opposition that can be marshalled may be *inversely* related to the size of the firm. Below a certain hierarchical level in the firm, employees are not likely to be politically identified with a policy favored by the company; nor are stockholders, unless their overall well-being is tied to that one firm. On the other hand, executives of smaller companies who have a direct stake in a policy outcome can act decisively. Since independent refineries are widely dispersed, their owners can influence many elected officials. So can oil jobbers, for the same reason. Executives of major oil companies may get invited to testify before Congress, but local gasoline dealers can deliver votes. They get heard in Washington too.

The operation of the mandatory oil import program of 1959 to 1973 provides a concrete example of how different economic interest groups fared in affecting policy:

> A roll call of the special interests in energy policy would find most of them the recipient of at least some favored treatment: small refiners, inland refiners, Northern Tier refiners, major oil companies, oil producers, petrochemical companies, Northeastern utilities and other identifiable and isolatable consuming interests, deep water terminal operators, island interests, West Coast consumers ... The major large group which *paid* for the benefits of the greater energy security achieved under oil import controls was the undifferentiated portion of the consuming public. With one possible exception, other identifiable functional groups were either unaffected, or actually made better off. The "possible exception" is the international operations of major oil companies and, perhaps, those oil companies themselves ... Virtually every controversy was resolved against the best interests of the original major company importers, a fact with important implications when the political economy of oil is examined. The political power of oil may be great, but based on the record of the mandatory quota program, this power is not found in the international giants of the industry.[12]

An important force in the politics of energy has been the perceived effect on the distribution of income. Oil and gas producers are considered rich, energy consumers poor. Measures favoring producers (whether or not affecting consumers directly) have been widely viewed as making income distribution more unequal, though the facts are uncertain.[13] The emphasis on "equity" rather than "efficiency" in energy decisions was clear in the defense of price controls on natural gas. Ironically, it was a concern for equity that ultimately led to the relaxation of these controls when it was recognized that they resulted in some persons getting all the gas they wanted at low prices while others had to do without. It then became politically possible for members of Congress from consuming states to support decontrol, and a bill passed. Also arousing egalitarian wrath were tax policies such as the percentage depletion allowance that favored the petroleum industry.

Finally, a streak of populism—the favoring of small over large economic units—has permeated public discussion of energy issues and affected policy outcomes. When the percentage depletion allowance was modified in the Tax Reduction Act of 1975, integrated petroleum firms lost the use of this tax reduction device, while independents and royalty owners continued to benefit, albeit at a scaled-down level.[14] Refineries classified as small (even though they were large firms by the standards of other industries), were given extra allocations of cheap foreign oil under the Mandatory Oil Import Program (MOIP)[15] and the subsequent crude oil entitlements program.[16]

The political sensitivity of energy is reflected in government's continuing involvement, traced out in a series of legislative, regulatory, and judicial actions, as outlined in Table 2-1.

Conflicting political and economic interests have spawned energy studies in great number. In briefly reviewing these studies we will try to discern the motivations and positions of the parties involved while paying special attention to the questions asked and how the nature of these questions and their analysis changed over time.

LEGACY OF DEPRESSION AND WAR

Social planning for energy received major impetus from both the Great Depression and World War II. During the Depression, the Tennessee Valley Authority (TVA), the Bonneville Power Administration (BPA), and the Columbia, Missouri, Colorado, and Central California Valley river basin projects brought energy planning together with regional development. The goals of the massive federal water power developments under TVA and BPA included irrigation, land reclamation, and industrialization, as well as electrification. In David E. Lilienthal's graphic vision, TVA was "Democracy on the March."[17] The Depression produced many trained, experienced professionals eager to devote their analytical talents and new cost/benefit techniques to a range of national issues, one of which was resource planning.

Table 2–1.   Major Energy Legislation, Regulations, and Judicial Decisions.

1920   *Mineral Leasing Act* (P.L. 66-146, 41 STAT. 437) Established the modern policies of issuing prospecting permits and leases for development of coal, oil, and gas resources of public lands.

1935   *Federal Power Act* (P.L. 49-333, 49 STAT. 847) Granted authority to the FPC to regulate interstate transmission of electricity and to regulate wholesale sales in interstate commerce.

1938   *Natural Gas Act* (P.L. 75-688, 52 STAT. 821) Granted authority to the FPC to regulate the interstate sale of natural gas.

1946   *Atomic Energy Act* (P.L. 79-585, 60 STAT. 755) Set up a civilian Atomic Energy Commission with authority over nuclear R&D. Amended 1954.

1954   *Phillips Petroleum Co. v. Wisconsin* (347 U.S. 672) Ordered the FPC to regulate the price of natural gas sold in interstate commerce by independent (nonpipeline) producers.

1955   *Trade Agreement Extension Act* (P.L. 69-86, 69 STAT. 162) Granted authority to the president to establish import quotas and authorized the Office of Defense Mobilization to advise him whenever imports may impair national security.

1957   *Price-Anderson Act* (P.L. 85-256, 71 STAT. 576) Set up a combined federal/private insurance program to cover nuclear power plant accidents and limited the liability of industry.

1959   *Adjusting Imports of Petroleum and Petroleum Products* (Presidential Proclamation #3279) Established the Mandatory Oil Import Program and the Oil Import Appeals Board.

1963   *Clean Air Act* (P.L. 88-206, 77 STAT. 392) Established programs for the prevention and abatement of air pollution.

1969   *Federal Coal Mine Health and Safety Act* (P.L. 91-173, 83 STAT. 742) Provided federal authority to promulgate and enforce health and safety regulations in coal mines.

1970   *National Environmental Policy Act* (P.L. 91-190, 83 STAT. 852) Established the Council on Environmental Quality and introduced the requirement for Environmental Impact Statements.

1970   *Clean Air Act Amendments* (P.L. 91-604, 84 STAT. 1676) Amended the Clean Air Act of 1963 by establishing ambient air quality and emissions standards and provided for research on the effects of air pollutants on public health and welfare. Also amended in 1977 (P.L. 95-95, 91 STAT. 685).

1971   *Calvert Cliffs v. AEC* (449 F.2d 1109 DC Cir.) Decided the AEC must examine environmental impacts of nuclear power facilities.

1973   *Trans-Alaska Pipeline Authority Act* (P.L. 93-153, 87 STAT. 584) Facilitated Alaska pipeline construction by barring court challenges and relaxing limitations on pipeline rights-of-way.

Table 2–1.   Continued.

1973   *Emergency Petroleum Allocation Act* (P.L. 93-159, 87 STAT. 627) Authorized the president to establish a comprehensive allocation program for
oil and oil products and to set prices for crude oil and refined petroleum
products.

1974   *Energy Reorganization Act* (P.L. 93-438, 88 STAT. 1233) Created ERDA to
assume nonregulatory functions of the AEC and other energy research
programs in government and set up the NRC to perform nuclear regulatory functions.

1975   *Energy Policy and Conservation Act* (P.L. 94-163, 89 STAT. 871) Established standby authority for an energy emergency, created a strategic petroleum reserve, mandated fuel efficiency standards for automobiles, and
continued oil price controls.

1977   *Surface Mining Control and Reclamation Act* (P.L.95-87, 91 STAT. 445) Set
environmental standards for surface mining coal.

1978   *Public Utility Regulatory Policies Act* (P.L.95-617, 92 STAT. 3117) Encouraged conservation of energy supplied by electric utilities, efficient
use of their facilities, and rate structures reflecting the actual marginal
cost of providing service.

1978   *Energy Tax Act* (P.L. 98-618, 92 STAT. 3174) Allowed tax credits for install-
ing solar, geothermal, and energy-saving equipment and imposed tax on
gas-guzzling cars.

1978   *National Energy Conservation Policy Act* (P.L. 95-619, 92 STAT. 3206) Required utilities to promote energy conservation, mandated efficiency
standards, and authorized conservation grants.

1978   *Powerplant and Industrial Fuel Use Act* (P.L.95-620, 92 STAT. 3289) Pro-
hibited new utility plants from burning oil or natural gas and existing
plants from using those fuels after 1990.

1978   *Natural Gas Policy Act* (P.L. 95-621, 92 STAT. 335) Established prices for
various classes of natural gas, set a schedule for decontrol of some gas
prices by 1985, and gave the president allocation authority in an emergency.

1980   *Windfall Profit Tax Act* (P.L. 96-223, 94 STAT. 229) Provided for an excise
tax on decontrolled domestic crude oil.

1980   *Energy Security Act* (P.L. 96-294, 94 STAT. 932) Created a quasi-independ-
ent Synthetic Fuels Corporation to provide loans, loan guarantees,
purchase and price agreements, and joint ventures to stimulate production of synfuels.

World War II demonstrated the critical role of liquid fuels in
war as well as peace. With the end of the war, the worry was
whether the depleted supply of fuels was sufficient to sustain
prosperity.[18] Was this to be the only generation upon which fate

had smiled? The eventual depletion of liquid and gaseous fuels, or (as the optimists put it) the necessary long-term energy transition, was an issue calling for integrated analysis of energy.[19]

The critical policy questions during the Truman Administration were whether to involve government, and if so, how? Some wanted government to serve as protector against rapacious business; others feared its becoming a diabolical meddler in private affairs. Most attention was directed to specific fuels. Yet a concept of national energy planning also began to emerge. Some believed that national security demanded resource independence; others were convinced that resource planning for energy was now an international concern. A coherent view was needed of prospective energy supplies and requirements.

The Interior Department was conducting fuel studies in response to temporary shortages and a worry that the Marshall Plan would threaten U.S. growth by draining energy supplies. These studies, not part of a systematic or sustained effort to provide a foundation for long-term decisions, suffered from association with particular policy controversies. A program staff formed in 1948 to improve the department's analytical capability initiated a study to appraise the adequacy of the department's programs in the energy field.[20] Harold J. Barnett led the study, applying novel analytical approaches to explore the implications of energy trends then on the horizon.[21] The influential methodology and conclusions of the study went with Barnett to the Paley Commission staff and later appeared in his seminal work on resource availability.[22] Barnett's report was referred to as the basis for most of the Interior Department's public statements about energy policy at the time.[23]

President Truman had come to believe that a comprehensive examination of a broad array of matters relating to materials and resources was essential. In an act with far-reaching consequences, he appointed a distinguished commission headed by William S. Paley, chairman of the Columbia Broadcasting System (CBS), to examine "the broader and longer range aspects of the nation's materials problem."[24] Federally funded, unprecedented in scope, the Paley Commission was authorized to call upon all government agencies as well as industry and academic experts. Results of the Commission's study were widely disseminated and influential, partly because of the skillful writing and effective graphics of public information specialists assigned to the project, including talent Paley could call upon from CBS.

The study, exercising broad vision, rose above facts and forecasts to reflect upon the processes by which materials are transformed into goods and services. Examining the policies and institutions required to ensure materials availability in the face of technological change and shifts in output mix, its analysis gave context to the disagreements and identified areas where more work was needed.

The Paley Commission advanced the state of resource analysis. Building upon the pioneering efforts of other groups, the Commission promoted the idea of a unified and evolving energy system.

> What is important above all else is that the energy supply must be viewed as a whole and the interrelations of its various sources and forms kept in the foreground in the formulation of policy. The central task of energy policy is not simply to solve the problems and seize the opportunities relating to each energy source; it is to fashion from all of them in combination a flexible pattern of energy supply which will grow to the expanding demand of the United States and other free nations.[25]

The Commission's study developed an explicit framework for tracing out price effects and projecting energy balances. In the process, it achieved reasonable estimates for the 1970s.

The effort gave rise to a new research institution, Resources for the Future (RFF), funded by the Ford Foundation. The intent was for RFF to pursue broad-gauged resource studies investigating materials, environment, and energy problems, and contribute to building a data base for analyzing these issues.

The Paley Commission, generally optimistic on the outlook for energy, considered oil imports, coal, oil shale, solar power, and nuclear energy to be potential alternatives to which the United States could turn as the need arose. It concluded that while conventional fuels would be adequate to 1975 and beyond, the nation should long before then begin its transition toward nonconventional sources.[26] The Commission expected costs to rise but not by enough to choke economic growth. Energy should be obtained from the lowest cost source, domestic or imported, with reliance upon private market forces for the required transitions. Government's role would be to monitor prospects far enough ahead to take action should supply problems develop or national security be threatened. The Commission did not explicitly anticipate oil interruptions of the sort that erupted in the 1970s but did consider military interdiction of supplies and explosive increases in demand due to war. It downplayed conventional oil stockpiles on the basis of cost but urged policies to foster standby capacity.[27]

The Commission's optimistic findings on long-term energy supply were widely accepted without adequate notice being taken of the preconditions laid down for achieving these outcomes. Rather than becoming less specialized, the economy became more resistant to fuel switching, and a smooth transition to nonconventional fuels did not occur. One reason was the unanticipated fall in energy prices during the 1950s and 1960s. The Commission had expected stable prices, or at most a modest increase if "opportunities to strengthen the United States energy position are vigorously pursued."[28] Instead, the composite fossil fuel price fell 29 percent from 1953 to 1969 and rose only slightly from then until 1974. The real price of fossil fuels even in 1973 was 21 percent below the level two decades before.[29] There was little incentive to get started on the production of synthetic fuels.

The Commission expected major synthetic fuels production by 1975.[30] Observing that "for years some people have been predicting the nation's crude oil supply was going to be exhausted within ten to twenty years," it left open the question of whether the 1975 level of conventional oil production would be lower or higher than that of 1950.[31] The Commission's sanguine view of energy availability came not from exaggerated estimates of oil and gas resources (its estimates were consistently too low) but from an underlying faith in technological progress. Oil and natural gas were short-term expedients to be used up as the world prepared for coal and new energy sources. Truman's Point Four Program, after all, called for U.S. technology to unlock prosperity in the Third World, and Eisenhower's Atoms for Peace initiative suggested that shared nuclear power would guarantee energy supplies for all.[32]

It was easy to forget the energy worries of the 1940s. From 1950 to 1960, estimated proven crude oil reserves in the Mideast soared by 460 percent (from 32.4 to 181.4 billion barrels), production expanded by 200 percent, and U.S. imports from the region grew by 172 percent.[33] Domestic oil reserves increased by 25 percent and output by about 30 percent. Although natural gas consumption doubled, gas reserves were being added at such a rapid pace (going from 185 to 262 trillion cubic feet between 1950 and 1960), that the price of gas rose only 68 percent from give-away levels at the wellhead. The real price of domestic oil fell 10 percent over the decade and electricity costs dropped by over one-quarter.[34] Further helping to reduce consumer costs was the shift to less expensive, more convenient oil and natural gas and to new and more efficient delivery systems and fuel-using equipment. The coal industry was beset by excess productive capacity, and low-cost nuclear power was thought to be just beyond the horizon.

This image of plenty allayed fears of exhausting energy sources. There was no government institution with the political status and technical competence to maintain the weather eye urged by the Paley Commission. Policymakers were unconcerned about nonproblems, and energy studies tended to be optimistic about supply.

In 1960 a study team at RFF under the direction of Sam H. Schurr (Bureau of Mines's first chief economist) examined data on energy production and use to make forecasts to 1975. The study concluded that at constant prices and costs, estimated demand could be covered by available supplies. "Fears about the adequacy of domestic energy resources should be relegated to a secondary position in public discussion and eventual policy decision ... The United States is *not* compelled to adopt particular policies because of impending resource exhaustion in any of the mineral fuels, or the threat of steeply rising costs."[35]

Somewhat later, Warren E. Morrison at the Department of the Interior developed an integrated set of energy balances with energy content (BTUs) as the fundamental unit of analysis.[36] In subsequent work, Morrison made projections to 1980 and 2000 using a scenario and formal modeling approach.[37] This research improved insights and was well known among the select band of energy analysts of that period. Yet, such efforts tended to be more reassuring than galvanizing in their policy effect. They did not prepare the nation for the 1970s.

Several haunting questions occur. Would it have made any difference had there been Paley-type commissions (with some claim on public attention and skill at analyzing data as it evolved) reporting on energy each five years from 1952 on? Given the information available, could better analysis have reached different conclusions? When, if ever, could warning of what was to occur in the 1970s have been issued? Would anyone have listened? Would a message have gotten through to policymakers and the general public? Most importantly, would it have mattered? Could any additional actions in 1957, 1962, or 1967 have reduced the shocks of 1973/74 and those that followed? If these shocks could have been lessened, would the delayed benefits have outweighed the substantial costs?[38]

There are no easy answers. The kinds of shocks that hit the energy system in the late 1960s and early 1970s were not easy to anticipate. How could one predict the full force of environmental restrictions, the rise of OPEC, the sharp increase in oil prices, and the dramatic slowdown in nuclear power development? With benefit of hindsight, it seems one might have foreseen some of the supply constraints that emerged in individual fuel sectors. Domestic oil and gas discovery per foot drilled started falling in the 1940s, and some very astute forecasts were in fact made.[39] But not much note was taken of them. The signals—garbled by oil production limits, import controls, and gas price regulations— were hard to interpret. It was difficult to spot the trend until it was well along.

Different analysis might have suggested that the conditions lowering energy costs could not last. It should have been clear that the astounding 7 to 8 percent rates of growth in energy use would soon swamp any supply system. A regular process that brought supply and demand trends together in a common framework might conceivably have pointed up the fragility of the energy system and given notice of its vulnerability to disruptions. But despite the studies that were made and the concern expressed, nothing happened. The *energy* problem retreated from public view, not to reappear until the 1970s. Instead, *fuel* issues occupied center stage politically and analytically.

ISSUES AND ANALYSIS PRE-EMBARGO

Dominating the energy policy picture in the two decades before the Arab oil embargo were nuclear power (particularly the breeder reactor), price controls on natural gas, and price supports for domestic oil, including restrictions on oil imports. Also of concern was the plight of the coal industry and Appalachia. All of these issues involved large amounts of money, strong emotions, conflicting regional interests, and powerful political antagonists. Yet an examination of the debates and analyses of the time reveals that with the exception of coal, the issues were discussed mostly in isolation from one another.[40] Considering this period will provide background for our discussion of the energy studies of the seventies in the following chapters.

Nuclear Policy

Public discussion of the possible use of nuclear power to generate electricity began almost immediately after Hiroshima. The first power-generating reactor in use in the United States was of the light-water reactor (LWR) design. It is asserted that this technology—derived from the submarine program—was chosen not because it was innately superior to alternative reactor concepts, but because it was relatively proven and available.[41] Despite a slow start, U.S. reactors using the LWR design had captured virtually all of the world market by the mid-1960s, and nuclear power was firmly established as a potential source of electricity to be owned and operated by private utilities.

Progress with reactors came only after resolution of three major controversies.[42] First, on the question of secrecy, the risks of helping the Soviet nuclear weapons program were deemed less serious than missing the potential benefits of nuclear power. Second, the New Deal ideology that favored public development of nuclear technology (and opposed turning taxpayer-financed achievements over to private firms) was rejected; private firms were allowed to enter the nuclear field on a for-profit basis. Finally, government promotion of nuclear power was accepted. Pushing the technology faster than dictated by market forces was justified to maintain the momentum of the program, benefit other countries, and build an export market for U.S. reactors. Thus, nuclear technology was released to the private sector under progressively relaxed secrecy restrictions. Federal funds went to speed development of advanced reactors while private firms assumed major responsibility for developing LWRs, the reactors they hoped to sell.

The LWR, Uranium Availability, and the Breeder.   Central to the nuclear power debate was the question of the uranium resource base and the breeder reactor, a technology that can extend that base.[43]

The fuel consumed in a LWR must contain approximately 3.5 percent of a special isotope of uranium, U-235, a fissile material that constitutes about 0.7 percent of natural uranium (the rest is nonfissile U-238). Natural uranium must be "enriched"—the U-235 isotope concentrated—for use as fuel in LWRs. The uranium "tails" left over from this process, mostly U-238 but including about 0.2 percent U-235, has no use in conventional LWR power production.

High-grade, low-cost deposits of uranium are known in relatively few places and are limited in size so that prospective uranium costs rise as more uranium is consumed. Most countries do not have effective access to uranium within their borders. The cost of enrichment is high, partly because the commonly used process of gaseous diffusion consumes enormous quantities of energy.[44] These factors make a technology that could use all the uranium (the U-238 as well as the U-235) very attractive. The breeder reactor is such a technology.

A portion of the U-238 within an LWR core is transmuted into plutonium P-239 when exposed to high levels of radioactivity. P-239, like U-235, is fissile and can be used to fuel reactors. The conversion ratio measures the amount of P-239 formed from U-238. The ratio, relatively low in LWRs, can be increased in breeder reactors to the point where more fissile material is formed from U-238 than is consumed. The P-239 can then be separated from the rest of the fuel in a procedure known as "reprocessing" to fuel other reactors and to replace the spent U-235. Gaining early favor was the liquid-metal fast-breeder reactor (LMFBR), a design that uses liquid sodium (instead of water) as a coolant and medium for heat transfer.

By turning otherwise useless uranium tails into fuel, and by using nearly all newly mined uranium to produce power, breeders effectively remove the specter of resource exhaustion and rising fuel prices. Breeders can also free nations without indigenous uranium supplies from perpetual dependence on foreign sources of fuel. Once the reactors are initially fueled, the much smaller amounts of uranium required in operation can be stockpiled at reasonable cost.

But breeders are more expensive to build than LWRs and the technology is more demanding. Moreover, breeders require widespread handling, transportation, and possession of plutonium, a highly toxic material that can be used for nuclear weapons. The danger is that weapons-grade nuclear materials could become available to any nation with a power program, and could be diverted to terrorist groups. Such issues, of central concern in the nuclear policy analyses of the 1970s, were not prominent in discussions of breeder development during the 1950s and 1960s. The issues considered important at that time were:

1. The difference in capital costs between breeders and LWRs;
2. The trajectory of differences in their fuel costs as determined by the cost of raw uranium, relative costs of enrichment and reprocessing, the demand for electricity, and the rate of growth in the use of LWRs;
3. The costs of breeder research and development, and the prospects for making electricity cheaper, while assuring safety;
4. The hazards and environmental risks of the breeder compared to alternative future sources of electricity, including the LWR;
5. The expected cost of energy from breeders compared to non-nuclear alternatives;
6. The discount rate appropriate in all these comparisons for reconciling current and future expenditures with future benefits.

Policy Analysis of the Breeder.   The breeder reactor had strong support.[45] Nuclear proponents were pleased that a frontier enterprise was underway. Even ideological opponents of government involvement in the private sector recognized potential social benefits of breeder technology unlikely to be produced by the normal workings of the market. The breeder provided a mission for otherwise underutilized national laboratories and intriguing scientific problems for cadres of trained professionals in government and academe. Breeder research offered political rewards for members of Congress whose districts contained facilities and contractors and challenging assignments for administrators in whose domains the work was accomplished.

The breeder was a natural way to bypass the fuel constraint, but the research and development program it required was expensive. From early on, questions were raised about the safety, environmental impact, and economic justification of the breeder.[46] Numerous analyses were launched to evaluate different levels of effort, examine alternative concepts, and sketch out a supply strategy into the twenty-first century. We mention two such analyses here.

The Atomic Energy Commission (AEC) conducted a review of the breeder program for the Joint Committee on Atomic Energy in 1966. The AEC study examined alternative nuclear power strategies on their own terms; that is, without considering other possibilities for supplying energy or reducing consumption. Benefits factored into the AEC calculation included "low-cost electrical energy, reductions in uranium ore requirements and in separative work demand, increases in plutonium production and use of uranium tailings." Costs were those "expected to be incurred by the government in the development of the breeder."[47] The AEC found a benefit-to-cost ratio significantly greater than 1 for cases examined where a discount rate of no more than 7 percent was assumed. It argued in favor of continued commitment to the LMFBR program.

Thomas B. Cochran challenged the AEC analysis and conclusions in a 1974 study in which he recalculated the benefit-to-cost figures under different assumptions and critiqued the health and safety implications of the breeder program. His results suggested much lower projected LMFBR benefits, especially when the standard government discount rate of 10 percent was applied. Cochran called the AEC's uranium supply projections too pessimistic, and he faulted the AEC study for its failure to investigate alternative means for generating electricity—including other reactor designs. "The revised data imply that commercial-size LMFBRs will not be competitive for several decades beyond the AEC target commercial entry date of 1986," wrote Cochran. Serious safety problems make it prudent "to delay commercial introduction of LMFBRs until there is a strong overriding economic justification for the program."[48]

The nuclear analyses of the 1960s and early 1970s did not notably affect the course of the nuclear enterprise. They tended to ignore or treat as technically solvable, the issues of waste disposal, reactor safety, and weapons proliferation. In their technoeconomic approach, the analyses tended to neglect the emotional connotation of nuclear weapons and the organizational, institutional, and moral problems the technology posed. Alvin Weinberg, former director of the Oak Ridge National Laboratory, put it thus:

> We nuclear people have made a Faustian bargain with society. On the one hand, we offer—in the catalytic nuclear burner—an inexhaustible source of energy. Even in the short range, when we use ordinary reactors, we offer energy that is cheaper than energy from fossil fuel. Moreover, this source of energy, when properly handled, is almost nonpolluting ... But the price that we demand of society for this magical energy source is both a vigilance and a longevity of our social institutions that we are quite unaccustomed to ... We have established a military priesthood which guards against inadvertent use of nuclear weapons, which maintains what a priori seems to be precarious balance between readiness to go to war and vigilance against human errors that would precipitate war ... Peaceful nuclear energy probably will make demands of the same sort on our society, and possibly of even longer duration ... Is mankind prepared to exert the eternal vigilance needed to ensure proper and safe operation of its nuclear energy system? This admittedly is a significant commitment that we ask of society.[49]

To Weinberg, it was "well worth the price." To others, it was not. The difference in views relates largely to what Weinberg called "transscientific" elements. It was values much more than facts that were at the heart of the debate over the breeder.

Natural Gas Price Control

Natural gas became important after World War II when development of large-diameter, high-pressure pipelines made possible its cheap transportation from the southwestern producing regions to major consuming centers. Between 1950 and 1960, natural gas consumption in the residential and commercial sectors rose about 160 percent, compared to an 80 percent increase in oil consumption and a fall of almost two-thirds in the consumption of coal. The proportion of homes heated by natural gas rose from one-fourth to over two-fifths, and the use of gas for electricity generation almost tripled.[50]

In an era of cheap energy, natural gas was cheaper still, and cleaner, more convenient, and more efficient as well. But until it could be economically transported to eastern and northern markets, it was only a byproduct of oil exploration and production. As demand grew, the price of gas started edging up, and by 1960 gas began to be sought for its own sake. This transition from a local to a national source of energy was an important development. Coal producers were dismayed at their loss of markets. Consumers wanted to know about prospects for long-term supplies at reasonable prices to justify the capital investment required to convert to gas-using equipment.

The Coming of Regulation.   Gas distributors and pipelines were regulated as public utilities. Was gas production to be regulated too? This question was resolved in 1954 in the *Phillips* decision wherein regulation was imposed on the field price of most natural gas flowing in interstate commerce by a Supreme Court interpretation of the Natural Gas Act of 1938.[51]

The *Phillips* case began in 1948 with a request by consumer representatives to limit prices charged at the wellhead by Phillips Petroleum Company for gas delivered across state lines. The consumer's concern was that if excessive charges were levied at the field level, regulation at subsequent stages could not protect them. As a matter of law, however, the primary question was what the statute meant. Whether protection was necessary because (for example) of monopoly power in the field markets was for Congress to decide.

The Federal Power Commission (FPC) held that the Natural Gas Act did not apply.[52] After a successful appeal by consumer representatives,[53] the Supreme Court agreed to hear the case and decided that sales at the wellhead by independent producers after resale in interstate commerce were indeed covered under the Natural Gas Act.[54] Gas sold in intrastate markets, and direct sales to ultimate consumers interstate, were left unaffected.

Thus was a resistant FPC thrust into the role of regulating the price of a depletable commodity—a task far different from traditional public utility regulation of wholesale electric power and interstate pipeline sales. The FPC had inadequate analytical capacity to determine how to operate in this new context and what price to set. While the responsibility for setting regulatory policy rested ultimately with Congress, "both the House and the Senate repeatedly showed themselves to be reluctant to take a position on this highly sensitive political issue, which affected millions of voters ... Congress left the problem of policy formation to the FPC."[55]

The *Phillips* decision and the regulation it mandated inspired a spurt of analyses that continued through the 1970s.[56] The central issues were:

1. Is there monopoly in the field markets for natural gas that justifies its regulation?
2. To what extent does gas supply respond to price?
3. Can regulatory techniques restrict the revenues of natural gas producers without seriously affecting gas supply and distorting its allocation to most productive uses?

Sophisticated econometric and other analyses of the first two questions[57] indicated that competition did exist in field markets and supply response to price was significant—findings referred to in formal FPC proceedings along with the FPC staff's own econometric testimony.[58] These were among the first times such analyses were used in a legal setting subject to cross-examination. Analysts not only had to do good work, they had to try to communicate to lay persons. Their efforts to do so were of only limited success.

Many of the studies performed during the 1960s were aimed at the regulatory process: how to set a price, what price to allow, and the effect different prices would have on different classes of gas. The FPC initially sought to regulate independent producers using the "public utility" concept or "cost of service" principle which allows the firm a reasonable rate of return on its historical investment or rate base. Charges to consumers are designed to cover both capital costs and out-of-pocket expenditures, with a rate schedule that specifies which consumers pay how much for each service rendered or unit of product delivered.

The effort to use this public utility approach was stymied because of the difficulty of ascertaining the "cost of service." It was also impossible to demonstrate a dependable relationship (on a company-by-company, investment-by-investment basis) between cost of service and output. The approach simply was not applicable to a risky industry whose resource was depleting. "Producers of natural gas cannot, by any stretch of the imagination, be properly classified as traditional public utilities."[59] In addition, administrative problems were nightmarish. Extensive studies would be required for each of the 5,600 independent producers. There were 11,000 rates and 33,000 supplements to these rates on file in 1960, with more coming in every day.[60] All would have to be perused. The FPC's solution was to abandon the individual company-based public utility approach in favor of an "area rate" concept which sought a reasonable rate for all the gas produced within an area whatever the cost incurred by an individual producing company.[61] Thus the regulatory focus was changed from company to commodity. With the area rate concept, the work of accountants designed to illuminate what a company had done became less important. The role of policy analysts, and especially economists concerned with the industry's aggregate price response, grew.

Regulation on the Defensive.   Gas producers and their allies consistently opposed regulation and sought to have it removed. They had little success, however, until a prospective gas shortage developed in the late 1960s. Positions changed and support for deregulation and higher prices started to come from gas transmission and the distribution companies (who wanted to maintain gas supplies) and some industrial consumers who feared (correctly, as it turned out) that political decisions would direct gas to small users and deny their firms the right to bid for it. Analyses were launched to estimate the price gas would reach if deregulated, the cost to consumers, and the effect of higher prices on gas production and consumption.[62] A core question was how to lessen efficiency losses while still meeting the most pressing income distribution goals.[63]

Whereas estimated supply and demand responses were emphasized in the formal debate, votes in Congress seemed to rest more on ideological factors and economic interests.[64] Thus the tide of political opinion started to change with the shifting regional and sectoral interests that arose with shortages in the regulated interstate market. This began around the turn of the decade.

Restrictions on interstate gas consumers with noninterruptible contracts were initiated in 1968 as Transcontinential Pipeline became the first company unable to supply enough gas to meet the needs of its distribution company's customers. (Intrastate consumers could get all the gas they wanted by bidding slightly higher than the regulated price; they paid a little more than the interstate consumers but faced no shortages.) By 1973, curtailments were widespread; 35 percent of all residential customers were served by utilities that restricted access to gas in some fashion. Near 0 in 1970/71, curtailments grew to almost 1.5 trillion cubic feet by 1975/76, about 7.5 percent of annual consumption.[65]

With gas shortages focused on interstate consumers, nonproducing states became more likely to support decontrol. Their citizens would then at least be able to get gas, and it would still be a bargain compared to other fuels. Producing states, on the other hand, became less supportive of decontrol. They now had to balance producer benefits against losses to consumers who would face higher prices as out-of-state customers sought to bid gas away.

The oil supply cutbacks of 1973/74, the sharp increase in oil prices, and the desire to reduce oil imports, added to the arguments for gas decontrol.[66] The growing recognition that certain consumers were getting all the gas they wanted at low prices while others had to either do without or substitute more expensive energy sources made it possible for members of Congress previously opposed to deregulation to support it. Deregulation would later fail in close votes in both the 93rd and 94th Congresses, only to reappear in the tortured Natural Gas Policy Act (NGPA) passed in 1978. The NGPA contained a crazy-quilt mixture of widened regulation (intrastate gas), deregulation of deep gas, an eventual free market, and immediate higher prices.[67] The act worked better than many expected, but did not end the conflict.[68]

The thirty-year controversy over natural gas regulation shows how difficult it is for analytical results to affect political decisions if they clash with strong feelings and dominant regional and sectoral interests. When economic analyses demonstrated the politically counterintuitive result that what was good for gas producers could also be good for (many) energy consumers,[69]

politicians with consumer constituencies were able to support decontrol measures. Appeals to parochial consumer interests succeeded where earlier efficiency arguments had failed.

Oil Price Policy

From the mid-1930s until about 1972, government effectively
raised the price of domestically produced crude oil above the
short-run free-market level. Then, in an about-face, it held the
price below market by imposing price restraints. We discuss this
turn of events later. First, we examine the series of actions by
which government elevated oil prices and, by means of tax provisions, simultaneously raised net revenues, domestic output, and
consumption.

How could this program have been sold to the electorate? First,
analysis of the economic impacts of the price-raising actions was
relatively scanty for some time. Second, emphasis on conservation
and national security probably kept the price effects of these actions from public consciousness. Third, the level of prices seemed
so low until late in the period that small increases (or reductions
in decreases) were tolerable and appeared fair, especially considering that the overall price trend was not upward but downward in
real terms. Finally, proindustry regional interests were deeply entrenched in Congress and had substantial political power. Well-
placed legislators were able to sustain programs that, whatever
their merits, were against the short-run economic interests of the
majority of the electorate.

The specific effects of the various price-raising instruments now
to be discussed were often in dispute. Yet, from an analytical perspective, it is surprising how little effort was made to estimate
their impact on income distribution by region, income class, energy consumption, economic growth, and competing energy industries. Until the 1969 report of the Cabinet Task Force on Oil
Import Control,[70] these questions were submerged by concerns
about physical waste and fears of shortages threatening national
security.[71]

Production Controls and Import Restrictions.   Under U.S. law,
the right to produce petroleum belongs to the owner of the sur-

face of the land (or his designee). But as with the "rule of capture" applied to wild animals (from the legal treatment of which oil and gas law arose), the petroleum itself belongs to the person who can bring it out of the ground (capture it).[72] Thus if a reservoir exists beneath land owned by more than one person, each will have an incentive to produce as much as possible as fast as possible so as to get a larger share of the total potential output. Producers act as if oil in the ground were effectively worthless to them (someone else might get it first); they produce more rapidly than they otherwise would, bringing more oil to the market and depressing its price. This leads to waste. More rapid production often means less of the original oil in place being recovered, and a lowered price reduces the inducement to prevent leaks and avoid evaporation or fire.[73] Joint operation of the reservoirs would remove incentives to competitive overproduction but are extremely difficult to arrange without coercion by government. Without government action, self-interest leads to production practices that partially ignore the future social utility of oil in the ground.

To make matters worse, giant oil discoveries and huge increases in production capacity occur erratically. In an uncontrolled environment, oil prices can first mount as production from established fields declines, only to plummet as a massive new field is brought into production. The discovery of the East Texas field in 1930 resulted in the prototypical boom. Wells were drilled platform-to-platform on city lots, and oil was stored behind earthen dikes because transportation facilities were inadequate to move it to markets. Prices in East Texas fell to as low as 10 cents per barrel in April 1933. The resulting chaos caused the declaration of martial law enforced by Texas Rangers to close down production, restore order, and lessen strife in the East Texas field. Prices recovered, but not until the end of September did they return to the $1.00 per barrel level of December 1932.[74]

Once the initial crisis passed, Texas and other producing states introduced orderly procedures and restricted output among producers on the basis of their potential production. The price stayed roughly even in constant dollars until the late 1940s. It then rose to a level from which it slowly eroded during the 1950s, and continued to decline between 1960 and 1973.[75] But price was not the only issue.

Security, resource independence, and the proper role of the United States in the international petroleum market became critical issues in the 1940s. They involved simultaneously the defense establishment, the State Department, and the Department of Interior. The United States could be self-sufficient in oil. Was it necessary to exploit foreign sources? Was it desirable? Could peacetime and emergency needs be met through enhanced production, conservation, and synfuels? How much should government intervene? Such questions were studied in 1948 by the staff of the National Security Resources Board (NSRB), an organization established by Truman. The report by Wallace E. Pratt on A National Liquid Fuels Policy "contained the first full development of the national security implications of petroleum policy as perceived especially by the oil majors, the State Department, and the NSRB."[76] The report recommended that the United States serve as middleman to secure oil from the Middle East for most of the world and designated rapid development of South and North American resources within a "minimum strategic defense area" for the estimated military and civilian needs of the United States.[77] These recommendations, questioned by those who saw no need for a *national* petroleum policy, failed to produce new directions in policy toward the oil industry.

Security-related concerns resurfaced in the mid-1950s when an influx of cheap oil from the Middle East and elsewhere threatened domestic prices and production. A descending spiral was in prospect: greater imports meant lower prices, which meant less domestic exploration and development and increased consumption, which led to still more imports. Heavy reliance on imported oil risked military interdiction of needed supplies and compromised national security. The desire for excess domestic supply capacity for use at home and by allies abroad in case of emergency was thwarted by the reduced profit levels. Government's response was to limit oil imports, first ineffectively through a voluntary program, then through the Mandatory Oil Import Program (MOIP) of 1959.[78]

When in the 1970s, fear of inflation overrode the fear of import dependence, import quotas were first relaxed and then removed. The price controls imposed by the Nixon Administration in August 1971 to keep domestic oil prices below world prices continued through the decade.[79] They were instrumental in causing and prolonging the energy crisis.

Prorationing.   Efforts to stabilize the market by limiting production conflicted with the rights of producers and inflicted higher costs on consumers. Innovative legislation and the political skill to pass it, along with the administrative resourcefulness to enforce it, were needed.[80] There were four objectives:

1. Limit the capital going into production to prevent wasteful duplication of facilities;
2. Limit the rate of production in each reservoir to optimize between ultimate physical recovery and efficient use of capital;
3. Provide for an equitable sharing of the right to produce oil among potential producers;
4. Limit the total amount of oil produced to maintain its price.

These objectives were fulfilled by an interrelated package of statutes and regulations, especially comprehensive in Texas.[81] Minimum well-spacing requirements limited the number of wells drilled. To establish a fair basis for restricting production from each well, an estimate was made of potential output—the maximum efficient rate (MER)—for each property.[82] Then an "allowable" based on the MER was determined monthly, implementing a proportionate reduction of the production potential of the state consistent with "market demand," the output goal. That goal, in turn, was set by administrative decision reached after considering how much oil purchasers said they wished to buy in the succeeding months. Output over and above market demand was presumed to be wasteful and this could be suppressed under the legal theory of the police power of the state. This overall program came to be called "market demand prorationing."

Regulations governing output restraint had to be artfully designed to avoid conflict with state laws and common-law holdings protecting the right of individuals to do as they saw fit with their property. They also had to avoid "burdening commerce" under the Interstate Commerce clause of the Constitution. Then they had to be enforced. Producers were numerous, scattered, and motivated to violate the restraints since the supported price was above the cost of production. Inasmuch as buyers were much fewer and easier to monitor, regulations were enforced through oil purchasers.

The federal government backed state efforts to avoid waste and support prices. The 1933 National Recovery Administration Petroleum Code provided "for import restrictions, minimum prices, administrative approval of development plans in new reservoirs, limitation of domestic production to total demand less imports, and allocation of domestic production among states, reservoirs and wells."[83] The enabling legislation, the National Industrial Recovery Act, was declared unconstitutional,[84] but Congress arranged another vehicle, the Connally "Hot Oil" Act.[85] This act stipulated that no oil produced in violation of a state conservation statute could pass into interstate commerce. In other legislation, formation of an Interstate Oil Compact was authorized; membership obligated states to pursue conservation, and formal information exchange facilitated coordination of their separate production control programs.[86] Federal involvement made it very difficult for oil production to exceed the levels set by states.

The national system depended on the Texas market demand prorationing program, which was remarkably effective and free of scandal during its almost three decades of influence. The Texas Railroad Commission managed the program under the rationale of reducing physical waste. There was little official recognition that price was even affected. Said Col. Ernest O. Thompson, powerful Chairman of the Texas Railroad Commission in its formative years, "After all, only so much oil can be sold in the marketplace, so what good is accomplished by crowding the market?"[87]

When narrowly drawn, the conservation argument on which the program was based was constitutionally sound. Less oil was left in the ground when reservoirs were abandoned, and less oil was lost to evaporation, leakage, and fires on the surface. "The best place to store oil until you need it," so it went, was "in God's reservoir."[88] But critics asked broader questions: Was it worthwhile to save the "conserved" oil? Were there not less-costly ways of preventing waste? From the efficiency perspective, the program was clearly flawed.[89] Yet the main political force of the criticism came from antioil, populist, and consumer interests who simply wanted lower prices.

Having stabilized oil prices, the system threatened to come apart when cheap foreign oil began to be imported in quantity in the mid-1950s. Each barrel of imported oil displaced domestic production and reduced the allowable percentage of operating capacity, thus lowering both the return from existing wells and the incentive to develop new ones. The capital cost was the same whatever the level of production, so that a reduction in allowed output reduced profits as though it were a reduction in price. In Texas, the average annual production allowable was more than 65 percent of capacity before 1954, but had dropped to 28 percent at the time the Mandatory Oil Import Program was instituted, and stayed at about that level until the late 1960s when the world oil market began to tighten.[90] Exploration and development tracked these changes. Drilling peaked in 1956 and went into a long-term decline that did not bottom out until 1971 and did not recover to the 1956 level until 1980.[91]

Since the MOIP limited imports based on domestic production, imports rose with consumption and shared the expanding market. Behind the protection of import controls, production limitations were maintained and domestic prices supported. But there was conflict over the level of imports to be allowed and over who would benefit from the cheaper foreign oil.[92] Domestic oil was produced more and used more. Less was left in the ground for the future.

The MOIP

reduced oil imports during 1959–73 and lessened the impact of the 1973 embargo. But it also imposed higher energy costs on the American public, encouraged depletion of domestic energy resources, altered the location and type of refinery capacity, and distorted oil exploration and development decisions. In doing so, its provisions directly and obviously improved the position of some individuals and firms at the expense of others, affected the regional distribution of income and wealth, and led to government influence over a multitude of apparently unrelated decisions.[93]

Oil Taxes and Oil Prices.   More vociferous than the conflicts over import controls and market demand prorationing have been those over oil industry taxation. The effects on income distribution are similar, but relatively more transparent in the case of tax preferences. At issue is the treatment of the return of capital to the industry through depletion and depreciation deductions and the timing of the deduction of expenses.[94] It is settled policy that income tax is to be levied only on income, the return over and above the investment originally made, not on the repayment of the investment itself. Thus, in principle, an oil company should not pay income tax on the liquidation of its capital asset—the value of the oil originally in place—but on the difference between that value and the price the oil was sold for, minus costs of production. Following this theory, the oil company should be able to replace its asset—oil in the ground—with untaxed funds via the depletion allowance. This treatment is consistent with that accorded a manufacturing firm that sets aside untaxed funds via a depreciation reserve to replace its worn-out factory. If capital rather than income is taxed, investment is discouraged and oil and gas production will be lower than the socially efficient level. Conversely, untaxed income encourages too much investing and is unfair to the taxpayers.

These conceptually straightforward taxing principles were the source of an enduring and corrosive political battle. President Truman avowed that he knew "of no loophole in the tax laws so inequitable as the excessive depletion exemptions now enjoyed by oil and mining interests."[95] Charged another critic:

> On the theory that the more you take out of an oil well the less you have left, an ingenious tax theory has been evolved to assure the production of the biggest and most blatant crop of millionaires that the nation has seen since the "robber barons" flourished in the late nineteenth century.[96]

Attitudes on tax preferences and depletion became a litmus test by which politicians were judged to be either "proindustry" or "proeverybody else."

It was the application, not the principle, that was critical in the tax treatment of a depleting asset. The central questions were what value to place on the oil in the ground at the time of its discovery, and whether the combination of the depletion allowance and other distinctive tax provisions provided appropriate treatment of that value. Legislation was passed in 1926 to use a percentage of the value of annual output (27.5 percent at that time) as the measure of original value.[97] (The same treatment was later accorded other mineral deposits, but at lower rates.) Each year, the oil company was allowed to deduct from gross revenues 27.5 percent of its income (up to 50 percent of its net revenues) as a return of capital. Expenses incurred in unsuccessful petroleum exploration and development, and "intangible drilling expenses" (those not resulting in a physical asset), were allowed to be deducted in the year incurred rather than having to be capitalized.

These provisions had complex effects. They left oil and gas relatively undertaxed but did not unduly enrich the oil companies because the greater after-tax revenues were dissipated in the extra investment made profitable. This enhanced investment added to excess capacity. There was increased domestic oil production and greater overall oil consumption. Consumers had to pay less under import limitations than they would have otherwise, and more of the U.S. resource base was drawn down, leaving a greater quantity of foreign oil in the ground. Thus, the tax system took on some of the political strain induced by production and import controls and made it easier for the controls to be maintained.

Oil Prices in Transition. Market demand prorationing was sufficient to stabilize oil prices above short-run, market-clearing levels until imports threatened the system. Then oil import controls were imposed. These instruments, along with special tax provisions, formed the triad that assured the continued prosperity of the petroleum industry for almost four decades.

Changing conditions in the early 1970s shifted the balance of political forces. As domestic production moved toward capacity levels, the import quota program became too expensive to maintain; sharply higher prices would have been required to equilibrate demand and domestic supply. The program's demise in 1973 was hastened by "the overwhelming concern with rising prices that became the preoccupation of national economic policy after 1969."[98] World oil prices inched upward, first to meet and then exceed controlled domestic prices controlled under the 1971 program. Supply restrictive mechanisms, so important to the industry before, lost their relevance and fell into disuse. The problem now was how to deal with shortages and climbing prices rather than gluts and falling prices. A stream of studies was launched to investigate the issues over which conflicting interests battled. These studies are the subject of subsequent chapters.

Appalachia in the Wake of a Declining Industry

Coal policy has been concerned largely with the depressed condition of Appalachia, the human misery associated with mining coal, and the unemployment incident to not mining it. Consumption of coal fell from 53 to 31 percent of the nation's mineral fuels mix in the decade after World War II.[99] Use peaked in 1947 at a level not regained until 1973.[100] Coal lost out to increasingly available natural gas and oil, whose low price, relative convenience, and cleanliness proved tough competition. It also suffered from such technological changes as conversion of rail transport to diesel. The impact on employment was magnified by mechanization of the mines. Manual loading went from 50 percent as the war ended to less than 15 percent by 1960. Decreasing labor intensity meant smaller portions of shrinking revenues went for labor services.[101] The coal industry was a mass employer of men until World War II. After the war, "investment capital and advancing technology threw 100,000 American coal miners into obsolescence."[102]

The double shock of depressed demand and technological change left Appalachia financially and socially debilitated. The area kept falling further behind. Unemployment in the coal-producing regions at the height of the coal depression was substantially higher than for the nation as a whole, and it was to get better only when enough people left for jobs in industrial cities such as Akron and Detroit.[103] Social services deteriorated, and a higher proportion of the population became dependent on a declining tax and employment base. Real median family incomes rose about as fast as for the nation as a whole during the 1950s, whereas in other poor areas it rose faster. In coal-dominated Appalachia, incomes were still only 56.5 percent of the national average in 1959.[104] Fifteen years after the war, the people of Appalachia were no closer than before to enjoying higher relative living standards.

Was there a national interest in shoring up the coal industry as a matter of energy policy? Was support to the industry an effective or efficient way to help the people of Appalachia? These were the questions analysts were called on to address. Those advocating government support for the coal industry made three arguments. The first, advanced by Eisenhower in setting up a committee to study coal's difficulties in 1954, had to do with national security.

> The soft coal industry is of course a very important part of our defense mobilization base. Despite the ever increasing demand for mechanical energy, the utilization of coal has actually dropped over the last several years and there is danger of a serious loss in our capacity to produce coal in the event of a major emergency in the future.[105]

The weakness of this contention was that much excess capacity existed, mining operations could be expanded rapidly, and the need for extraordinary expansion of coal usage due to war seemed remote. In a later study the director of the Office of Emergency Planning concluded:

> I do not question the desirability of continuing measures to remedy hardships in the coal region. Of course these measures should be continued. But I do not believe that the national security section of the Trade Expansion Act can under the showing of this investigation be the medium for government assistance.[106]

The second argument claimed that since progressive exhaustion of oil and gas would in a few decades force up their prices and lead to a return to coal, it was in the national interest to protect the coal industry and so avoid the necessity for later rebuilding. It was widely accepted that coal would someday again be the primary hydrocarbon source. But there was disagreement over what this meant for policy.[107] To the Paley Commission, "The negative approach which seeks to cure the coal industry's past ills by imposing artificial restrictions upon competing fuels is unacceptable." In the interlude of low-cost alternatives, the coal industry could be positioned for the resurgence to come by developing new means for using coal, methods for converting it to liquids and gases, and techniques for mining it more safely and at lower cost.[108]

The third argument was that oil and gas were premium fuels that should be limited to specialized uses best matching their characteristics, leaving bulk-heat markets to coal. The future value of the oil and gas thus saved, according to this position, would be sufficient to pay society (if not individual producers) to forego the benefits of cheaper fuel in the interim.[109] The signals given by the price system were flawed, it was contended, and did not reflect the true comparative value of the fuels. The conclusion, based on inclusive evidence, was that government intervention was therefore appropriate.

None of these arguments won many votes, and the politics of coal continued to revolve around the plight of the people of Appalachia.

> The Eisenhower administration treated the coal industry as an unemployment problem not as an energy supply matter. As a contributor to unemployment it was by no means a trivial source of worry. Nevertheless it was not self evident that the distress of jobless miners could be relieved by giving artificial respiration to a declining industry. Other social programs were available for this purpose.[110]

Presidents Kennedy and Johnson were especially sympathetic to the people problem, carrying out programs targeted at Appalachia and working to mitigate poverty generally. On the coal front, however, few actions were taken. The critical competitor to coal for generating electricity was residual fuel oil. Controls on its importation were progressively relaxed from 1963 forward and were effectively eliminated by 1966—"a victory of the consuming eastern states over the coal interests."[111] To make matters worse for coal, expanded concern for the environment led to passage under Kennedy of the Clean Air Act,[112] further depressing coal markets.

The progressive decline of the coal industry was finally reversed in the 1970s when competitive fuel prices began to rise as controls begot shortages in natural gas and oil. Nixon became the first president since Roosevelt not to be pressured throughout his administration to do something for coal as a means of relieving the poverty of the depressed Appalachian region.

The Federal Response.   Federal action to help coal never got much beyond the study stage. Eisenhower's Cabinet Committee on Energy Supplies and Resources Policy, for example, had suggested "steps to remove any handicap to coal's competitiveness arising from rail freight charges," and promotion of "coal markets abroad by urging foreign governments to reduce restrictions on the access of U.S. suppliers to their markets and by counseling the Export-Import Bank to regard the financing of coal exports sympathetically."[113] But by downgrading the committee from cabinet to advisory status because of misgivings about its recommendations on oil imports, Eisenhower effectively avoided giving presidential endorsement to its conclusions about coal.[114]

During the Kennedy years, a Senate National Fuels and Energy Study Group was formed to investigate the wisdom of a national fuels policy that would provide end-use controls and fuel preferences—an idea fostered and strongly supported by the coal industry but opposed by oil interests. To the discomfort of the coal interests, the Senate study found the nation's resource base adequate to meet projected requirements for *each* fuel, making unnecessary the special efforts to use coal in order to save oil and gas. According to the study, the nation had the "ability to be self sufficient in oil."[115]

In 1963 Kennedy established an antecedent committee to the Appalachian Regional Commission.[116] Ever mindful of his political debt to the region that gave his campaign its real start, Kennedy instructed the committee to "make a careful study of the woes of Appalachia and recommend remedial legislation for submission to Congress."[117] The Commission's 1965 report to President Johnson suggested a policy that would rehabilitate Appalachia through capital investment in education, highways, and the reforestation of private lands. But the report did not provide for utilization of the massive coal beds—the area's "sole industrial resource of any real consequence."[118]

Finally, President Johnson appointed a special secret Cabinet Committee on Appalachian Coal during a flurry over residual oil import controls, directing the committee to "determine what steps can and should be taken now by the executive departments, to carry forward the policy of developing, within the framework of private enterprise, new and additional markets for the Appalachian coal industry and assuring its participation in the national energy market on a fair competitive basis." The committee called false the claims by the industry that a reduction in residual oil import quotas would create 33,000 additional jobs and opposed increased restrictions.[119]

Coal Policy and Analysis in Retrospect. The economic and social problems of Appalachia predisposed successive administrations to use coal policy as a tool for regional assistance. But analysis showed that helping coal would be an expensive and ineffective way to relieve poverty and alleviate the dreariness of life in the coal belt. Politically, major legislation benefiting coal could not be passed on the basis of special interests alone. And analysis would not support the contention that a national interest such as security or the long-term cost of energy was at stake. Resurgence of coal markets awaited the squeeze on oil and gas supplies and the price rises of the early 1970s.

As for the coal industry, it came through this period leaner and more efficient. With the demand growth of the 1970s, the industry was able to expand without the overhang of obsolete facilities and ineffectual management that would have existed had continuous rescue efforts been mounted. In retrospect, outcomes for the industry and the people of the region may have been as good as circumstances allowed. The social service programs of the Great Society were disproportionately directed at Appalachia. Yet even by 1969 the median family income of the coal-producing part of the region was only 50 percent of the national average, *down* from 56.5 percent ten years earlier.[120] The region remained poor, but many of its people moved out to find opportunities elsewhere. The better lives many found can be counted as a benefit of a coal policy that refused to use extraordinary means to sustain firms that could not compete. Migration that was in the best interests of those leaving might have been discouraged had stronger protection measures fostered false hopes of burgeoning jobs in the coal industry.

CONCLUSIONS

Thus we see that energy did not first become a policy issue with the Arab oil embargo. There had been policy controversy over nuclear power, gas, oil, and coal for at least two decades before. And there were many analyses. Yet, at no time did energy issues capture the attention of the nation as they would after 1973. The industry expertise and energy data base that had been steadily growing would be drawn upon for the comprehensive energy studies of the seventies.

Although specific energy problems were not the major themes on which elections were won or lost during the 1950s and 1960s, no president could ignore energy. Powerful special interests demanded notice. As for analysis, it did seem to make a difference. Studies like the one by the Paley Commission reflected and affected the way the nation thought about its energy and resource problems.

3 REPORT ON THE ENERGY STUDIES

A profusion of energy studies, triggered by growing concern about future sources and needs for energy, ran through the decade of the seventies. The studies provide a rich laboratory for investigating policy analysis, a mid-twentieth-century "politechnic" art form whose use in national problemsolving is still evolving. In this chapter, we outline fourteen of the energy studies, sketch their objectives and outcomes, and explore their meaning both to the understanding of energy problems and to the practice of policy analysis in general. Subsequent chapters deal with the studies individually. As backdrop for the later discussion, we present the results of a questionnaire given to a group of energy experts concerning the quality, attention, and influence of the fourteen studies. The responses corroborate impressions from interviews, foreshadow later sections of the book, and raise some central questions.

Like a recurrent bad dream, there was repeated fitfulness from energy problems in the United States during the 1970s. Many energy studies were conducted. We examined fourteen in detail. They covered the principal issues being argued and had a decided influence—directly or indirectly—on the form and content of the debate. The analyses might not always have been as helpful as one might have wished, but the fault often lay with how the analyses were used. We will try to be clear on the differences between content and use, and between objectives and effects, as we proceed to consider the studies individually.

FOURTEEN VIGNETTES

Table 3-1 lists the fourteen studies. Half of them were undertaken specifically at the behest of government; another four were conducted at the initiative or with the primary sponsorship of private foundations. Staffing of the studies was varied in nature, ranging from the use of analysts internal to the responsible organization, to the assembling of teams of researchers and expert panels from outside, to the employment of external consultants and research contracts. Issues covered by the studies included

Table 3–1.   Classification of the Fourteen Studies.

	Origin[a]	Staff[b]	Subject[c]	Level[d]
Ford Energy Policy Project	F	C,T	D,E	1
Project Independence Report	G	S	P	1
Rasmussen Reactor Safety Study	G	T	N	1
ERDA-48 and 76-1	G	C,S	A	1
Synfuels Commercialization	G	C	A	3
MOPPS	G	C,S,T	A	1
Ford-MITRE Study	F	X	E,N	1
Lovins's Analysis	I	S	A,D	1
WAES Study	I	T,X	W	2
CIA Assessment	G	S	W	1
CONAES	G	T,X	A,D,E,N	1
Stobaugh and Yergin	U	T	A,D,W	2
RFF-Mellon Study	F	S	A,D,E,P	3
Ford-RFF Study	F	X	A,D,E,P	3

[a] Origin: F = Foundation sponsored; G = Government commissioned; I = Individual initiative; U = University based.

[b] Staffing: C = Consultants or Contracts; S = Staff members or Self; T = Team assembled Temporarily; X = eXpert panel.

[c] Subjects: A = Alternate technologies; D = Demand and conservation; E = Energy/Economy relationship; P = Prices and markets; N = Nuclear power; W = World oil situation.

[d] Level of Politicization and Controversy: 1 = highest, 2 = moderate, 3 = lowest.

conservation and demand, nuclear power, the prospects for alternative sources of energy, the relationship between energy and the economy, the workings of energy prices and markets, and the future of world oil.

Ford Energy Policy Project

We have noted in Table 3-1 that no fewer than nine of the fourteen studies were highly controversial and politicized in their execution, reception, or use. The tone was set by the first, the clairvoyant Energy Policy Project of the Ford Foundation. Begun in 1972 in response to environmental concerns then at a peak (energy was not yet front page news), the project released its final report in the fall of 1974 as the public was reeling from the first pangs of the energy crisis.

A mixture of plaudits and protest greeted publication of the report. Irate supporters of business-as-usual received the document as an advocacy piece for energy conservation, which they equated with misguided restraint on economic growth. Perhaps some of the antagonism and resentment aroused by the study could have been avoided without compromising the central message. The project caused much trouble for itself, for example, in its emphasis on regulatory rather than market measures to enforce conservation. Still, in promoting the need for moderation in energy use, the project was prophetic. Its heresy became the new orthodoxy within four years.

Project Independence Report

The United States in 1974 was outraged at what was widely regarded as blackmail and strong-arm tactics by Arab oil-exporting countries. President Nixon vowed to make the nation energy self-sufficient by 1980! To help plan the drive to independence, the White House approved a major in-house analysis by the newly formed Federal Energy Administration (FEA). The FEA effort, the second study we review, revolved around development and use of the Project Independence Evaluation System (PIES), a large collection of data and computer models.

The FEA used PIES in 1974 to prepare the Project Independence Report, which pointedly disavowed Nixon's original self-sufficiency goals. The report focused instead on alternate strategies for dealing with the problem of energy security and drew attention to the effects of energy prices. The Republican administration employed PIES repeatedly over the next two years to justify and buttress White House positions on price deregulation and other issues, greatly irritating Democratic members of Congress and consumer advocates.

Rasmussen Reactor Safety Study

Nuclear power, the bright hope of the sixties, might have been expected to lead the march in the attack on energy problems. But public concerns about nuclear safety were already mounting by 1972. The Atomic Energy Commission (AEC), eliminated in a bureaucratic reorganization a few years later, was at the time responsible for overseeing the civilian nuclear energy program. With the hope of easing safety concerns, the AEC created an ambitious research project to demystify the complex and emotional subject of risk in nuclear reactors. The project was placed under the direction of M.I.T. nuclear engineer Norman Rasmussen. Released in final form in the fall of 1975, it was used extensively thereafter in campaigns to reassure the public. Some believe it should have been used instead to reduce the chance of an accident like that at Three Mile Island. The Rasmussen effort is the third of our studies—one of the most controversial.

ERDA-48 and 76-1

In the bureaucratic response to the energy crisis, the government created the Energy Research and Development Administration (ERDA) as a successor to the AEC. By statutory requirement, the new agency was under pressure to produce an R&D plan in the second quarter of 1975. Rushing to meet the deadline, drafters of the plan (known as ERDA-48) slighted conservation options and adopted a supply focus, thereby upsetting environmentalists and advocates of energy moderation. Stunned by the criticism this first report evoked, ERDA produced in the following year the much more conservation-oriented and politically wise ERDA 76-1 report. The two plans, together, constitute our fourth study.

Synfuels Commercialization

A supply alternative favored by ERDA and enjoying strong political support within Congress and industry was synthetic production of oil and gas from coal and oil shale. This was an energy source the United States could itself control, although the cost was uncertain. President Ford set an exuberant Project-Independence style goal for synfuels in his 1975 State of the Union address. A White House task force was set up to examine the goal and recommend a realistic program for congressional approval. The task force commissioned a cost-benefit analysis. It is our fifth study. The study's conclusion that "no program" would be the best program was not the recommendation forwarded to Congress.

MOPPS

ERDA went out of business in 1977, the year the incoming Carter Administration created the Energy Department. Just before making its exit, the agency organized its most comprehensive study yet, the Market Oriented Program Planning Study (MOPPS). Funds were unavailable to finance all of ERDA's increasingly expensive technological programs. MOPPS was intended to winnow the list of alternatives and mediate among competing program managers. It was not supposed to become the center of a highly publicized conflict with the new administration over estimates of future gas supply. But that is what happened. MOPPS is our sixth study.

Ford-MITRE Study

Nuclear anxiety soared to new heights in 1974 with the startling announcement that India had exploded an atomic bomb with plutonium from the spent fuel of a research reactor supplied by the Canadians. The hazards of plutonium production and the breeder program quickly became urgent issues for those concerned about nuclear proliferation, including presidential candidate Jimmy Carter. Carter's views conflicted with those of certain American allies abroad and many members of industry at home. He was glad to have his positions reinforced by the Ford-MITRE nuclear energy policy study, a second Ford Foundation effort, this one focusing on nuclear matters. Several contributors to the Ford-MITRE study joined the Carter team in key nuclear policy positions. The Ford-MITRE study is seventh on our list.

Lovins's Analysis

For understandable reasons, those who opposed nuclear development typically favored conservation and renewable sources of energy. This school of thought was called the "soft path" by the lively Amory Lovins, *enfant terrible* to the nuclear and electric utility industries. Lovins was an articulate analyst/polemicist who preached widely to audiences in many countries. His writings and arguments form our eighth study.

WAES Study

The 1970s brought the United States to the realization that domestic oil production would never again be able to meet domestic consumption. It was alarming to see much of the world's oil located in some of the most politically volatile and unfriendly regions of the globe. With many of the advanced economies considerably less energy self-sufficient than the United States, people worried about the problems ahead and wondered what could be done about them. How long would the free world's available oil hold out? Carroll Wilson of M.I.T. brought together a group of people from the industrialized countries to address these concerns and pool national statistics. Wilson's international Workshop on Alternative Energy Strategies (WAES) is the ninth of our studies.

CIA Assessment

WAES assumed that the Soviet bloc, on net, would be neither importers nor exporters of oil. An assessment by the CIA disagreed. CIA economists estimated that Communist countries in the aggregate could be transformed within eight years from a body exporting one million barrels of oil a day to one *importing* several times that amount. These dire predictions, originally presented to Carter in a classified briefing, were publicized by the president to underscore the need for his National Energy Plan. The political ploy backfired. The CIA study is tenth on our list.

CONAES

Over and over again the energy studies illustrated the difference between objectives and outcomes—between the sponsor's initial intentions and the study's subsequent course. Nowhere was the contrast sharper than in the work of the Committee on Nuclear and Alternative Energy Systems (CONAES), set up by the National Academy of Sciences at the request of ERDA. Asked to focus on the breeder program, CONAES ranged across almost the entire scope of the nation's energy problems, bringing together hundreds of experts of varied disciplinary backgrounds and ideological views. Over-budget and ridden with strife, CONAES had a terrible time obtaining closure. Yet it provided a valuable training ground for very talented people and helped significantly in the rethinking of attitudes on energy supply, energy use, and their relationship to the economy. CONAES is study eleven. Policy analysis in the seventies would not have been the same without it.

Stobaugh and Yergin

Study twelve appeared just as gasoline lines were beginning to form once more in the aftermath of the Iranian revolution. Well-written, skillfully communicated and marketed, and favored by fortuity, the study report became a national best seller to the great surprise of its coeditors Robert Stobaugh and Daniel Yergin. In their advocacy of conservation, Stobaugh and Yergin echoed points made by the Ford Energy Policy Project and a CONAES panel (in an article in *Science*) as well as by Amory Lovins. In their espousal of solar energy, they again lined up with Lovins. But there was a difference. Stobaugh, leader of the project, was an expert on the oil industry. He was also professor at the Harvard Business School, an institution honored by a community not known to be enthusiastic about either conservation, solar, or Mr. Lovins. *Report of the Energy Project at the Harvard Business School* was the subtitle of the book. This implied to some readers approval of the report's positions by business and industry, despite a statement by the authors that the book represented only their own views. Although commendation for the report did come from a broader cross-section of the political spectrum than for any previous energy study, many remained unpersuaded. The clarion call for conservation and solar—music to the ears of the soft-path constituency—did not avert a massive cutback in solar support by the Reagan Administration. It may have helped prevent the administration from severely emasculating the conservation programs.

RFF-Mellon Study

Resources for the Future (RFF) is a research organization with a carefully cultivated reputation for scholarly and objective analysis. Unlike some of their Washington colleagues next door at the Brookings Institution or those a few blocks away at the American Enterprise Institute, the Senior Fellows at RFF have generally avoided taking sides in partisan political disputes. So it was no surprise to find an RFF energy study (funded by the Mellon Foundation) declining to propose any one policy plan that might be interpreted as a prescription for solving the energy problem. Instead, the study presented information and analysis for assessing a wide range of policy options. RFF-Mellon, thirteenth on our list of studies, is among the most factually comprehensive of the group.

Ford-RFF Study

The last of the fourteen studies, the third energy project of the Ford Foundation, was certainly not the "least" of the lot, according to a sampling we made of informed opinions on these studies. The project, headquartered at RFF, was conducted as a panel of author/experts in the pattern of the earlier Ford-MITRE effort. Codifying and articulating much of what had been learned about the energy problem in the previous studies, Ford-RFF was a product of the times. Bringing together some of the best and clearest thinking at the end of the decade, it fits well as anchor analysis for the energy studies of the seventies.

TRADITIONALISTS AND REFORMISTS

In doing the research for this book, we interviewed hundreds of energy policy analysts and policymakers about the energy studies and about the role of analysis in policy. We gave the "attitude" questionnaire reported on in Appendix A to 150 of these people to explore their perspectives on key energy issues. The results help us understand the nature of the energy debate. Two core viewpoints emerged: "traditionalist" and "reformist." We will say a bit about these two viewpoints here. Details are left to the appendix.

Traditionalists were growth oriented. They favored nuclear power and were skeptical about the near-term promise of solar energy. Quoting from Appendix A,

> The traditionalists were convinced that the marketplace and higher energy prices could help solve the energy problem and thought deregulation of oil and natural gas was the key to efficient allocation ... Reformists had great sensitivity to environmental concerns. They believed that environmental protection laws should be enforced vigorously ... Reformists favored a resource-conserving ethic and were troubled by the profound implications of today's energy decisions for future generations. They did not regard the American standard of living as dependent upon substantially more energy in the future. They were against primary reliance on nuclear power and were for greater emphasis on renewable sources of energy such as solar and biomass.

Traditionalists and reformists, not surprisingly, had very different opinions of many of the studies.

REPORT CARD ON THE STUDIES

To obtain opinions on the studies in a systematic way, we designed a second questionnaire—a "study" questionnaire—that listed the studies by name in the order their final reports were released. For each, we asked about the *quality* of the analysis, the *attention* it attracted in the public media and government circles, and the *influence* it had (or was expected to have) in public policy considerations. Of the 150 people who took the attitude questionnaire, 135 also took the study questionnaire. This was sample I. To form a control group, we gave the study questionnaire to 54 additional policy analysts and policymakers in the energy field. This was sample II. Both samples were taken early in 1980.

The ratings were averaged within each sample separately and converted to letter grades. The two samples were significantly different in organizational profile. For example, whereas 30 percent of the respondents in sample I were government workers and 7 percent were from industry, in sample II, 17 percent were from government and 17 percent were from industry. We expected the results from the two samples to show noticeable disparities. They did not. Although there were wide variations from person to person, the two sets of averages were very similar. We took the near-identity of results as an indication that the ratings were reasonably representative of the impressions of energy experts at the time.

Reassured, we combined the results from the two samples to obtain a single set of letter grades for the fourteen studies. The grades, as set out in Table 3-2, suggest that attention and influence tend to be linked in people's minds. The only noticeable exception was the Ford-MITRE study, which scored low for attention but high for influence. The reason may be found in the fact that Ford-MITRE meshed well with Carter's position on nuclear energy. Several of the study's participants assumed important roles in the administration and were able to put into effect some of the study's main recommendations. On the other hand, the study did not receive much coverage in the press, and those seeking copies complained that the report was hard to come by.

Table 3–2. Grades for the Fourteen Studies.[a]

	Quality	Attention	Influence
Ford Energy Policy Project	C	A	A
Project Independence Report	D	B	C
Rasmussen Reactor Safety Study	B	A	A
ERDA-48 and 76-1	D	D	D
Synfuels Commercialization	C	D	D
MOPPS	C	D	E
Ford-MITRE Study	B	C	A
Lovins's Analysis	D	A	A
WAES Study	C	C	C
CIA Assessment	C	B	B
CONAES	B	C	C
Stobaugh and Yergin	C	A	A
RFF-Mellon Study	A	D	D
Ford-RFF Study	A	D	C

[a] Based on the results of samples I and II combined. Only in a small number of instances did these two sets of results show any significant differences. Grades run from a high of A to a low of E.

Table 3-2 bears out a suspicion we had before sending out the questionnaire. Impressions of influence and quality are often inversely related, resembling the inverse relation between mass popularity and elitist views of quality in TV programming, but for different reasons. Studies generally rated high in quality tend to be noncontroversial and integrative in nature. In reflecting ideas already known and accepted, they are not as likely to attract attention and exert influence (other things equal) as studies with striking and fiery conclusions. In no case did a study scoring an A for quality also score an A for influence. The only two studies receiving an A for quality scored C and D for influence. Conversely, the five studies scoring A for influence obtained two B's, two C's, and one D for quality.

On quality, Ford-RFF was the top choice of both samples I and II. It was the only study to receive an A by all occupational groups and was rated first by both traditionalists and reformists, ahead even of the studies most naturally akin to these two viewpoints. In second place on the quality dimension was RFF-Mellon. Again, its position was the same in the replies of both samples I and II. The ranking of the two RFF studies was not affected by removal of the replies from respondents associated with RFF.

With respect to attention received, Stobaugh-Yergin and Rasmussen shared top honors. Also high in this category were Lovins and Ford EPP. As for influence, it was a close race among five of the studies: Lovins, Ford EPP, Stobaugh-Yergin, Rasmussen, and Ford-MITRE. *Energy Future*, the Stobaugh-Yergin book, having just been out a few months at the time of the questionnaire, had already attracted a great deal of attention. People seemed to feel it was going to have a strong influence on government policy. Perhaps it would have, had Carter been reelected.

The low quality ratings received by a few of the studies were unexpected and, at first, perplexing. They seem to have been significantly affected by the perspective of the scorer. Lowest in both samples I and II was Lovins; next lowest were ERDA and Project Independence. There were 45 traditionalists and 15 reformists among the 135 respondents in sample I. They rated Lovins and Project Independence very differently. Traditionalists gave Project Independence a middle ranking and Lovins a definite last. Reformists, on the other hand, placed Lovins fairly high and put Project Independence at the bottom. Both played down ERDA. Table 3-3 summarizes the grades by viewpoint.

In last place for attention received was the Synfuels Commercialization study. It was the least known of the analyses. Next lowest were ERDA, MOPPS, and RFF-Mellon. These same four studies were also considered to have had the smallest influence— MOPPS, never officially released, the least of all.

We partitioned responses according to whether respondents came from government, universities, or industry. Since 12 of the 135 respondents in sample I were engaged in electric utility research and 9 were employed by RFF, these two groups were further singled out to see if any special slants might have been introduced by their disproportionately high numbers.

RFF replies were generally consistent with the overall results. They showed a slight tendency to rate the quality and attention of all studies higher and the influence lower than other respondents, and they displayed predictable self-esteem on the quality of their own two studies. Respondents engaging in electric utility research were generally negative on the quality of Stobaugh-Yer-

Table 3–3. The Studies as Graded by Traditionalists and Reformists.[a]

	Tr	Re	Tr	Re	Tr	Re
	Quality		Attention		Influence	
Ford Energy Policy Project	D	A−	A	B	A	A−
Project Independence Report	C	E	B	B	C	D
Rasmussen Reactor Safety Study	A	D	A	A	A	A
ERDA-48 and 76-1	D	E	D	C	D	D
Synfuels Commercialization	C	D	D	D	D	B
MOPPS	C	D	D	D	E	E
Ford-MITRE Study	B	B	C	D	A	A−
Lovins's Analysis	E	A−	A	A	A	A−
WAES Study	C	B	C	C	C	B
CIA Assessment	C	B	B	B	B	A
CONAES	B	C	C	C	D	D
Stobaugh and Yergin	D	A	A	A	A	A
RFF-Mellon Study	A	B−	D	D	D	E
Ford-RFF Study	A	A	D	D	D	C

[a] The first grade under each category is based on the average scores given by Traditionalists (Tr), the second by Reformists (Re). Grades run from a high of A to a low of E.

gin and positive on Rasmussen. Their negative rating of Lovins went beyond that of traditionalists overall, which is understandable since Lovins is not enthusiastic about the customary operations of conventional electric utilities.

The largest occupational groupings in sample I were government (with 40 respondents) and universities (with 41 respondents). Government responses conformed quite closely to the overall results and reflected no noticeable biases. Somewhat more surprising, the same was true of the university group—an indication, perhaps, of the increasing secularization of academe.

Of the 189 people in the two samples, about half had been associated with at least one of the fourteen studies, over 10 percent with two or more of the studies, and about 4 percent with as many as three of the studies. We were curious to see how study-group members felt about the studies in which they participated.

Participants in every study without exception clearly demonstrated pride in their work and, on average, scored their study higher in quality than did others. The same tendency (except for

Ford-RFF) showed up in ratings of influence, but to a lesser degree. Self-evaluations of attention displayed a different pattern. Participants in MOPPS and the two RFF studies gave their studies no higher an attention score than did nonparticipants.

In general, respondents exhibited less divergence of opinion in rating attention than in rating quality or influence. Attention is a more concrete concept. It can be measured by column-inches in the *New York Times*, coverage on the evening news, references by colleagues, and requests for interviews. When a report is commercially published, the attention it receives can be measured by sales of the book. Table 3-4 shows that the attention scores correlated almost perfectly with sales. No such tangible indices exist for quality and influence.

IN PURSUIT OF INFLUENCE

There is a natural tendency in policy circles to judge the value of a study according to the direct and immediate influence it seems to have in political debate and the drafting of legislation. By one means or another, policy studies aspire to exert influence. This common yardstick has two problems. First, it uses a standard that has no reliable measure. Second, it neglects a great deal of

Table 3–4.   Sales and Attention Scores of Published Studies.[a]

Report	Published	Hard	Soft	Total	Score
Ford-EPP	10/74	5,000	65,000	70,000	A
Ford-MITRE	3/77	1,800	22,000	23,800	C
Lovins	6/77	1,600	15,000		
	2/79		41,000	57,600	A
WAES	5/77	9,800	18,900	28,700	C
CONAES	4/80	1,300	24,800	26,100	C
Stobaugh-Yergin	7/79	100,000			
	10/80		80,000	180,000	A
RFF-Mellon	9/79	2,000	13,000	15,000	D
Ford-RFF	9/79	2,300	16,000	18,300	D

[a] Sales figures given are approximate through May 1982.

what in the last analysis is a study's real worth: its indirect and longer term impact.

We held up the influence yardstick during the course of interviews with policymakers by asking how important the various energy studies had been to their thinking. We were not able to place much credence in the answers obtained. For one thing, it was exceedingly difficult for individuals to identify the factors that had brought them to a particular position. For another, in their attempts to recall or discern the uncertain factors, policymakers were swayed by other unperceived factors—such as ego and agreeableness to the psyche. Some seemed reluctant to share credit or give attribution for their views, as if admitting that they could be persuaded would detract from the image they held of themselves as independent thinkers. Politicians are not alone in this vanity.

It seems that impressions of influence were generally shaped— in the absence of hard evidence—by the attention a study received rather than by consideration of its merits. In rating influence, respondents appeared to minimize or ignore how they personally reacted to the study. If it had been the indirect or longer term impact of the study that they were assessing, influence ratings might have been less negatively (and perhaps positively) related to ratings of quality.

Emphasis on influence, narrowly considered, can blind one to other functions served by policy analysis. Table 3-2 indicates that four of the fourteen studies received average influence grades of D or lower, and another four received grades of C. Yet all contributed to policy development: in training analysts, helping officials learn their job, creating expertise, providing a sounding board, educating outsiders, challenging positions, developing insights, and in testing, reforming, regurgitating, and digesting ideas.

One close observer involved in five of the fourteen studies believes that interaction of the analyses and participants with the political world helped frame the debate, improved the understanding of energy problems by decisionmakers (Ford EPP and Ford-MITRE are cited as examples), and led ultimately to a measure of consensus.

Some studies added important new insights to the debate, even though their quality might have been uneven. Lovins and the Stobaugh-Yergin project certainly fell into this category. MOPPS helped break open the natural gas issue, and contributed to a major change in the Carter Administration policy on the subject. The sequence of ERDA studies were themselves a process of learning within the agency, which is an important role analysis can play ... The synthetic fuels study had its real impact during extensive testimony before Congress ... The kind of program we might have had rested heavily on the data and insights developed during the work. The Rasmussen study could have had a similar role in program development, but did not. Also of central importance was the overlap of the persons participating in these studies ... The cast of characters kept changing, and the cross-pollination of ideas was enormous. By the time we got to Ford-RFF, the study group was composed of participants from virtually all the prior studies. It was largely the insights developed from five years of analytical effort and political response (and not the considerable wisdom of the individuals) that gave this study its character.[1]

This notably positive perspective was not shared by many whom we interviewed. Some thought policy analysis largely irrelevant. We record their cynicism in later chapters.

Policy analysis is varied in its functions and the expectations held for it. There is a positive way of looking at it, a negative way, and much room in between. We try not to be judgmental or take sides ourselves, but neither are we entirely neutral. In particular, we believe in the potential value of policy analysis in a democratic society.

Our major assumption is that there is a place and need in such a society for experts whose primary professional allegiance is to the systematic and dispassionate examination of complex problems facing policymakers. But the experts must not work in isolation. They should be in open and effective communication with those having other interests, and be cognizant of the wide range of viewpoints in society-at-large. Improving this communication is a first order of business in ensuring that policy analysis will achieve the potential of which it is capable.

Those are our assumptions. We state them at the outset. Our concluding chapter notes the significance of assumptions in determining the results of an analysis. Our analysis of analysis is no exception to the rule.

4 AN UNCERTAIN BEGINNING

Attempts to alert the nation and assess the critical state of energy started before the oil embargo and quickly spread, yet had little immediate success in mobilizing policy or changing deep-seated attitudes. The initial energy studies, quickly politicized, led to controversy more than resolution, producing skepticism instead of constructive action. But the studies did identify the issues and map the battleground. In this chapter, with discussion of the Ford Foundation's Energy Policy Project and the White House's Project Independence, we examine energy policy analysis inside and outside of government around the time of the first sharp rises in the price of oil.

A major purpose of the Energy Policy Project is to help interest the public in a national energy policy ... Should a crisis occur ... the Project should be ready to respond.[1]

Thus did the first large energy study of the 1970s chart its course. This extract from an early information plan for the Energy Policy Project (EPP) of the Ford Foundation forebodes the outbreak of the oil crisis the following year. With its initiation in 1972, the project's timing was excellent; its perception of impending problems, prophetic. In retrospect, the need for the review of policy and the rethinking of energy use in the United States was critical at this point. By the early seventies, the nation had come to take cheap and plentiful energy for granted. Greater use of electricity was being promoted. Yet, domestic oil production was turning down, regulated natural gas was in short supply, the use of coal was in disfavor for environmental reasons, and the expectations for nuclear power were unrealistically high. The fact is that despite a general complacency, the energy imbalance was growing dangerously. Energy consumption was outpacing domestic supply more with each passing month.

As the decade began, the price of crude on the world market was about $2.00 a barrel. In the United States, where oil import quotas had been in effect for over ten years, the price was higher: $3.30 a barrel at the wellhead. Prominent economists believed that oil prices would not increase and were likely in fact to fall if quotas were removed. After all, the cost of lifting oil in the Mideast was just pennies per barrel. President Nixon's Cabinet Task Force on Oil Import Control wrote on February 2, 1970:

> We do not predict a substantial price rise in world oil markets ... No projected change in our imports over the next decade will be large enough in proportion to total world demand to affect significantly the incentive or ability to raise world prices ... The world market seems likely to be more competitive in the future than in the past because the growing number and diversity of producing countries and companies make it more difficult to organize and enforce a cartel ... The landed price of foreign crude by 1980 may well decline.[2]

The task force, reflecting the thinking of the times, underestimated the oil-exporting countries and overestimated the prospects for domestic oil. With the advantages of hindsight, the task force should have given more weight to the uncannily accurate forecast of a topping out of domestic production beginning in 1970. The forecast, known as "Hubbert's bubble," was made by geologist M. King Hubbert twenty-two years earlier.[3] Conventional assumptions were more reassuring. Warnings about the risks of growing dependence on foreign oil were not effectively heeded.

The task force recommended phased replacement of import controls with a tariff system. But the tariff proposal met stiff opposition. As a compromise, the White House adopted a system of import licenses and fees: 63 cents a barrel for petroleum products and 21 cents a barrel for crude oil.[4] It was a token gesture. The allure of foreign oil did not fade. Imports from the Mideast climbed sharply, and the subsequent jump in crude prices signaled the onset of the crisis.

The crisis of 1973/74 was a decisive moment in history, a turning point, a critical time when visions of prosperity gave way to feelings of constraint and a mindset of energy abundance turned to the possibility of energy limitations. Like the calm before a storm, the period prior to the crisis was an opportune time for analysis—for taking stock, assessing the hazard, and reexamining policy. The Energy Policy Project was in an excellent position to do this. When the storm hit, the project staff did its best to "respond." It briefed Congress and the press and put out a widely quoted preliminary report. But the pressure was intense. EPP analysts were caught up in the swirl along with increasing numbers of other research investigators—engineers, economists, lawyers, and scientists from government, industry, universities, consulting firms, and research organizations—all offering or seeking explanations of what was happening and what could be done to restore order. Energy became a frontier subject of analysis. In these next chapters, we review some of the major studies which resulted.

A number of observations emerge from this examination. The outcome of the studies seldom matched their initial intent. Sharp divisions of opinion attended both their creation and execution. The early studies, begun on an uncertain tack, were immersed in the swell of outrage and hysteria induced by the gasoline lines and price hikes. In this politically turbulent environment, the studies themselves became objects of contention. How they were perceived varied widely depending upon whether the findings conformed or clashed with the preconceptions, loyalties, and interests of readers. Interpretations of findings were varied and contradictory.[5]

THE FORD FOUNDATION'S ENERGY POLICY PROJECT (EPP)

The EPP was the first major study to give visibility to the issue of energy conservation and to question the assumption that historic rates of growth in energy consumption were required to maintain a dynamic economy. These ideas were very controversial in October 1974 when the EPP's final report, *A Time to Choose*, appeared.[6]

The report recommended that energy growth be reduced from its historic annual long-term average of 3.4 percent to under 2 percent, and possibly to 0 before 1990. It argued for government action and industry reform to spur greater efficiency of energy use in buildings, automobiles, and industrial processes. The authors condemned tax subsidies to the petroleum industry and promotional rates in the utility industry. They took the position that the market was "ineffective" for dealing with problems of environmental degradation, international security, consumer protection, and building design. These conclusions did not endear them to staunch growth advocates and defenders of the free enterprise system.

The report was severely criticized—and lavishly praised. Some damned it as a partisan document whose analysis did not support its findings. The *Wall Street Journal* charged it with "excessive environmentalism, bad economics, and authoritarian politics."[7] Others extolled its clear warnings, persuasive statement of the problems, and ringing call to action.

Publication of *A Time to Choose* was a major media event. Publicity was extensive and well orchestrated. The report was presented at simultaneous press conferences in New York and Washington. Over 6,000 copies were distributed free-of-charge to members of Congress, the governors of the fifty states (including Georgia's Governor Jimmy Carter), the federal bureaucracy, and the press; tens of thousands more copies were sold. The three scenarios of possible futures developed in the report—historical growth, technical fix, and zero energy growth—attracted much attention in public discussions of the energy crisis. A preliminary report published six months before the conclusion of the project under the title *Exploring Energy Choices*[8] was very influential because of its timing and received even wider dissemination than *A Time to Choose*, including 300,000 copies through the Book-of-the-Month Club. In addition, a score of specially commissioned research volumes were published, several of considerable importance to later energy work.[9] If outreach were the measure, the EPP would have to be considered a resounding success.

But the attention a report receives will not by itself determine whether it exerts a constructive influence on policy. The overall persuasiveness and soundness of its arguments, the credibility of its assumptions, and its skill in weighing and putting into perspective divergent points of view are also important. Although *A Time to Choose* attracted attention, it also antagonized. Some found its prescriptions too strong and its economic analysis too weak. It received wide and sympathetic coverage by the media and probably had an influence on the fuel economy legislation later passed by Congress. Yet it won few friends within the business community and among free-market economists. Some sharing its bias questioned its basis.[10] One economist, Armen Alchian, attacking it in a countertract, was moved to vitriolic phrases.

> *A Time to Choose*—better titled *A Time to Confuse*—enters the Guinness Book of World Records for most errors of economic analysis and fact in one book, is arrogant in assertions of waste and inefficiency, is paternalist in its conception of energy consumption management, is politically naive, and uses demagoguery. That is a shocking indictment of a final report of a $4 million project financed by the Ford Foundation.[11]

What were the aims of this much-debated project? A hint of incongruity is discernible in the initial goals of director S. David Freeman and sponsor McGeorge Bundy. Freeman, later an energy advisor to President Carter and subsequently chairman of the Tennessee Valley Authority (TVA), wanted to educate the members of Congress and the press. He did not shy from controversy. McGeorge Bundy, president of the Ford Foundation and former Special Assistant for National Security Affairs to Presidents Kennedy and Johnson,[12] wanted to draw attention to "issues of great national interest" without "lobbying for a particular point of view."[13]

The director and sponsor both wanted to educate and alert, publicize the energy problem, and promote public debate. Bundy was "determined to make the Ford Foundation a relevant force in American life."[14] The project was initially set to run for just a year and culminate in a dramatic policy statement. Plans were for broad public dissemination of the findings through an ambitious outreach program, including a film for wide distribution.

Freeman expressed his views on energy in an article published in *The Bulletin of the Atomic Scientists* shortly before his selection to head the project. "What I am suggesting is that we take a harder look at the growing demand for energy to determine whether the rate of growth really needs to be as high as we are currently experiencing and projecting for the future."[15] This was the focus the Ford Foundation was looking for; in that respect, it was not to be disappointed.

Upon his retirement from the Foundation eight years later, Bundy put aside his other feelings about the EPP and looked back with pride on the "right and timely" message it had delivered: "That this country could and should get along with less energy than historic patterns of growth suggested."[16] The low energy demand estimates that had been heretical to many in the early 1970s had by the end of the decade become commonplace.[17] Conservation was no longer considered inconsistent with economic and technological progress.

But Freeman's strong advocacy mission did not coincide with the Foundation's nonpartisan aims. The differences became apparent after the project got underway. Contention developed between Freeman and certain members of the project's advisory board. It was not what Bundy and his staff had in mind when they launched the enterprise, and it would cause them to reconsider their approach in later Ford Foundation studies. Midway along, Bundy asked Foundation officer Marshall Robinson, an economist, to oversee the project and mediate with the board.

Bundy had alluded to his hopes for the role of the advisory board in a foreword to the preliminary report:

> Most public analyses address a limited segment of the problem or argue from the standpoint of a particular interested party. Both kinds of study are important, but neither is a substitute for a broader effort to set the issues of national policy in a general framework which can assist citizens and their representatives in reaching balanced judgments ... The Energy Policy Project has been carefully designed ... to avoid control by any special interest.[18]

The advisory board was set up to establish a diversified base of opinion and balance points of view. But it was *advisory only*, emphasizes Freeman. That was "clearly spelled out in writing from the beginning."[19] Freeman had the power and made the decisions. He kept the board informed but was not obliged to follow its advice. Statements in opposition to his views were "largely ignored," claims William Tavoulareas, president of Mobil Oil

Corporation. Tavoulareas, who was Freeman's chief critic on the board and a constant thorn in Bundy's side, charged that the results of the study were "preordained."[20] Other board members maintained that *A Time to Choose*, whatever its faults, provided a "useful framework for public discussion."[21]

Indications are that the EPP did have an important political impact. Some insist that it exerted great influence on members of Congress and made a big impression on Governor Jimmy Carter. Freeman helped the Carter campaign with energy policy in 1976 and became a member of the team that wrote Carter's National Energy Plan (NEP). As full-time consultant to the Senate Commerce Committe, he was active in efforts to get legislation passed on fuel economy and building standards.

One critic labeled *A Time to Choose* a "No Growth report on energy," and charged Freeman with having translated its ideas "right into government policy."[22] But Freeman was only one member of the team that drafted the conservation-oriented NEP. The issues that caused it to run aground in Senate debate—namely, its stand on oil taxes and natural gas pricing—were not the main issues of *A Time to Choose*.

From Environment to Energy

Freeman, with a background in engineering and law, had served with the Tennessee Valley Authority (TVA), the Federal Power Commission (FPC), and the White House Office of Science and Technology (OST). Concerned with the impact of energy growth on the environment, he argued ardently for moderation in energy use to avoid further multiplication of power plants. An admirer of New Deal reform as practiced at TVA under the late David Lilienthal, his allegiance was to government, not business; his expertise was in the world of consumer advocacy and public regulation, not research projects and economic analysis. Assistant to the head of the FPC when natural gas price controls were being put into effect, Freeman doubted that controls would induce shortages by stimulating demand and discouraging supply, given the imperfect market conditions in the gas industry.

Why did the Ford Foundation commission the EPP and select Freeman to head it? The Foundation's interest in energy dates back to 1952 and its support for creation of Resources for the Future (RFF). When the idea for an energy policy project was being discussed, the Foundation gave thought to housing it at RFF but dismissed the notion. Gordon Harrison, who established and headed Ford's Office of Resources and the Environment, feared that RFF "would in its usual style produce a thorough but essentially academic analysis ... rather than a guide for policy action." Says Harrison, who wanted the study "to make a difference," RFF had built its reputation as a scholarly research organization on its "scrupulous avoidance of being caught on any one side of any question." Harrison believed it could only maintain its objectivity in controversial matters of policy by studiously balancing the pros and cons of every position analyzed so that no partisan could take offense and no policymaker feel swayed.[23]

Harrison, with a strong ecological sense, had been looking into the possibility of cooperating with other foundations and the government to form a nonpartisan, nonpolitical institute to carry out policy analysis of environmental issues. These issues had risen to the top of the political agenda by 1970, the year the Council on Environmental Quality (CEQ) and the Environmental Protection Agency (EPA) were launched. Yet there was little systematic study of environmental policy and no mechanism for carrying it out efficiently. Harrison proposed an Environmental Institute to fill the need.

Ford's Board of Trustees approved the appropriateness of this initiative in the fall of 1970. A budget was set up to provide beginning capital for the institute contingent on receiving matching funds from the National Science Foundation (NSF). Bundy, considering federal commitment to the undertaking as crucial to its success, entered into discussions with White House staff—discussions that for a while seemed to be bearing fruit. In Nixon's environmental message to Congress on February 8, 1971, the President pledged federal support for the institute and anticipated that it "would provide new and alternative strategies for dealing with the whole spectrum of environmental problems."

Despite the presidential sanction, White House politics cast a pall over the proposal. Some speculate that the president's staff did not want an independent body outside of its control developing public policy recommendations. In any case, the nomination of Alain Enthoven as director of the institute was vetoed by

Charles Colson and other White House assistants. Enthoven, a registered Democrat who had served as assistant secretary of defense under Robert McNamara (a close Bundy associate during the Kennedy and Johnson years) alarmed some influential Republicans. Others approached for the position, such as Howard W. Johnson of M.I.T., were "put off" by the "demonstrated indifference if not hostility of the Nixon Administration," and by the way it was "intruding politics into a personnel selection that was supposed to be on the merits."[24]

Discouraged about a cooperative venture with the administration, Harrison proposed that part of the funds budgeted for the institute be applied to an "out-of-town tryout" of policy analysis—one that might demonstrate the value of an Environmental Policy Institute and possibly revive interest in it at a later time. This more modest proposal became the EPP.

Where did energy enter the picture? Harrison and several colleagues at the Ford Foundation were concerned about the environmental implications of rapid increases in energy production and use. They felt that energy urgently needed sophisticated management to protect the long-term stability of the environment. Environmental impacts, not uncertainties about the availability of oil, were at the root of their interest in conservation.

The Foundation was also concerned about impending disruptions to the supply of energy. Projections of electricity demand were still on an exponential curve upward. Loud public outcries about the pollution associated with electricity generation had caused new power plant construction to be stalled, and David Lilienthal had come to Bundy to seek Foundation help in easing the deadlock. Electrical brownouts were growing in number, the New York City blackout was fresh in people's minds, and warnings of a national energy crisis appeared ominously in the press.

> For the third straight summer, Americans by the millions are living under the daily threat of power brownouts, blackouts and possible electricity rationing. But it is more than a seasonal shortage of power. It is part of a national crisis that won't go away—the energy crisis.[25]

According to Edward Ames, who was to take over project management of the EPP under Harrison, the Foundation was stymied in its attempts to locate learned people in the customary places who were questioning how the nation was using and producing energy. "We realized that we would have to step out of the mainstream of thinking on energy matters," says Ames. "Energy conservation needed to be explored, but it was anathema to government and industry." And normal centers of research "were in lockstep with the government-industry position."[26]

Ames and Harrison believed that Freeman was the most qualified nonindustry person available to do the job they felt needed doing, and had sufficient visibility and national prominence to command a hearing.

Having left his White House post as OST director of energy policy, Freeman was working on a book sponsored by the Twentieth Century Fund dealing with the conflict between environmental values and increasing demand for electric power,[27]—a near-perfect match, it would seem, with the Foundation's interests.

Critics have faulted the sponsors for their selection of a man who on the record could not have been expected to view the issues impartially. The Foundation, aware of the potential for controversy in making the appointment, took an "all but unprecedented step in staff procedure" of bringing Freeman before the board of trustees for examination.[28]

> Freeman was asked if he could conduct a study with an open mind, whether he was capable of restraining his bias while constructing an objective analysis ... The trustees were very particular on this point; they hoped for a comprehensive review of the alternatives rather than a passionate declaration of principles ... Freeman said something to the effect that, although he had an open mind, he didn't have an empty one.[29]

The trustees gave their approval.

Flutterings in the Coop

Before the project was half-way along, there were "flutterings in the coop."[30] On January 25, 1973, in a widely publicized speech to the Consumer Federation of America, Freeman vigorously opposed removal of price controls on natural gas and the increase in oil prices that the administration had recommended to alleviate fuel shortages and encourage exploration for new domestic sources. Without taking care to make clear that he was stating only his personal views, Freeman charged that the oil shortages of the winter of 1972/73 were "manufactured" in Washington and "could have been averted with a stroke of the President's pen" by scrapping oil import restrictions.[31] After several incensed members of the project's advisory board threatened to resign, Freeman issued a statement to the effect that his speech did not reflect judgments or conclusions of the EPP. But to an interviewer afterward he maintained, "I've been in the habit of expressing my opinions in government and out, and shall continue to do so."[32] And he did—in congressional testimony and continued communications with legislative assistants and the press.

Before final publication of *A Time to Choose*, the Foundation brought the advisory board together with the project staff at a Colorado mountain retreat to discuss the draft. The board insisted that judgments be clearly separated from analysis in a Conclusions and Recommendations chapter,[33] with a section of dissenting opinions and clarifying comments by individual board members to follow. The board's limit on the length of these individual comments was firmly enforced by board chairman, Gilbert White, over the strenuous objections of board member Tavoulareas. Not to be thwarted, Tavoulareas proceeded to issue portions of his over-sized dissent in newspaper advertisements and a booklet financed and widely distributed by his company.[34] He also maintained pressure on the Foundation.

Under duress, the Foundation eventually agreed to publish Tavoulareas's full critique in a separate volume, along with a reply by economist Carl Kaysen, a board member whose views were considerably more sympathetic to the report's position than to Tavoulareas's, thus qualifying him to give the reader "the benefit of more than one perspective."[35]

As Kaysen saw it, the root issue between Tavoulareas and himself was whether energy conservation could achieve real savings and benefit the economy without impairing economic growth. But the exchange had more the characteristics of a political confrontation than a scholarly debate. "It took 18 months of negotiation," says Kaysen. "Every time I submitted a draft, the Mobil staff went to work with the figures again."[36]

On the level of national policy, the difference between Kaysen and Tavoulareas was one of priorities. Tavoulareas gave priority to expanding supply rather than restricting demand and to pro-

viding consumers with maximum choice. He believed that inducing automobile companies "to orient their manufacturing
operations towards the smaller cars" in the interests of saving energy was contrary to true customer preferences. It could only result in "unsold inventories of small cars and shortages of larger
cars."[37] (This imbalance did in fact exist at the time of Tavoulareas's critique. Later, however, in the wake of the 1978/79 crude
oil price rises and the decontrol of domestic oil, the situation reversed itself, causing major problems for the U.S. automobile industry.)

Tavoulareas's position might be stated as, "I don't oppose conservation, but let's start by increasing supply and not reduce consumption any more than we have to." In contrast, the
conservationist's logic would be, "let's start by moderating demand and not *produce* any more than we have to." Free-market
economists have a different point of view; they want the price of
energy to rise to its natural level so that supply and demand can
take care of themselves.

Kaysen is a market economist with qualifications. He doubted
that the unaided market could balance implied threats (to security and environment) with economic benefits (of more and cheaper
energy).[38] He and fellow board member Harvey Brooks saw this
market failing as justification for recommending conservation but
feared it might be missed or dismissed because of the book's "rhetorical excesses" and "populist" speechmaking.[39] Here is an example:

> The basic problem is that the energy industry, particularly the oil
> industry, possesses a unique combination of political advantages
> which has enabled it to exert considerable influence on public policy.
> This influence is manifested in a variety of energy policies that are
> highly favorable to the industry. A necessary first step in reforming
> energy policies is to remove the main sources of the oil industry's
> disproportionate political strength.[40]

To its credit, the EPP did succeed in placing energy conservation on the policy agenda as a legitimate subject for serious discussion—making what was a political nonissue into an important
item for public debate and further analysis. In the process, it is
accused of having identified conservation unnecessarily with government control. To economist Walter J. Mead, the proposed reg-

ulatory measures were uncomfortably similar to ones that had failed in the past.[41] These measures seemed likely to undermine the efficiency of a market already weakened by government intervention.

One reason the market was not working effectively to restrain energy demand was because of price ceilings and a gasoline tax much lower than in Europe or Japan. Although *A Time to Choose* recommended rescinding oil industry subsidies and the preferential block rates of utilities, it did not question low energy prices in the United States. It did not connect these prices to what it referred to as "the national habit of extravagance in energy use."[42] Instead, it viewed price controls as necessary for the poor—a protection against high energy costs and excessive profits. It did not examine the possibility that controls might seriously distort the marketplace, sapping economic efficiency in the name of economic welfare. It did not explore whether the regulation of natural gas was really serving the interests of the lower income groups. It missed the chance to investigate these important issues and thereby gave the appearance of being one-sided.

Of course, any study seeking to raise public consciousness on the need for energy conservation in the early 1970s would have had to have been controversial and, at least to some degree, polarizing in effect. Yet the question remains: Was some of the hostility that the EPP engendered—and from which it suffered—avoidable?

Some argue that for the EPP to have been a more cogent influence in the debate on energy policy, it would have had to have been credible to leaders of the energy industries. They believe industry representatives on the advisory board could then have helped foster a sympathetic audience for the report among businessmen. Advisory board chairman White had hoped the board would serve in such a participatory function, but, he says, "Freeman could not recognize it as an educational device."[43] Would the project have gained greater favor with a director received as an impartial manager of research rather than as a partisan with a cause? Would it have had the same punch and would its message have had a sustained long-term effect?

"No," replies Harrison. "The point is that *no* analysis would have been credible to leaders of the energy industries except one that adopted their prejudices." Had Freeman been perceived as an impartial manager of research, asserts Harrison, "he would no doubt have got less flak"—and also "a good deal less attention." "To make people mad, especially among the power establishments, is no fault; it is an indispensable part of being effective."[44]

So the argument goes on the virtues of dispassionate analysis versus the power of a strong advocacy position. The issue has been debated repeatedly and will probably continue to be with each new policy study.

The angry reaction by industry to *A Time to Choose* draws attention to the real conflict triggered by the Ford EPP. It was a battle more of clashing doctrines than technical substance: a battle between those committed to economic growth and those distressed by abuses to the environment and the depletion of precious resources, between those who see material progress as the salvation of society and those who regard it as a "destructive engine" in need of control.[45] Later studies would take up the banner of conservation with increasing success, yet the polarization and hostility that surfaced in the EPP would continue to appear throughout most of the analyses of the seventies.

Spawning Ground and Initiator of Research

Going into the seventies, most energy expertise was in fuel-specific areas. Not many experts were well versed in energy matters outside of the context of a single industry. Walter J. Mead, senior economist for the EPP when it was initiated in 1972, conjectures that before the start of the study there were fewer than two dozen professionally qualified, nonindustry energy economists in the United States. By the time the final report was published in 1974, Mead estimates, the number had grown to several hundred.

The work force for the project was of modest size. The project "opted for hiring a somewhat larger number of young, bright people rather than a smaller staff made up entirely of more experienced people." In all, the core group consisted of fifteen to twenty persons, including editors and administrators, aided by approximately one hundred consultants, some doing special pieces of research.[46] There were many contracts written for outside studies, leading eventually to a score of published research reports. All told, the project involved many able people and provided a valuable training ground for analysis of problems not previously experienced or thought through.

In the opinion of staff member Steven C. Carhart, there was not one but two quite separate projects, and they did not have much to do with each other. The first was a "political dog fight," with Freeman and Tavoulareas as principals. The second was a serious technical study to produce the scenarios and provide effective documentation for the conservation recommendation.

According to Carhart, the EPP never had a research director who kept all of the conceptual pieces together. Monte Canfield, Jr., deputy director of the project, managed the budget and day-to-day operations and participated in development of scenarios.[47] Freeman initiated many of the outside contracts, some, he says, "by people that I knew had views that I disagreed with."[48] Several of the outsiders reached conclusions markedly at variance with those presented in A Time to Choose. Their results were published independently.

In some instances, outside work would be cited or incorporated to provide support for positions internally developed. The macroeconomic modeling work of Edward Hudson and Dale Jorgenson questioned the popular idea of a tight linkage or "lockstep relation" between energy consumption and economic growth. An important conclusion of A Time to Choose is that energy demand can be moderated without upsetting the economy.

> The general conclusion of the analyses, stated simply, is that neither the economy nor employment necessarily suffers from lower growth in overall energy use ... The United States can grow and prosper and have plenty of jobs—and still conserve energy.[49]

This thesis, based on the project's engineering analyses but not generally accepted at the time, appeared to gain support from the findings of Hudson and Jorgenson. Their report, cited as confirmation, was made an appendix to A Time to Choose. Yet, says Jorgenson, "I was convinced by our EPP research that higher energy prices reduced economic growth." He adds, Freeman chose to regard the effect as "small."[50]

In their simulations, Hudson and Jorgenson had introduced an energy tax as high as 15 percent in 2000 A.D. to reduce energy demand to the levels postulated in the project's zero-energy-growth scenario, redistributing the substantial revenues resulting from the tax (131 billion current dollars in 2000 A.D.) into government spending on health, education, and transport services. In this "dual mechanism: energy use is directly discouraged by taxes, and demand is further redirected by a change in spending patterns towards nonenergy-intensive production."[51] The "dual mechanism," with its assumption that revenue from higher energy prices does not leave the country, led to dampened energy consumption with only a modest adverse impact on GNP.

In later investigations, the modelers took account of the financial drain from the United States to oil-exporting countries, and they added a mechanism to represent the effects on technical change and capital accumulation.[52] The results strengthened their conviction that higher energy prices and limited energy availability do adversely affect productivity and economic growth, although not as much as the conventional wisdom of the time had led people to believe.

The controversial nature of the EPP overshadowed much of its technical work, though several outside research reports earned acclaim and recognition in their own right. The Brannon study of energy taxes and subsidies and the Willrich/Taylor investigation of nuclear risk are two of the most respected. The full array of reports is impressive. Many of the authors continued to contribute to the policy analysis of energy problems, as did EPP staff members. Their developing careers and maturing views through later energy studies resembled bees carrying pollen. Ideas considered hasty or outlandish in one study would get implanted and reshaped in a later study, gradually gaining adherents and respectability.

All told, the project took two and a half years instead of one to complete and consumed over four times as much money as originally allocated to it from the prospective budget of the aborted Environmental Institute. In fact, it was a valuable source of funds and an initiator of expanded energy research at a time when the nation was just beginning to awaken to its worsening energy problem. The need for such work was still not generally understood. The government appeared to be heading in precisely the opposite direction.

ENERGY POLICY ANALYSIS ELSEWHERE

Early in his second term in office, President Nixon terminated two White House offices doing work in energy, transferring their activities to other parts of the bureaucracy. The first was the Office of Emergency Preparedness, which had among its responsibilities the monitoring of the nation's energy situation and the tracking of oil imports. The second was the Office of Science and Technology, where Freeman, up until his departure in mid-1971, had a group working on a range of energy and environmental studies.

Reorganization of the executive branch was much discussed in the early 1970s. Termination of the two offices came about not from a desire to deemphasize energy research, says Roy Ash, then chairman of Litton Industries and head of Nixon's Council on Government Reorganization, but from an attempt to reduce the excessive number of reporting lines to the president and move functions out of the White House that would be better served elsewhere. The Ash Council had recommended a sweeping consolidation of cabinet operations, including creation of four major executive departments headed by senior cabinet officers—the corporate equivalent of executive vice presidents. The new departments were "designed to force competing advocacy groups to resolve their differences at the departmental rather than the presidential level and to lead cabinet members rather than Executive Office assistants to make the crucial evaluations of the relative costs and benefits of competing policies."[53] One of the new departments, a Department of Natural Resources, would contain the Interior Department, several other government agencies dealing with natural resources, and a centralized body for energy policy.

The proposal for a Department of Natural Resources appeared in Nixon's June 1971 energy message, publicized as the first on energy policies ever submitted by an American president. Despite the billing, the message failed to rouse the Congress.

Plans for reorganizing the cabinet never passed, having been "smothered" by groups benefiting from the advocacy system, according to George Shultz, Nixon's secretary of the treasury.[54] The proposed reorganization would have required changing the congressional committee structure, which is patterned along lines reflecting the division of responsibilities within the executive branch. This threatened the power network in Congress. There was also fear that a strengthened presidency with consolidated lines of authority could become an "imperial presidency." "That was a red flag," says Ash.[55]

By the beginning of his second term of office, Nixon knew the Departmental Reorganization Plan would not get Congressional approval, but he was still determined to make some personnel and organizational changes. His request for pro forma, written resignations from all cabinet officers and senior White House staff, intended to show he meant business, created a stir. Executive Reorganization Plan No. 1 of 1973 transferred the oil import responsibilities of the Office of Emergency Preparedness to the Treasury Department. It also transferred the policy functions of the Office of Science and Technology to the National Science Foundation, a temporary locus for the analysis of energy policy. George Shultz was assigned the additional role of assistant to the president for economic affairs (including oil prices and imports). Shultz's deputy in Treasury, William E. Simon, was appointed chairman of an interdepartmental Oil Policy Committee with responsibility for the oil import program. Simon was "an activist from the business world" who did not "know a thing about oil."[56] He would learn on the job.

Scientists regarded abolition of the Office of Science and Technology and its (extragovernmental) affiliate, the President's Science Advisory Committee (PSAC) as a serious downgrading of their status on the national scene.[57] These organizations, and the science advisor who headed both, were important symbols to the scientific community. PSAC and the science advisor originated during the 1950s as means for giving public attention to the importance of science in government. The Office of Science and Technology was established in 1962.

By the 1970s, according to Ash, PSAC was losing the ear of the president. It went from having "50 percent of its work load assigned by the President" originally, to "10 percent under Johnson" and only "5 percent under Nixon." For the most part, says Ash, "PSAC was assigning itself projects of little relevance to the President's interests."

Dismissed presidential assistant John Ehrlichman suggests that demotion of the science advisor was "pure Richard Nixon." Lee DuBridge (who served under Nixon as science advisor) was "out of step on the ABM" (antiballistic missile), says Ehrlichman. "From then on, OST was on its way out; it took time, but Nixon finally moved it."[58]

There was a touch of anti-intellectualism in the air and it affected the climate for analysis. It appeared that science, linked with policy analysis in the public eye, was losing attention and support in government's upper reaches. Scientists were out of favor with an administration that considered them political adversaries. With energy problems looming, it was an unfortunate time for a breach in communications between science and government.

Crude oil prices began shooting up in the fall of 1973 and energy research started to surge. One important fountain of analysis appeared within the executive branch itself. To provide background for this development, we pause to review energy policy operations in the White House at the time: a montage of starts and stops, changing personalities, and shifting assignments.

The White House In Disarray

Richard Nixon was under great stress from the intensifying Watergate investigation throughout 1973. The vacuum of leadership contributed to organizational turmoil in the White House and heavy political infighting. Three years before his Watergate-related departure in April 1973, John Ehrlichman, then executive director of the White House Domestic Council, formed a subcommittee (described as "the first effective energy policy body of the Nixon administration"[59]) at the urging of science advisor DuBridge. The subcommittee, including Freeman, provided material for the president's energy message of June 1971. It was headed initially by Paul McCracken, chairman of the Council of Economic Advisers.

In light of later developments, it is interesting to observe that energy policy was the domain of a White House subcommittee at the start of the decade. The subcommittee devised steps to meet the fuel shortages expected in the winter of 1970/71 and set up working groups to address longer term problems. Energy was not a priority issue at the time. More important were environmental concerns, the opening of negotiations with mainland China, worsening inflation, and the weakening of the dollar.

In an attempt to stabilize the inflation-racked economy, Nixon adopted a series of economic measures in August 1971 that included a freeze on oil prices, thus disregarding the stimulative effect that price ceilings would exert on energy demand, and the distortions among products that would result.[60] Natural gas prices had been regulated for some time. Controls would become an increasing anomaly and would complicate the energy problem through the decade.

Controls were the wrong medicine. McCracken summed it up well when he said, "it's hard to think of a more effective way of creating a fuel crisis than to decree U.S. price ceilings ... below those prevailing in the world market."[61] Oil demand continued to increase beyond levels of domestic production, the pressure to remove import restrictions became uncontainable, and the quota system, atrophying slowly but surely, was finally dismantled. Still, major price hikes were not anticipated. A reassuring view held that with the entrance of the United States as a powerful customer in the world oil market overseas, competition in the production of oil would increase. Given the extremely low costs of producing petroleum, it was theorized, the price of oil would decline over the long term.[62]

After the June 1971 energy message, McCracken's subcommittee languished and energy policy slipped a notch in White House status. When McCracken left government in January 1972, presidential assistant for international economic policy Peter M. Flanigan took over the subcommittee and initiated another set of energy task force studies. Flanigan, whose background was in finance, came out in favor of gas deregulation, drawing support for his position from a number of government and industry studies and from the results of an M.I.T. computer model of the oil and gas industry built by economists Paul MacAvoy and Robert Pindyck. The MacAvoy-Pindyck model was one of the first of a growing number of energy models to be used in the emerging energy policy debate.[63]

One indication of the low profile of energy policy in the White House during this period was the broad portfolio of projects Flanigan carried. Energy was but one. Used extensively by Nixon as an emissary to the business community, Flanigan was subject to charges of conflict of interest. His having been president of a tanker company did not endear him to environmental and consumer groups, and he was unable to take an active role in the formation of policy.

It was politically adept Interior Secretary Rogers C.B. Morton who nominally held the energy subcommittee chairmanship for much of this period. But Morton is said to have been "out of the room" during many of the energy policy meetings. Flanigan took the initiative and led the energy planning effort, assisted by, among others, James E. Akins, a controversial State Department expert on Mideast oil. Akin's strong views on the dangers of rising oil consumption and the need for aggressive energy R&D were, according to Ehrlichman, generally "discounted," though Simon credits Akins with having contributed to his education on oil.[64]

After Nixon's reelection, energy began to loom as fertile political ground. The Democrats, led by Senator Henry Jackson, were sowing political seeds, having earlier initiated an influential several-year effort of energy hearings and analysis known as the National Fuels and Energy Policy Study.[65] "Everyone wanted to get involved," notes Ash, who was now to join the administration fulltime as head of OMB.[66] Ehrlichman was appointed with Shultz and Henry Kissinger to a Special Energy Committee. The committee recruited systems analyst Charles J. DiBona, head of the Center for Naval Analysis, as its special consultant on energy. DiBona, a former Navy officer, had no industrial associations at the time.[67] His instructions were to put together a small National Energy Office that would prepare itself over the course of the year to play a coordinating role in energy analysis for the White House.

Before DiBona's appointment was official, Ehrlichman had him reviewing a set of energy policy option papers based on the work of the Flanigan committee. Soon DiBona was preparing a second energy message for Nixon to present. The message came on April 18, 1973, just days before Ehrlichman's departure.

DiBona, innocent in energy policy but seasoned in his concern for national security, was a staunch believer in unregulated free enterprise.[68] He strongly opposed price and allocation controls on oil, went along with the recommendation of the Oil Policy Committee that import quotas be scrapped, and favored aggressive domestic energy production and development. Shultz credited DiBona several years afterward with having made "massive and creative contributions to the analysis of policy issues and to the April 1973 energy message, a document whose recommendations—including the elimination of price controls on natural gas—hold up well today."[69] Yet with its low profile on research and development, it was found wanting at the time. "The original version of the energy message called for a much enlarged program," explains DiBona. "This was taken out by OMB."[70]

DiBona negotiated on the energy message until 2:00 a.m. on the morning of April 18, victim of a collapse in White House communications at what the Watergate tapes later revealed was a critical point in the thickening intrigue. Ehrlichman, who under other circumstances may have been able to get OMB's decision reversed, was unreachable. Shultz, who favored the enlarged program, was out of town. Ash, who opposed it, felt the boost in energy R&D should be financed primarily by private industry, not by government. He did not want the president coming out in favor of any major increase in government spending during a period when he was impounding funds Congress had allocated.

The April message removed by proclamation all existing tariffs on foreign crude oil and products, suspended limitations on the amount that could be imported, and adopted license requirements in place of controls.[71] Oil imports continued to rise. With the production cuts of oil-exporting countries later that year, crude oil prices started to climb.

The energy message outlined organizational steps that would begin to activate players in subsequent events. The Interior Department was directed to develop a capacity for gathering and analyzing energy data, and for coordinating energy conservation programs throughout the federal establishment. Recruited for the purpose was Eric Zausner, an energetic twenty-eight-year-old with the Council of Environmental Quality.[72] Also cited in the energy message were the Oil Policy Committee in the Treasury Department under Deputy Secretary William E. Simon and a new division of energy and science within OMB under John Sawhill, associate director for energy and natural resources. Simon and Sawhill would assume prominent roles in energy policy as people jockeyed for position in the wake of Ehrlichman's exit and Nixon's increasing absorption with Watergate.

On June 29, 1973 (still pre-embargo), energy policy received a sudden promotion in the White House. While the recommendation for a Department of Energy and Natural Resources was bogged down by jurisdictional rivalries in Congress, the Senate approved Henry Jackson's idea for an energy council to function like the Council on Environmental Quality. Nixon reacted quickly, determined not to be upstaged. At the urging of Roy Ash, he issued an executive order restructuring the energy bureaucracy and creating a White House Energy Policy Office. He proposed an independent Energy Research and Development Administration (ERDA) to incorporate the research activities of the AEC and Interior's Office of Coal Research, while delegating to a separate body the AEC's licensing and regulatory functions.

Nixon regarded energy as primarily a political problem and sought someone with political clout to head the Energy Policy Office. Turned down by William Scranton of Pennsylvania, the president appointed Colorado Governor John A. Love to the position and made DiBona deputy head. The analyst's job he had innocently taken on, recalls DiBona, turned out to be "90 percent politics and only 10 percent analysis."[73]

Love, the first of the energy czars, never had a statutory base. He was very vulnerable in the intense political infighting that erupted. Responsibility for the oil import program, transferred from Simon to Love in October 1973, was soon back in Simon's hands. Love, with little prior experience in the energy field and seemingly indecisive, resigned rather than remain as a "superfluous" advisor when Simon was elevated above him in still another reorganization of the energy bureaucracy after the embargo. DiBona also resigned.

"To this day, I don't know why they offered me the job," Love said afterward.[74] Charged with not moving fast enough on crucial issues, Love claims that the Nixon Administration did not realize the depth of the problem and was not receptive to the actions he thought necessary. "To be honest, its been difficult to try to do anything meaningful and even to get the attention of the President."[75] Most telling was the power struggle that Love and DiBona waged with Simon over how to deal with the prospect of a mandatory allocation program emerging from Congress. DiBona, who opposed the program, believes Simon considered allocation inevitable and pushed it as a preemptive maneuver despite strong free-market leanings. Simon later called the centralized allocation process over which he ruled a "disaster."[76]

Nixon was by now fighting to retain credibility with the electorate. An issue hotly disputed during this turbulent period was whether a gasoline tax or rationing was the best way to curtail energy demand. To Simon and Shultz, rationing was an anathema. To Nixon, so was a gasoline tax or anything else that might further alienate the nation. It was not a time for forceful policy initiative of that kind.

PROJECT INDEPENDENCE

When the Arab countries announced their embargo and production cutbacks in October 1973, the president, preoccupied with Watergate but under pressure to respond aggressively to what was widely regarded as blackmail from abroad, acted hastily rather than in concert with America's allies. On November 7, 1973, Nixon delivered an address outlining his steps for dealing with the energy emergency. The concluding words were designed to stir feelings of patriotism in a nation approaching its bicentennial anniversary.

> Let us unite in committing the resources of the Nation to a major new endeavor, an endeavor that in the Bicentennial Era we can appropriately call "Project Independence."
> Let us set as our national goal, in the spirit of Apollo, with the determination of the Manhattan Project, that by the end of this decade we will have developed the potential to meet our energy needs without depending on any foreign energy sources.
> Let us pledge that by 1980, under Project Independence, we shall be able to meet America's energy needs from America's own energy resources.[77]

Project Independence, as thus formulated, was both vain and in vain. It was a desperate act by a man stubbornly trying to reverse the tide of opinion that would finally drive him from office nine months later. Apollo and the Manhattan Project notwithstanding, Nixon's concept of self-sufficiency—zero oil imports by 1980—was unachievable. Energy experts pointed out the impossibility of the goal straightaway.[78]

Undeterred, Nixon called for a "blueprint" to provide a strategy and course of action for energy independence. To work out the blueprint and handle the growing shortage in petroleum supplies, Nixon issued an executive order in December 1973 creating a temporary Federal Energy Office (FEO), pending congressional approval of a Federal Energy Administration (FEA) that would assume some of the functions planned for the now shelved Department of Energy and Natural Resources. The bill establishing the FEA was signed into law the following May. FEA gained statutory authority on July 1, 1974.

With this impetus, a federal energy bureaucracy and in-house analytical capability finally began to take shape. Shultz arranged for Simon to be appointed FEO administrator and Ash arranged for Sawhill to be named deputy administrator. Treasury and OMB were in this way both factored into the new operation.

Simon insisted on keeping his Treasury post in order to maintain a bureaucratic base—and thus remained in line to become secretary when Shultz left government a few months later. Sawhill, taking over from Simon at the new agency, assumed an active role in fashioning its research strategy. He appeared "committed to a rational planning process" for long-term decisions.[79]

Despite the respect he commanded in many circles, Sawhill did not reign over the energy bureaucracy much longer than had either Love or Simon. He never established effective rapport with Gerald Ford. Within two months of Ford's having taken over for Nixon (August 1974), Sawhill was out, the loser in a power struggle with the president's friend, Interior Secretary Rogers Morton.

Morton, whom Ford knighted "overall boss" of the national energy program,[80] may have felt the FEA was poaching on his dominion. To Morton, Sawhill was not a "team player." Sawhill's strenuous promotion of a gasoline tax just prior to the 1974 congressional elections was contrary to Ford's political judgment, and his outspoken opposition to a crash program to develop synthetic fuels was jarring too. Some said Sawhill's proclivity for giving speeches and granting interviews was his political undoing. It conflicted head-on with the president's desire to avoid public discussion of the issues prior to an official decision.

Morton was Ford's advisor on organization. To wonder how he came to be named "overall boss" of energy at the same time that

Sawhill was presumably "energy czar" would be an admission of political innocence. It is reminiscent of Morton's serving as nominal chairman of the energy subcommittee of the White House Domestic Council when Flanigan was leading the subcommittee in energy planning, or of Simon's being inserted as FEO administrator above supposed "energy czar" John Love. The number one spot in energy was about as tranquil a position to occupy in the 1970s as chief of programming for a TV network. It was not like directing the Mint.

The fact is that if we include Ehrlichman, McCracken, Flanigan, and DiBona, as well as Love, Simon, Sawhill, and Morton, and if we continue with Zarb, O'Leary, Schlesinger, and Duncan (who enter the narrative further on), there was, all told, a succession of twelve energy principals during the decade—a disconcerting commentary on the political turbulence of energy policy during this period.

It is remarkable that any in-house analysis at all was possible under the circumstances. Yet an impressive model-building effort was achieved right within the FEA. Among those most responsible for bringing it about and getting an audience for it were Sawhill, who launched and supported the work, Frank G. Zarb, who as Sawhill's successor at the FEA brought the results of the analysis to the personal attention of President Ford (with whom he enjoyed a good working relationship), and Eric Zausner, who oversaw the team analysis and formed a bridge to Zarb.

Zarb had served at OMB as acting administrator of the allocation program. His background, like that of both Simon and Sawhill, was in finance. Congress set up an Energy Resources Council in October 1974 at the same time it created ERDA. When Morton became chairman of the council, Zarb was named its executive director. Soon afterward, he was appointed FEA administrator as well, replacing Sawhill. With Zarb in this dual capacity, communication with President Ford was no longer a problem. The FEA became the lead body in conducting analysis and in influencing White House energy policy. The FEA dominated the Energy Resources Council because the FEA dominated the analysis. For a time, it may have seemed as though some measure of stability had finally come to energy policy activities at the White House.

PIES and the Analytical Development

According to Genesis (1:1), "in the beginning of creation ... the earth was without form and void." The description fits the early days of the FEO. But amidst the turmoil, a formidable and youthfully aggressive capability for energy policy analysis developed.

As deputy assistant secretary in the Interior Department, Eric Zausner had created three new offices in mid-1973 for energy research, conservation, and analysis, staffed with analysts who went on to distinguish themselves in the energy field. One was John H. Gibbons, later to become head of the influential CONAES Demand and Conservation Panel and subsequently director of the Congressional Office of Technology Assessment. Others included Bart Holaday, assigned to set up the data and analysis operation, and Bruce Pasternack, whom Zarb singles out as "key," along with Zausner, in getting the president and Congress "to focus on a total new data base."[81]

In putting together the analysis team, Holaday recruited David Wood who had worked in energy conservation studies at the Office of Emergency Preparedness. Together they hired former Air Force Academy instructor William W. Hogan, who launched the model-building effort that would come to be known as the Project Independence Evaluation System (PIES). In recruiting, the practice was to acquire the best people that could be found. As if to challenge the brash image he projects, Zausner says, "I made a success by hiring people much smarter than me."

Zausner's analytical experience prior to joining Interior had been in the financial analysis of solid waste management and pollution control. Here again, as in the case of the Ford EPP, concern with environmental problems led to work on energy problems. But the FEA's 781-page Project Independence Report was very different in purpose and content from *A Time to Choose*. By coincidence, the two reports were released within a month of each other in late 1974.

Sawhill acknowledged FEA's debt to the EPP. We were desperate, he said, for whatever information and assistance was available. FEA used the EPP scenario approach in making its projections of energy futures. FEA complemented EPP, according to J. Frederick Weinhold, who had been responsible for much of the EPP's engineering analysis before taking charge of long-term R&D forecasting for the FEA. Weinhold believes the FEA work met most of the objections raised about EPP's limited analysis, provided the quantitative basis that EPP lacked, and thereby legitimized its concepts.

Project Independence underwent a significant change of direction during its execution. Sawhill and Zausner effectively redesigned the project, making it into an overall evaluation of the nation's energy problems and an exploration of policy options.[82] Zausner did not share what he called the "simple view of the world" held by the initiators of Project Independence. "We dumped the word blueprint," he says. "If I had my way, we would have dumped independence, too."[83]

The Project Independence report evaluated a wide range of conceivable means for increasing supplies and reducing demand based on assumed future oil prices of $7 and $11 per barrel (in constant dollars). It held out little hope of achieving zero energy imports. To reach such a goal by 1985, it declared, would require "consistently higher domestic energy prices, a reduction in real Gross National Product, inflation, and possible local or sectoral economic disruptions," with depletion of domestic reserves and "substantial environmental impact."[84]

The attempt to deemphasize the political quest for self-sufficiency was only partially successful. Project Independence was still "first and foremost a political statement," Zausner acknowledges, "even though we tried not to make it that." The project was accused of starting with the bottom line. Even as late as October 1974, Zausner was allowing for the possibility of cutting import requirements to zero by 1985 based on optimistic projections for domestic oil and natural gas production from the Outer Continental Shelf.[85] Zausner's optimism supported Sawhill's contention that a crash program to develop synthetic fuels was unnecessary.

The ghost of independence was indeed not easy to give up. President Ford would replay the theme the very next year with a scheme proposed by his vice president, Domestic Council Chairman Nelson Rockefeller, for a $100 billion quasi-public Energy Independence Authority to provide government support for energy projects. The proposal split the Administration. FEA chief Zarb backed it as a fighting response to the OPEC cartel, while chairman of the Council of Economic Advisers Alan Greenspan and Treasury Secretary Simon strongly opposed it as interference with the market. Also in opposition was an unusual coalition of environmentalists, political conservatives, and oil-state producers. Congress voted down the proposal.[86]

In recharting Project Independence, Sawhill and Zausner decided that what the FEA needed was its own integrated data base and a continuing process for doing policy analysis and comparing alternatives. A strong data and analytical capability could help bring a sense of order and mission to the internal operations of the FEA and at the same time forcefully establish its political legitimacy within the bureaucracy. Machinery for analysis, after all, can confer status and authority like that associated with more material technologies. Information is power, and many did not want that power concentrated in the oil companies.

Thus, the scope of Project Independence was broadened, the target date extended, and a heroic effort mounted to develop a data base and assemble an integrated set of computer models for projecting energy/economy futures and evaluating presidential initiatives in energy policy. The Project Independence team, eventually numbering hundreds of people from inside and outside FEA, was assembled into a score of working groups and task forces. It was a singularly ambitious exertion—a tour de force. A tight deadline was established to have results ready for consideration in President Ford's upcoming State of the Union address. This sense of urgency, amplified by the sense of history the project inspired, resulted in a deep commitment by staff members. It was a perilous period for more than one staff member's marriage.

By November 1974, a year from Nixon's call for independence, and half that time from the initiation of serious work on system development, the models were run and the final report prepared. Just two months earlier, Zausner had called a "go, no-go" meeting of his top staff to decide whether to continue with the models or seek another way of producing a Project Independence report.[87] Recalls David Wood, "It was like going down an icy slope. There was no turning back."[88]

The PIES models contained representations of alternative sources of energy supply and energy demand. Also developed were models focusing on the macroeconomy, economic impact, vulnerability, and environmental factors. Existing models were adapted to the system and new models were built, as appropriate. An integrating model constructed by system architect William Hogan brought estimated energy supplies together with projected energy demand to effect a "partial equilibrium" that produced price and quantity forecasts.[89]

The builders of PIES were confronted with two schools of thought. First were estimates of future supplies of coal, oil, gas, and electricity, largely by engineers and geologists. Second were estimates of future energy demand by economists and econometricians. The langauges and methodologies of these two sources of expertise were very different, as were their points of view. Economists predicted significant increases in production and decreases in consumption resulting from rising energy prices.[90] Engineers, on the other hand, tended to ignore or discount the effect of price. The divergent conclusions from these polar positions were not easy to reconcile. PIES tried to bring together the best of both perspectives.

Sawhill was proud of the PIES achievement. In composing his preface to the final report, he gave the models star billing.

> Perhaps more important than this report are the framework and analytical tools developed during the last several months ... to evaluate changing world and domestic conditions and the impacts of alternative policy actions.[91]

Rogers Morton was unimpressed. He dismissed the FEA analysis as "a lot of fancy footwork with computers,"[92] and rebuffed Sawhill's attempt to provide FEA input for a presidential address in October 1974. Sawhill's recommendation of a gasoline tax and his warnings against establishing a crash program in synthetic fuels met a similar fate.

At Its Heyday

The situation changed when Zarb took over the FEA at the end of November. PIES was used actively throughout President Ford's term of office. Its structure, scope, and data base grew. For the first time, the government appeared to have its own source of energy information and analysis—reasonably integrated, increasingly regionalized, and operated and maintained in-house.

Zarb and Zausner were among the advisers who assembled with President Ford at Camp David in December 1974 to review energy policy options in preparation for the State of the Union message in January. Briefing books outlining the options and tracing out their consequences were prepared for the meeting with information generated by PIES. There was a follow-up meeting the same month in Washington and a third meeting with the vacationing president in Vail, Colorado over the Christmas holidays.

Ford played an active role in the give-and-take. He was described as being much more involved and interested in energy policy than Nixon had been after the Arab oil embargo.[93] Ford decided to move for complete decontrol of oil and gas and to set a production goal of 1 million barrels a day of shale oil and synthetic fuels by 1985. This bullish goal was restrained compared to a proposal for producing 4.1 million barrels a day by 1982 floated by the secretary of commerce the previous January.

A dominant political consideration in fashioning energy policy during the seventies was to avoid actions that would benefit the oil companies excessively as the value of petroleum soared. A temporary tariff on imported oil was installed for a time, but that turned out to be almost as much of a political hot potato as decontrol. Raising prices, in effect, was left to OPEC.

President Ford recommended decontrol in his State of the Union message on January 15, 1975. The proposal was opposed vigorously by the Democratic Congress. The debate lasted a full year, with PIES playing an active role. Personalities and politics took over where the model left off, according to a principal player at the time.

Many members of Congress were concerned about the effect further oil price rises might have on the inflation-prone economy as it was recovering from recession. Gasoline prices were viewed as inflation inducing, and there was doubt that an effective windfall profits tax could be passed to offset the large profits that would accrue to the oil companies under decontrol. The Democratic opposition tended to regard the use of PIES as a blatant political power play by the Republican Administration to advance its position. Resentment and suspicion of PIES built up as the resistance to decontrol increased.

To appease the opposition, President Ford submitted a substitute 30-month plan to Congress on July 17, 1975, for phased price decontrol. Accompanying the plan were computer evaluations of the estimated impact of the plan on the economy. The FEA claimed the impact would be minimal. Others disagreed.

Walter W. (Chip) Schroeder was a modeling specialist hired from M.I.T. as a "defensive weapon against PIES" by the influential House Subcommittee on Energy and Power. Schroeder, mistrustful of the FEA results, reviewed the simulations that FEA had run based on a Data Resources Inc. (DRI) model of the economy. He discovered a crucial technical flaw. The "base case" (the standard against which the implications of the decontrol program were compared) already had decontrol assumed within it. The error was innocent, but important.

When the comparison was redone against a simulated future that assumed continuation of price controls, much larger effects on the economy were obtained than those reported by the FEA.[94] But even with the mistake corrected, the economic costs of decontrol calculated by the FEA were consistently less than costs estimated by a number of congressional and independent analyses. Schroeder argued that the FEA's downward bias was due to its assuming away the possibility of future price increases in OPEC oil, and neglecting sympathetic price rises in coal and natural gas.[95]

President Ford actually submitted two decontrol programs. The second had phasing extended to thirty-nine months. Both proposals were defeated. Perhaps a different order of analysis or research strategy could have saved the day. There is no way to know. "I believe the political forces were just too dominant," says James L. Sweeney, PIES analyst at the time, who did follow-up work on the impacts jointly with Schroeder.[96]

The political skirmishes and FEA testimony continued over the remaining months of the year. Then, just prior to the New Hampshire and Florida presidential primaries, the administration finally yielded to the Democratic Congress. On December 22, 1975, at Zarb's urging, President Ford reluctantly signed into law the Energy Policy and Conservation Act (EPCA) rolling back prices and providing for a forty-month continuation of controls under a complex system that Jimmy Carter finally phased out in April 1979. William Simon, who served under Ford, called EPCA the worst error of the Ford administration."[97]

"Somewhere along the line the purpose of PIES changed," asserts Schroeder. "Instead of being used as a forecasting model to investigate the interaction of current trends, it began being used as an instrument for advocating policy options." In Schroeder's opinion, Congress viewed PIES as a means for the White House to use numbers "to obscure the frequently weak logic underlying its policy proposals." Congress, says Schroeder, rejected PIES's "overly optimistic portrayal of the response of oil and gas supply to price increases."[98]

Another criticism of the FEA analysis comes from economist Neil de Marchi, who reviewed the energy policy papers of the Nixon and Ford administrations.

> The net result of FEA's efforts ... to quantify the inflationary impact of Ford's program ... was to discredit the Administration. FEA's early estimates of the impact ... were unrealistically low ... FEA, in making its numerical comparisons, was doing the Administration a disservice ... Once the defense of Ford's plan was altered from efficiency (with supplementary devices to safeguard equity) to that of least impact on prices, output, and unemployment, the way was opened to make ad hoc adjustments so as—supposedly—to lessen the impact on particular groups, industries, or regions.[99]

As the criteria began to move away from economic efficiency toward distributional equity, argues de Marchi, "the basis of the original program was weakened and the issues on which decisions affecting energy policy were to be made were much obscured." Energy became a "full-fledged political good," and the battle among special interests escalated.

Old Models Never Die

The PIES effort received a staffing setback in mid-1976 when Zausner, Hogan, and Sweeney departed from the FEA. Zausner went to a consulting firm; Hogan and Sweeney returned to academe. "When the good people left," laments an outside consultant who had worked closely with the FEA in the development and running of PIES, "there was no longer any reason to take PIES very seriously." But other "good" members of the modeling team did remain, and the system continued to be used, albeit under a spreading cloud.

Having provided the basis for the policy scenarios published in the 1976 *National Energy Outlook*, PIES was again put to work producing scenarios of gas deregulation for the 1977 *National Energy Outlook*.[100] Though published in draft form, the 1977 report was never disseminated. It had the misfortune of making its appearance at the same time as the incoming Carter Administration, whose members were determined to put distance between themselves and their predecessors. Yet they were not about to dispense with the imposing analytical and informational capability that the FEA had developed.

Congressional opponents to decontrol, meanwhile, were still smarting from the blows inflicted on them by Zausner's analytical arsenal. They recognized the need for energy information and analysis, but wanted it nonpolitical and accessible to them. In August 1976, Congress passed the Energy Conservation and Production Act (ECPA) creating a separate and independent office within the FEA to "develop a National Energy Information System containing accurate, coordinated, comparable, and credible energy information for energy-related policy decisions by the FEA, other government agencies, the Congress, the president, and the public." The act specified that the new office and the PIES analysts within it be "insulated from FEA's role in formulating and advocating national policy," and set up an auditing team from outside the FEA as an official watchdog.[101] The auditors charged that the PIES analysts were assigned a sensitive policy-relevant role contrary to the spirit of ECPA at the very beginning of the Carter Administration.[102]

What happened is that Carter's energy advisers under James Schlesinger were hurrying to put together a National Energy Plan that the president had promised to present to the nation within three months of taking office. Schlesinger's staff asked for a PIES impact analysis and specified scenario assumptions, it is alleged, that strengthened the case they wanted to make.[103] The PIES team, uncertain about its future in the soon-to-be-created Department of Energy, was ready and eager to oblige. It responded with a "crash" effort that reads almost like a replay of the original PIES operation.

> People worked night and day ... We did not document as we developed. We did not carefully examine each of the thousands of data

items, checking and double checking for accuracy. We did not validate the new model structures. We broke every rule of professional standards tied to flat notions of goodness. We did ask questions with the precision imposed by a model specification. The dialogue with the White House Energy Office surfaced real difficulties that in turn led to improved versions of the Plan . . . We did, in fact, deliver an analysis before April 20.[104]

The "collected wisdom" assimilated by the analysts during this exercise enabled them subsequently "to perform scores of analyses for Congress and the Administration." But Congress was not always appreciative, and the former irritation with PIES reappeared in full measure.[105] Targets for congressional displeasure now included Lincoln E. Moses, Stanford professor of statistics, who had become administrator of the newly formed Energy Information Administration (EIA), a separate and independent office established to carry out the provisions of ECPA.

Glad to shed unpleasant past connotations and having no personal loyalties to PIES, Moses observed with amusement if not relief when EIA analysts decided to retire the PIES label. A tongue-in-cheek obituary was distributed.

Project Independence Evaluation System (PIES), age 4, a complex evaluation, forecasting, and analysis system and beloved project . . . passed away on July 20, 1978, at the Department of Energy . . . PIES is survived by a number of offspring and relatives who will carry on the task of forecasting and analysis.[106]

EIA analysts acknowledge they were persuaded of the need for a new name because of the association of PIES with the "old administration." Also, they say, few of the original elements of PIES remained intact and the system was no longer being used for short-term projections, a use for which it had been much criticized. The name chosen for the survivor was Midterm Energy Forecasting System. But familiar and catchy acronyms, like celebrated generals, never die. Many persisted in calling the system "PIES."

A complaint about PIES from the start was that it was difficult to understand and its runs were impossible to reproduce. The charge was made repeatedly despite the record number of times the system was assessed. The complaints grew louder when PIES was run for the team formulating the National Energy Plan

(NEP), an operation cloaked from public view by the White House. The secrecy caused problems. It was never explained why PIES seemed to be looking much less favorably toward deregulation in 1978/79 under Carter than it had in 1975/76 under Ford.[107] Gas-producing states did not care at all for this change of perspective or for Carter's proposed incentive price ceiling of approximately $1.75 per thousand cubic feet on all new natural gas.[108]

There followed a stinging critique by the auditors[109] and the most comprehensive review of PIES yet—a lengthy assessment by the state of Texas with funding from the governor's office.[110]

> Attempts were made to identify the source of the NEP analysis. It was assumed that the source was PIES, an integrated set of computer models housed at the FEA. Numerous informal attempts were made to obtain documentation of the models and the specific parameters and assumptions used in preparing the NEP, with attention focused on the $1.75/mcf ceiling price ... It was confirmed that PIES was indeed the source of the technical analysis for the NEP.[111]

The busy analysts now responsible for PIES, suffering the pangs of reorganization and turnover of personnel, had trouble dealing with the request of the Texans for computer tapes and documentation of results. The Texans, after about a year of frustration, including resort to the Freedom of Information Act, finally managed to perform a review by transferring operation of the modeling system to their own computer. Administrator Moses, according to the Texans, encouraged his initially reluctant staff to search out the information needed for the transfer.

The total amount of attention accorded PIES is without precedent in the still brief history of energy policy modeling. The system stands as a worn symbol of the heroic attempt to apply analysis on a grand scale to the energy debate of the seventies. Although resisted and resented, PIES was nevertheless the beginning of a serious attempt to apply analysis to national energy policy on an integrated basis from within the federal bureaucracy—an attempt that met with constant opposition. Born at a time of tension and confusion, with an independence mission that was ill-conceived, PIES, through its ebbs and flows, stirred almost as much controversy as its contemporary, the Ford Energy Policy Project.[112]

With the EPP, the argument had been over energy consumption and economic growth. With Project Independence, it was more over energy prices, economic impacts, and oil imports. Energy policy analysis at the start of the seventies was not an arbiter of disagreement; if anything, it was a provocateur. It did help to define and sharpen the issues and expose their political complexity. It did raise questions; but it did not settle them. The EPP brought the subjects of energy consumption and resource limitations onto the political agenda and sparked an essential debate. Project Independence did the same for self-sufficiency and the role of energy prices. The EPP gave needed visibility to the conservation issue and questioned the linkage of energy with the economy. Project Independence drew attention to the influence of price and the need for strategic petroleum reserves. Both studies, controversies aside, were important steps the nation would take along the way from frenzied outrage in 1974 to more sober recognition and appraisal of the energy problem.

5 IN SEARCH OF POLICY FOR ENERGY R&D

The flow of energy analyses continued without interruption through the 1970s. An early one, the Rasmussen reactor safety study, was criticized for being overly friendly to industry (especially the nuclear industry), just as the Ford Energy Policy Project (EPP) was criticized for being overly hostile. The two offer interesting contrasts and parallels. Research and development was a repeated subject of study, including several efforts by the Energy Research and Development Administration (ERDA) for purposes of planning and evaluation. President Ford's ambitions for strong government-promoted development of a synthetic fuels industry were questioned by the useful but little-known Synfuels Commercialization Study, a cost-benefit analysis for the Office of Management and Budget (OMB). ERDA's last analytical hurrah, dubbed MOPPS (Market Oriented Program Planning Study), was initiated early in 1977 as the Republicans departed from the White House and the Carter team moved in. Non grata with the new administration, this unusual study generated optimistic natural gas estimates whose good tidings caused problems for promoters of President Carter's beleaguered National Energy Plan. MOPPS became widely publicized in the process.

One reason the [manager of fuels] was "still inclined toward nuclear" was the excellent operating experience his company had had with its

121

reactors. The Three Mile Island facility, for example, near Harrisburg Pa., generated 79.4 percent of the electricity it was designed to produce during the first six months of 1975.[1]

This enthusiasm by General Public Utilities for its Unit 1 reactor at Three Mile Island (TMI-1) was expressed unhesitatingly in November 1975 in the face of growing qualms about commercial nuclear energy. The safety record of the nuclear industry was, at this point, still essentially unblemished. A fourteen-volume Atomic Energy Commission (AEC) study of reactor safety (WASH-1400 in AEC terminology), led by M.I.T. professor Norman Rasmussen, had just been released and was being cited by nuclear advocates in an attempt to ease the public's concern about the safety of nuclear power.[2]

The nuclear resistance movement was not to be subdued. Within the next three years, the reassuring conclusions of WASH-1400 were being officially questioned and the government was criticizing aspects of the report in a highly publicized press release.[3] Then the dam burst. As if on cue, the newly operating Unit 2 at Three Mile Island (TMI-2) suffered serious damage to its core, leaking radioactive steam into its containment with radioactivity escaping into the surrounding air. Compounding human error with instrument inadequacy and component failure, "TMI," as the accident was to be known, came to be a symbol internationally for renewed anxiety about whether nuclear reactors were really safe.

The riveting display of drama, shock, outrage, and suspense produced by TMI on March 28, 1979, surpassed that of the Hollywood movie that eerily forshadowed its occurrence just shortly before.[4] The accident, the worst in the history of civilian nuclear power, severely tarnished the silver anniversary of the Atomic Energy Act of 1954, which marked the beginning of that history.[5] Resistance to nuclear power, starting in the mid-sixties just as light-water reactors were experiencing their first major commercial success, reached a crescendo in the aftermath of TMI. For energy consumers already suffering from a second round of sharply rising oil prices, and for electric utilities straining to perfect an alternate fuel source, TMI was a serious shock. General Public Utilities, forced by government to keep the power off at its undamaged TMI-1 reactor and facing cleanup costs at TMI-2 estimated at over \$1 billion, appealed for emergency aid to avoid bankruptcy.[6]

A PROCESSION OF STUDIES

Nuclear issues were high on the policy agenda throughout the seventies as the government struggled to fashion an all-encompassing energy policy. It should not be surprising, given the technical and political complexity of nuclear technology, that analysis played a key role. The Rasmussen study, in particular, was widely cited in the continuing nuclear debate.

Non-nuclear issues were important as well, especially as misgivings about nuclear energy grew. One set of questions had to do with how best to make use of the country's abundant coal reserves and whether the manufacture of synthetic oil and gas from coal was economically justifiable. ERDA addressed such questions in its effort to carry out its research and development mission; so did the White House's Synfuels Commercialization Study. This chapter discusses several studies having to do directly or indirectly with R&D.

Nuclear Reactor Safety

Growth in the use of electricity in the United States displayed steady annual increases prior to the 1970s. With the Atomic Energy Act of 1954, Congress endorsed the idea of using nuclear energy to help meet the nation's steadily growing electrical demand. To assist the fledgling nuclear industry in getting started, and to promote orders by electric utilities for water-cooled nuclear plants, Congress passed the Price-Anderson amendment to the Atomic Energy Act in 1957. This controversial insurance program indemnified utilities against liabilities of greater than \$60 million arising from a nuclear accident (plus \$5 million for each operating plant), and transferred to government the remaining liability up to a maximum of \$560 million. To provide supporting analysis for the Price-Anderson amendment, the AEC had its Brookhaven Laboratory make a study of the consequences of commercial nuclear mishaps.

The safety of nuclear reactors has been a topic of central concern since the beginning of civilian nuclear power. A major reason is that the heat generated in nuclear fuel continues at a high level long after the fuel has stopped fissioning. Unless the mechanisms for cooling a reactor are effective and reliable, this heat can melt the reactor core and release hazardous amounts of radioactivity. The Brookhaven study examined the damages such an accident could inflict, but had no methodology for calculating its likelihood.

The Price-Anderson amendment was approved. Yet the Brookhaven study, known as WASH-740, may have been used more by the opposition than the sponsors. The study estimated that a worst-case accident, which it considered highly unlikely, could cause 3,400 deaths, 43,000 injuries, and $7 billion in property damage. These alarming numbers became grist for the mill of nuclear critics, to the consternation of the AEC. In 1972, as the renewal date for the Price-Anderson Act approached, the AEC (with James Schlesinger as chairman) commissioned another study of reactor safety—one that would include the calculation of risks. This time, it is alleged, the commission took greater care to see that its study would not backfire.[7]

According to one commentary, the AEC first sought to have the study led by "a believer in the safety of nuclear plants," Manson Benedict, an Institute Professor at M.I.T.'s department of nuclear engineering,[8] but ultimately selected his colleague, Norman C. Rasmussen, whom Benedict suggested was better qualified for the assignment.[9] "Whereas Rasmussen initially proposed that the study be done at M.I.T., the commission chose to have it done at AEC headquarters where they could keep close watch as it progressed."[10]

In later defending its appointment, the AEC said it needed someone knowledgeable in the operation of nuclear power plants and in the analysis of industrial accidents, and also someone acceptable to the nuclear industry to ensure that utilities and manufacturers would "make essential data available to the study group."[11] Contrast this with the criteria used in the choice of David Freeman to head the EPP. Freeman was considered anti-industry and a consumer advocate; Rasmussen, proindustry and a nuclear advocate.[12] Ironically, Freeman's two and one-half year, 4 million dollar project started just months before Rasmussen's three-year effort of roughly the same cost.

Rasmussen maintained from the start that his study was neutral and that he and his sixty researchers were impartial in their investigations. Doubt was cast by a letter he wrote to the AEC before the selection. The letter was made public later under the Freedom of Information Act at the behest of the Union of Concerned Scientists. In the letter, Rasmussen referred to parts of the proposed study as "a manageable task that might have significant benefit for the nuclear industry." For "the report to be useful," he ventured, it "must have reasonable acceptance by people in industry."[13]

Rasmussen released his report, WASH-1400,[14] in draft form in August 1974, the month Richard Nixon resigned from the presidency. Dixie Lee Ray was then chairman of the AEC. The final version appeared on October 30, 1975. WASH-1400 endeavored to do what WASH-740 had not—estimate accident probabilities. The exceedingly low probabilities it imputed to serious accidents (whose consequences it agreed were enormous), made the report much more agreeable to the nuclear fraternity, a brotherhood dubbed by one cynic as "a religion in search of a bible."[15]

The report's popularly written executive summary minimized the reactor hazard. In a figure on its second page, it equated the results of a meteorite impacting the earth to expected fatalities from 100 nuclear plants: 100 or more fatalities once every 100,000 years; 1,000 or more fatalities once every million years.[16] The figure should have labeled these fatalities "early fatalities," as a footnote points out. Expected early fatalities from a reactor accident are small compared to probable longer term consequences such as cancers, for which comparison with a meteorite does not apply.[17] Rasmussen said later, "I wish we hadn't put it in," referring to the heavily quoted meteorite analogy. "It was grossly misused and misled the public."[18]

The Reactor Safety Study had a "soothing tone," commented Harold W. Lewis, who chaired an ad hoc review group that critiqued the study for the Nuclear Regulatory Commission (NRC) at the urging of Congress.[19] As Lewis saw it, WASH-1400 "was widely used as a propaganda item by AEC and NRC."[20] Timed to coincide with congressional review of the Price-Anderson extension in 1975, the study was accorded a key role in securing the act's renewal. Rasmussen denies that was its primary goal.[21] It was also used extensively by electric utilities in public relations programs fashioned to inform and calm customers about nuclear energy. In the words of one congressman, "Professional assurances about the low degree of risk were very important to the public perception of that risk ... [WASH-1400] was very, very central to the feeling that nuclear power is safe."[22]

So much for the image. What about the technical content? As with the EPP, beneath the political facade was a serious research study, one that contributed to methodology in the difficult and murky field of ultralow probability, ultrahigh consequence assessment. The study endeavored to apply decision analysis to reliability theory. It employed a concept of branching structures (fault trees and event trees) from the space program that traced out the likelihood of possible sequences leading to failure in complex systems. Reviewers such as Lewis remarked later that the risk assessment methodology that WASH-1400 developed was still "the best thing around," sound and deserving of wider use by the NRC.

Indeed, Lewis was disturbed that the commission's staff had not applied the methodology more to its regulatory and licensing decisions. "The licensing regulatory staff had never liked WASH-1400," said NRC Commissioner John Ahearn. "They had refused to participate in it because they didn't like/trust/understand probabilistic analysis." Those that did participate in WASH-1400 "were in the Research Division and stayed there."[23]

The NRC claimed it had scarcely ever used WASH-1400 in its licensing decisions,[24] while acknowledging that the study's executive summary had been unduly optimistic.[25] The NRC had been under considerable fire from Congress and antinuclear groups because of its identification with nuclear interests. It gave the Lewis critique wide publicity. Rasmussen, for one, believed it was attempting to gain an image of objectivity by adopting a critical stance toward the pronuclear study it had inherited from the AEC.

The Lewis critique considered WASH-1400's probability calculations opaque and "inscrutable," and faulted the study's treatment of system dependencies and common-cause failures; that is, causes of failure that could affect several parts of the system at once. The reviewers had a nagging suspicion that the analysis was incomplete. Overlooking certain statistically significant sequences of events leading to system failure would make the estimated accident probabilities too low. But choosing parameter values that were overly conservative, and giving insufficient weight to human adaptability (such as that displayed at a nuclear plant fire at Brown's Ferry in 1975) would make the estimated accident probabilities too high. The net effect, said the reviewers, was not that the probability estimates were necessarily wrong, but that the error bounds or confidence limits attributed to these estimates were grossly understated.

The subtlety of the distinction was missed in the reading and writing of press accounts, and thus an impression was created that reactors were much riskier than the Rasmussen report had allowed.[26] While Lewis, a self-confessed nuclear advocate, was confiding that he personally believed the estimates were very likely too high,[27] a story in the *Washington Post* was suggesting erroneously that the Lewis review had found the estimates too low.[28]

The NRC's criticism of the Rasmussen report was a signal for some that a massive reappraisal of nuclear power by the American public was underway. The Union of Concerned Scientists immediately called for sixteen nuclear plants to be closed because there was no longer any technical basis for judging them safe.[29] The NRC demurred but did later shut down five plants because of what it said was "deficient analysis of their ability to withstand earthquakes."[30] This came just two weeks before TMI.

With TMI, questions began to fly. Did TMI prove that nuclear power plants are a serious hazard after all? Or did it rather show that even in the most alarming of circumstances the danger still can be contained? Was TMI the beginning of the end for civilian nuclear power? Or, by bringing the spectre of nuclear disaster out of the realm of the feared unknown and suggesting needed corrective steps, had it, in fact, ushered in the age of commercial nuclear power?

There was an effective moratorium on new nuclear construction in the United States. Only two new orders for nuclear power units had been placed in the three years before TMI, and those by a single electric company.[31] The nuclear industry was in the doldrums. Right after the accident, the NRC temporarily stopped granting operating licenses for recently completed plants as well as construction permits for new ones. "This measure, designed to allow breathing space for assessment of the implications of TMI, immediately affected 9 of the 94 plants then in some stage of planning or construction."[32]

The time required to get a facility planned, approved, and built stretched to twelve years and more. Construction costs were soaring—several billion dollars per plant—and a price tag of $1 billion plus had been placed on the cleanup at TMI. The economics of nuclear power no longer justified the buoyant optimism of earlier decades. With electricity demand projections down substantially from prior years,[33] consultants advised utility executives to abandon plans for nuclear reactors in deference to their stockholders.[34] The private (financial) hazard of reactors was dominating the public (safety) hazard that had been the preoccupation of WASH-1400.

What meaning did TMI have for the Rasmussen analysis? Were the human errors and sequence of events leading to failure covered by the WASH-1400 calculations? Was the hydrogen bubble that developed at TMI foreseen by the analysts?[35] The presidential commission set up to investigate the accident maintained that the study had anticipated the very "loss of feedwater plus failure of a safety valve" that took place at TMI.[36] It also maintained that the probability of the failure sequence was "high enough, based on WASH-1400, that since there had been more than 400 reactor years of nuclear power plant operation in the United States, such an accident should have been expected."[37] Commission members "found themselves wondering why the NRC had not attended to WASH-1400 more closely."

> WASH-1400 contains three important messages. These involve expected frequency of accidents, methods for improving reactor safety, and the most likely types of accidents. Perhaps it is the fault of the report that these messages were not emphasized, because the conclusion most often associated with WASH-1400—reactors are safe—receives the primary emphasis ... Perhaps it is the fault of the NRC that more effort was dedicated to criticizing WASH-1400 than was applied to understanding its messages. One of the clearest messages in WASH-1400 is that reactor accident risk is dominated by small-break loss-of-coolant accidents and transient-initiated accidents—the small upsets that can make a mighty accident, as they did at Three Mile Island.[38]

In the Ford EPP, the important conservation conclusion was thought to have been undermined by the populist tone and rhetorical excesses of *A Time to Choose*. In the Rasmussen study, similarly, the report's valuable analysis of most likely failure sequences was vitiated by the public-relations tone of the executive summary. Such is the fate of analysis in a political setting.

ADDRESSING RESEARCH AND DEVELOPMENT POLICY

In discussions of the energy problem, the picture was gradually being accepted of long-term limitations on oil and gas, due both to their finite supply and to the change in proprietary rights exercised by producing countries. The United States needed to examine how much energy it really required and what it could do to promote and prepare for a transition to successor fuels. How much should the government be spending on energy R&D and how should these funds be allocated?

Attempts at formulating a comprehensive national strategy for the development of energy technologies go back at least to 1963 during the Kennedy years[39] and might have come ten years before that had the study of Truman's Paley Commission attracted more interest within the Eisenhower Administration.[40] But long-term energy supply was a political nonissue during the 1950s. There had been abundant oil finds in the Mideast, and there was Eisenhower's reassuring Atoms for Peace program. Expenditures on the development of nuclear power for civilian use rose steadily, and there seemed little reason to worry about an overall plan for energy R&D.

When political priorities shifted in 1973, work had to be done quickly to assess the costs and benefits of coal, synthetic fuels, tar sands, nuclear, solar, and geothermal. Charles DiBona, White House energy planner in the 1973/74 period, despaired at the unavailability of such comparative cost information when preparing the energy message for Nixon.[41]

On June 29, 1973, the president asked AEC Chairman Dixy Lee Ray to review federal and private energy research and development activities and formulate "an integrated energy research and development program for the nation." Nixon stipulated a figure of $10 billion for R&D to be spent over five years. Senator Henry M. Jackson, meanwhile, was promoting a program of twice that amount to free the United States from dependence on Arab oil.

In early December, Chairman Ray reported back with recommendations for an intensified program to develop a range of energy technologies, including processes to substitute coal for oil and gas "on a massive scale."[42] Also recommended was increased spending on nuclear fusion, solar, and geothermal energy. The report suggested a 50 percent increase in the energy research budget, without offering false hope about replacing all oil imports from the Mideast.[42]

Compiled quickly with a sterling roster of contributors, the report was an impressive job. Its identification with the AEC made some worry, but many felt the conclusions were "about right." "Though dismissed by the anti-nuclear crowd," says Dixy Lee Ray, the report "*did* achieve considerable credibility in international energy circles and in the nongovernmental energy community." Its being neglected, she explains, "had far more to do with the growing Watergate problem throughout the spring and summer of 1974 than with any credibility question."[44]

The idea for ERDA came before Congress the same month the Ray report was released. No resolution was in sight to the jurisdictional dispute blocking approval of the proposed Department of Energy and Natural Resources. ERDA, combining energy operations from several government agencies and separating research and development from regulatory functions, was "entirely political in its motivation," says Dixy Lee Ray. "Some of the impetus came from long smoldering, anti-nuclear, anti-AEC sentiments, part came from intra-executive branch squabbling."[45]

It took about a year for ERDA to gain congressional approval. Work commenced on January 20, 1975. By then, both *A Time to Choose* and the Project Independence report had been published. The influence of these reports was hinted at in the comments of ERDA administrator, Robert C. Seamans, Jr. at a news conference on ERDA's unveiling. Said Seamans, "There is no way we can become self-sufficient in 10 years or any time in the future if we keep using energy at the same rate we have been." Seamans pledged to undertake extensive planning and analysis of energy conservation strategies straightaway.[46]

The White House had been careful to select someone *not* from the nuclear fraternity to head ERDA. Seamans, a former secretary of the Air Force with a background in the National Aeronautics and Space Administration, had been serving as president of the National Academy of Engineering. Despite the small brass statue of Don Quixote on the table in his office, nuclear and coal, not windmills, were to be ERDA's main thrust.

To give its program balance, ERDA initiated a concerted R&D effort in non-nuclear fields. A significant portion of its funds went to support work on the conversion of coal, in line with President Ford's production goal of 1 million barrels of synthetic fuels a day by 1985. Seamans noted this goal would not be easy to attain. On nuclear energy he had reason for even greater caution. The AEC had been targeting for a thousand nuclear-powered generating plants by the year 2000. President Ford, more restrained, proposed 200 nuclear plants by 1985. Even that projection was beginning to seem unrealistic.

The single most important issue facing ERDA was what to do with the very controversial liquid-metal fast-breeder reactor (LMFBR) program that was consuming a major share of the agency's research dollars. *A Time to Choose* had deplored what it referred to as the "open-ended government funding commitment to the LMFBR demonstration project," and called for its immediate termination.[47] To assess the program, Seamans innocently commissioned a National Academy of Sciences review similar to one recommended by *A Time to Choose*. Thus began the grueling, protracted marathon of analysis described in Chapter 7.

ERDA's Two Plans

Congress decreed that ERDA, as a new agency, had to come up with a budget and research plan within six months. The agency's nebulous mission needed focus. Seamans assigned responsibility for producing the plan to his deputy, Robert Fri, and to his chief of planning and analysis, Roger W. A. LeGassie.

Fri says he and LeGassie were "both modelers at heart."[48] They decided on a modeling approach after "several false starts" that reduced the time available for analysis to about ninety days. They wanted "numbers" and a way of establishing some sense of organization and priorities among the R&D projects "thrown" at them. What they finally latched onto was the

Brookhaven model, operated within an accounting framework called the Reference Energy System by a group at the Brookhaven National Laboratory that ERDA had inherited from the AEC. The Reference Energy System traces back to work funded by Freeman when he was at the Office of Science and Technology in 1969. Used in the Dixy Lee Ray report, the system was well along in its development and detailed enough to permit ERDA to make use of its results. And it was available! With ninety days to go, that was the "clincher."

The Reference Energy System was an extremely useful, informative representation. It detailed the costs and capacities of alternate energy technologies and mapped out the flows of energy throughout the economy, from source through transportation, conversion, and distribution to end use. That is, from ground and tanker, to refinery and electrical generating station, to factory, home, office, and means of transportation. The illuminating graph of the system's interrelationships was referred to in so many talks and articles on the energy problem during the early and mid-1970s that it became a virtual emblem for energy analysis.

The Brookhaven model used the framework of the Reference Energy System to seek national energy choices that would minimize total overall cost. In its ERDA applications, the model had certain limitations that affected the way it was used and (some believe) the conclusions to which it led. It had an abruptness known as a "bing-bang" quality that is characteristic of programming models. A small change in input assumptions, such as in an operating parameter, could significantly alter the projected choice of future technologies: for example, from no synthetic fuel in 1985 to a million barrels a day, or from electricity in 2000 A.D. predominantly generated by nuclear plants to electricity produced mostly from coal. Avoiding such discontinuities required adding artificial constraints that made the model suspect.

Another problem with the model was that it assumed a fixed set of end-use energy demands independent of supply assumptions. Varying these demands from one run to the next in simulating different scenarios could not adequately reflect the effect of supply (via price) on consumption—nor, therefore, on conservation. Combining the Brookhaven model with a version of the

Hudson-Jorgenson model of the economy developed for the EPP would later correct this deficiency.

ERDA met the tight statutory deadline for reporting to Congress with a plan published on June 30, 1975[49] called ERDA-48. It was a hurried analysis, completed under pressure. Its scenario runs, unconvincing and poorly received, were considered more a justification of ERDA's R&D programs than a comparison of alternate R&D strategies. Negative reaction came especially from conservationists and environmentalists.

Critics such as Amory Lovins of the Friends of the Earth charged ERDA-48 with neglecting both the social impact of the technologies it recommended and the ethical issues underlying their use. A major assessment of the ERDA plan called it "a good start,"[50] but pointed out that it had not taken adequate account of nontechnical constraints such as transportation and the availability of resources, manpower, and capital. Nor had it paid enough attention to such factors as public acceptability and institutional, jurisdictional, economic, and environmental compatibility.

ERDA-48 was considered promotional rather than professional in tone. It was supply oriented, which is not surprising for the report of a technological R&D organization (although technology can be applied to reduce demand as well as increase supply). Most attention went to fossil and uranium-based technologies, which was consistent with the official position of the Ford Administration.

Some charged that ERDA downgraded conservation by running the Brookhaven model with preset demand numbers.[51] The Brookhaven modelers ran a conservation scenario reported in an appendix to ERDA-48. Although showing the largest short-run impact of any of the scenarios, it was not reflected in ERDA's policy recommendations. "All the numbers were there in ERDA-48," says one of the ERDA analysts. "As people got conservation religion, they *noticed* they were there."[52]

"We had the basic material to arrive at a conservation conclusion," admits LeGassie, but "we missed it."[53] By a "conservation conclusion," LeGassie meant a projected level of energy consumption in 2000 A.D. of about 100 quads (quadrillion BTUs). This was the projection given by the Zero Energy Growth scenario of the EPP. About 40 percent higher than U.S. energy consumption in 1975, it was among the lowest projections at the time.

Instead, the ERDA plan forecast a doubling of consumption to over 140 quads by the end of the century. Even this high estimate was well within the range of 124 to 187 quads projected in the Energy Policy Project's Technical Fix and Historical scenarios.

The principal message of ERDA-48 was that no single energy technology could guarantee filling long-term energy needs, so "as many options as possible should be nurtured."[54] ERDA did not want to foreclose possibilities or alienate constituencies unnecessarily. To the cynical reader, ERDA's strategy was the natural, self-interested position of an agency trying to expand its budget and staff. Says Fri, "ERDA-48 was in fact an attempt to determine if priority could clearly be given to one technology over another."[55] The lesson learned was that the type of analysis done for ERDA-48 was not "sufficiently discriminating" for this purpose. In April 1976, ERDA came out with a revised plan. Since the formation of the agency, domestic production of natural gas had fallen almost 7 percent and crude oil production had slipped by 4.5 percent. Petroleum imports were approaching 40 percent of domestic consumption. The quest for energy independence was not going very well.

The revised plan, called ERDA 76-1, elevated conservation to "the highest priority" category, previously reserved for favored supply programs. The report argued that it was cheaper to save a barrel of oil than to produce one, and that conservation was generally less damaging to the environment and less intensive in its capital requirements than was production. ERDA was responding to critiques of its initial plan and to points made at two-day public meetings held after publication of the plan. Additional analysis of research and development options was done for ERDA 76-1 based on linkage of the Brookhaven system with the Hudson-Jorgenson model.[56]

Although dismissed by some as a political placebo, ERDA 76-1 enjoyed a better reception than ERDA-48. It signaled the nation's increasing willingness to face realities. ERDA 76-1 was the first official statement coming from within government to put conservation at the top of the list of recommended energy measures. This action came only after vigorous fighting within the administration. In a period of two years, the idea of conservation had become accepted.

SYNFUELS COMMERCIALIZATION

Energy studies gradually began to give more emphasis to demand strategies in view of the rising cost and uncertainty of world oil, the leveling out of domestic oil and gas production, the unfavorable economics of advanced energy technologies, and the significant political resistance met by these technologies. The nuclear voyage was running into a head wind of public opposition. Besides, electricity did not provide a complete substitute for the liquid and gaseous fuels at the heart of the energy problem. A more direct replacement was possible from synfuels—oil and gas produced synthetically from either coal or oil shale, both of which the United States possessed in bountiful store. Synfuels could also be obtained from biomass and organic wastes, renewable sources that commanded political support from some of the same conservationists who, for environmental reasons, opposed massive build-ups in the utilization of coal and oil shale.

There has never been significant synfuels production in the United States. Projected production costs have always been higher than the price of oil and gas. But this has not discouraged synfuel optimists. In 1975, after the first rise in crude oil prices, the optimists argued vigorously that synfuels production could be brought to commercial scale and made economic within ten years (that is, by 1985, the revised time frame for Project Independence) given major government assistance and the launching of a new synfuels industry.

The highly successful lunar Apollo mission, a ten-year, $25 billion effort was a showcase for U.S. managerial expertise and technological prowess in the 1970s. President Ford announced another ten-year program in his January 1975 State of the Union address; this one was to develop an industry to produce 1 million barrels of synthetic fuels a day by 1985. Ford set the price of the Synfuels Commercialization program at $6 billion, a relative bargain compared to the cost of going to the moon.

Promoting synfuels in the White House were FEA chief Zarb and his replacement at OMB, William T. McCormick, Jr. McCormick, after receiving a Ph.D. in nuclear engineering from M.I.T., had served first with the AEC and then with Love's Energy Policy Office. The president's ambitious synfuels goal, without benefit of careful prior analysis, may have taken McCormick and Zarb by surprise. They quickly formed an interagency task force on synthetic fuels with McCormick as chairman to examine a variety of program alternatives. The task force initiated a Synfuels Commercialization Study under the aegis of the president's Energy Resources Council. The study's decision analysis of alternative synfuel production levels is the least known of the analyses we review.

The Synfuels Commercialization Study, like Project Independence and the Dixy Lee Ray and ERDA studies before it, reexamined the suggested path to energy independence. It was another example of how a presidential declaration can create the need for analysis after the fact. Edward Cazalet and a team of associates from the Stanford Research Institute (SRI) took about four months to complete the study using an energy supply-demand balancing model developed earlier for the Gulf Oil Corporation. (Gulf had been considering the possibility of participating in a coal-gasification venture in the Powder River Basin of Montana and Wyoming. Cazalet's team recommended against it.)

The SRI-Gulf model was comprehensive. It included in its portrayal of the energy system a wide assortment of energy forms, conversion technologies, transportation modes, demand sectors, and geographical regions.[57] One comparison of it with models being used by FEA and ERDA called it "the most useful" for the synfuels exercise because of its sophisticated method of matching supply and demand in its projections.[58] This won the synfuels study for SRI.[59]

The modelers worked within a decision-analysis framework. Decision analysis is a means for explicitly estimating future uncertainties in assessing the costs and benefits of decision alternatives. The model was asked to evaluate four possible program levels of synfuel development.

1. No program. Continuation of research and development and normal investment decisions by U.S. industry without federal involvement in commercialization.

2. Information program. Federal assistance designed to establish a 350,000 barrel a day capacity.
3. Medium program. (2), but for a 1,000,000 barrel a day capacity.
4. Maximum program. (2), but for a 1,700,000 barrel a day capacity.

For only the first of these alternatives were expected costs less than expected benefits (both measured in discounted 1975 dollars). The analysts concluded that the government should not intervene in synfuels commercialization.[60] Members of the task force, in light of the president's announced goal of 1 million barrels a day, were predisposed to favor the medium program. They pressed the analysts to find more benefit categories.

When the analysts held firm, McCormick proposed (2) as a compromise, contending that the net cost (over benefits) of the information program was a small price to pay given other potential benefits that he insisted could not be taken into account by the analysis. The task force's final report relegated the "no program" option to a technical section written by the analysts.[61] The executive summary did not mention it.

> In view of the relatively small risk and expected cost of the "information" option and the other potential benefits not quantified in the analysis, *it is recommended that a program be undertaken with a budgetary authority to install immediately a capacity of approximately 350,000 barrels a day.* This option does not preclude achieving the goal of 1,000,000 barrels a day by 1985 but defers the decision to commit firmly to that goal until later in the decade pending additional information on environmental factors, ongoing R&D programs, industry response, world oil price, and domestic supply and demand. This conditional two-phased approach could provide the opportunity to capture most of the potential benefits of a larger program without risking large potential costs.[62]

Every policy analysis has "contention points": assumptions and issues on which interested parties have significant differences of opinion.[63] "Critical" differences affect the policy results of the analysis. The most critical contention points in the Synfuels Commercialization Study were the projected strength of the OPEC cartel (and by implication, the future price of world oil), the future cost of synfuels technology, and the projected supply of domestic fuels relative to the total demand for energy.[64] The values assumed for these factors were decisive in determining the recommended program level.

The decision analysis gave 50 percent probability to the likelihood of a strong cartel in 1985 and 50 percent to the likelihood of a weak cartel. A probability split of 80/20, to reflect the prospect of rising oil prices, would have favored the information program, other factors held constant. But the cost of synfuels technology shot up faster than the analysis assumed. Increasing that cost would have militated against the information program. Attempts to reexamine the question with more recent data suggest that actual oil prices and synfuels costs, had they been foreseen, would not have changed the study's conclusions appreciably.

In late 1975, the House of Representatives, by a vote of 263 to 140, defeated an attempt to attach a $6 billion synfuels measure to the ERDA authorization bill. The Senate had earlier passed the measure by 82 to 10. The fact that it was phrased in terms of "loan guarantees" may have been its downfall since a bill to provide $2.7 billion in loan guarantees to financially troubled New York City had just been turned back. Congress was reluctant to give the appearance to incensed New York voters that it was granting to the oil industry what it was denying them. Representative John Dingell called the synfuels measure a $6 billion giveaway to the oil companies.[65]

A move to resurrect the program the following year without tying it to the ERDA authorization bill just missed succeeding. It lost on a 193 to 192 procedural vote.[66] A critical report by the General Accounting Office contributed to its defeat,[67] as did the mistaken impression that the synfuels program was linked with the $100 billion Energy Independence Authority promoted by Vice President Nelson Rockefeller. The Rockefeller plan eventually died in committee, after arousing fierce opposition within the top echelons of the administration. Liberal environmentalists joined with free-market conservatives in an uncommon coalition dedicated to keeping the federal government from getting more deeply into the energy business.

The many political aspects of the synfuels proposal complicate any attempt to evaluate the specific role played by the cost/benefit analysis in the final outcome. The analysis did restrain the task force in its recommendation, although not as much as the analysts would have liked. Their results supported the views of the staffs of the Council of Economic Advisers and the Subcommittee on Energy and Power, who opposed the synfuels measure. Some believe the content of the study was influential in the vote on loan guarantees. Clearly, there is no reason why policy analysis, when conducted fairly by people of firm conviction, must serve the political interests of the sponsor.

A CHANGE OF FOCUS

Nixon's approach to energy, enunciated in his June 1971 presidential message on national energy policy and reemphasized at the time of the Arab oil embargo, rested on an assumption that the economy needed cheap, abundant energy for continued growth. Project Independence called for an accelerated effort by industry to expand supplies and make the country energy self-sufficient. It was an appeal to national pride and technological prowess. As ambitions soared, Nixon's energy chief, William Simon, confidently predicted that the new energy sources developed would make the United States an exporter of fuels in the 1980s, even causing the value of petroleum in the ground to fall.[68]

But Project Independence did not become the Project Apollo it was to emulate. Its goal of total independence by 1980 "was revealed to be a costly, unrealistic illusion even before implementation plans could be formulated."[69] The Ford Administration did not renounce the project but it postponed the target date by five years and reduced the goals for nuclear energy, synthetic fuels, and new petroleum discoveries. Vigorous resistance by environmentalists and cost-benefit analyses such as the Synfuels Commercialization Study helped moderate expectations. As the world slipped into recession, the rising demand for energy abated and the cry to expand energy supplies grew fainter. It was no longer a primary issue in Ford's last year of office, causing what some considered to be "a major anomaly of the 1976 presidential campaign."[70]

Carter, victorious over Ford, saw energy come back into the news even before moving into 1600 Pennsylvania Avenue. The winter of 1976/77 was unusually severe in the northeastern and midwestern parts of the country, and a shortage of natural gas developed. Many consumers could not buy gas at any price even though it was still being flared in many parts of the world.

Carter, deciding to make energy a key issue, shifted emphasis from increasing energy supply to reducing energy demand. On television the day after taking office, he set the tone by asking the American people to lower their thermostats to 65 degrees.[71] Carter spoke of the need for "a coherent energy policy"[72] and pledged to offer a comprehensive plan within ninety days, leaving his energy team three months to get a document drafted. Once again, a presidential promise would set off a crash analytical effort.

Time seemed of the essence with Carter. By the end of his second week as president, he signed into law an emergency bill to move gas from areas with a surplus to those suffering a deficiency.[73] He took the occasion, in a televised "fireside chat," to remind the nation that the energy problem did not begin and would not end that winter. Appearing in a cardigan sweater, he struck a conservation theme reminiscent of the EPP and the views of S. David Freeman.[74] Freeman, serving Carter as energy adviser during the campaign, became a member of the president's energy team after the election.

Carter's approach to conservation focused on the disproportionately high use of energy by Americans. Under pressure from abroad, the president strove to exercise the moral leadership he felt was required to get consumption down to levels more in keeping with those of European countries. His advisers argued that "strong presidential leadership is essential in explaining to the public the necessity of sacrifices today to enable lower energy costs and security in the future."[75]

Carter spoke of profligate waste, the likelihood of continuing shortages, and the necessity for personal sacrifice. He could have built his case around the fact that energy economics were out of balance in the United States because of regulated prices, then asked for the nation's forbearance during the difficult times ahead when energy prices would have to be allowed to rise to discourage energy consumption and encourage the development of alternatives to oil and gas. He could have focused on the need for greater efficiency in the use of energy: lighter cars, better insulated buildings, energy-saving technological advances. Instead, he emphasized the need for self-restraint. His was a sermon as much as a program. Pricing energy more realistically and utilizing it more productively was secondary to using it less. Normal government incentives and the operation of the marketplace would not suffice. Cutbacks were in order and the American "dream of plenty" would have to be curtailed.

> Tonight I want to have an unpleasant talk with you about a problem unprecedented in our history. With the exception of preventing war, this is the greatest challenge that our country will face during our lifetimes.[76]

Choosing a phrase from Henry James that Nixon had used before him in a different context,[77] Carter called for a "moral equivalent of war" against spiraling energy use. He issued the call on April 18, 1977, two days before unveiling the National Energy Plan.[78]

The plan turned out to be less severe than Carter's somber tone suggested. The president was trying to convince an electorate whom polls called skeptical that there really was an energy problem.[79] He did not mention at first the provisions for tax rebates and other economic cushions built into the plan by the analysts who drafted it. These analysts did not see the plan as all that burdensome to the general population, but because of a failure of communication with Carter's speechwriter, their positive perspective was expurgated from the April 18 speech.

At a press conference a few days later, Carter drew attention to the more encouraging aspects of the program. Meanwhile, energy experts were interpreting the plan for reporters. The public was getting a confusing message. Successive headlines in the *Washington Post* over a five-day period tell the story: "Energy Outlook Grim" (April 19); then, "Future Called Not So Bleak As Depicted" (April 20); finally, "Energy Plan Now Pictured As Consumer Boon" (April 23).[80] The ambiguity lost the administration some precious credibility at a critical point in its effort to gain support for the program.[81]

James R. Schlesinger was Carter's energy program coordinator. Schlesinger became the first secretary of energy and presided over the department from the time of its formation until agreeing to resign as part of a cabinet reorganization in 1979. A Harvard economist with Republican sympathies, Schlesinger nevertheless shared values and had good rapport with the president. But his intellectually brusque, sometimes lecturing manner offended many members of Congress. It was said to have contributed to his abrupt dismissal as secretary of defense during the Ford Administration.[82] Schlesinger had also served under Nixon as head, successively, of the CIA and AEC. At the AEC, he led an effort to convert the agency from a role protective of the nuclear industry to one serving the public interest. Schlesinger appeared responsive to environmental concerns, yet his support for atomic energy and his launching of the Clinch River breeder reactor project aroused the suspicions of many Carter supporters. They wondered why the president had chosen him.

Schlesinger was an "unabashed, if qualified, defender of the value of analysis in policy formation."[83] Carter was too. Attracted to analysis, Carter was perceptive and detailed in his questioning. The president made liberal use of analysis in the design and promotion of his programs. But ironically, it was of no help in getting his energy plan accepted.

The plan was passed quickly as a single package by a House of Representatives under strong Democratic leadership. The victory was short-lived. The Senate mutilated the plan in almost endless debate. Influential senators were not as ready to vote along party lines as were their colleagues in the House. Many of them disagreed with the administration's reasoning and rejected its analysis.[84] Their staffs felt resentment at not having been consulted more broadly in creation of the plan. Carter's order midway in the drafting process to restrict external communication prior to official release was but one instance of a lack of political finesse in dealing with Congress during the year of Senate debate (late 1977 and most of 1978). The business of organizing and running a massive new department kept many members of the presidential energy team from being more attentive to political matters on Capitol Hill.

Natural gas deregulation was the plan's Achilles heel. Congress rejected the administration's analytical arguments and resorted to its own calculations. Chip Schroeder, critic of PIES, had a model suggesting that average-cost pricing under partial deregulation would drive the price of unregulated gas above long-run, market-clearing levels. Opponents of deregulation referred extensively to Schroeder's work and to an analysis by the Congressional Budget Office.[85] Those favoring deregulation, on the other hand, cited a hand-calculation by Republican Congressman David

A. Stockman of Michigan, an aggressive and articulate proderegulation member of the Energy and Power Subcommittee.[86] Stockman assumed a very favorable supply response to price rises and factored into his calculation substantial savings in replacement fuels and transportation costs which he attributed to the pipeline availability of greater quantities of natural gas. His conclusion was that deregulation would end up saving rather than costing the nation money.

The administration was calling for somewhat higher gas prices but not immediate deregulation. Its attempt at a politically palatable compromise between producer and consumer positions satisfied neither side. Congressmen representing producing states viewed the proposed legislation as a retreat from Carter's campaign promise to the Governor of Oklahoma to work for deregulation, while congressmen defending consumer interests saw the legislation as a costly burden to their constituents. The administration's calculations convinced no one. They appeared to seriously understate producer revenues by underestimating projected gas prices in the unregulated intrastate market. The debate was heated, engendering so much confusion and mistrust that the state of Texas finally launched its own assessment of the PIES model.

When the final energy package was passed in October 1978, it looked very different from the grand design submitted to Congress eighteen months before. Schlesinger's policy chief, Alvin Alm, was later to express regret at the decision to include the inflammatory natural gas issue in the energy plan that he and his policy team had painstakingly fashioned under great pressure of time. Said one observer, referring to the fine tuning and "elegant complexity" of the plan, it was a "case of good analysis and bad politics."[87]

Other analyses also played roles in the troubled history of the National Energy Plan. We conclude this chapter with discussion of ERDA's last study, conducted in the months immediately preceding the formation of the Department of Energy.

MOPPS GETS PRESS

When in January 1977 ERDA Acting Administrator Robert W. Fri set in motion MOPPS—the study whose acronym suggests a cleanup—he had no reason to suspect it would become grist for the mill of the editors of the *Wall Street Journal* in a crusade against regulated energy prices and Carter's National Energy Plan.

> The study estimated that at $2.25 per thousand cubic feet the nation would be awash in natural gas ... How could there be an energy crisis? How could the president go on television with ringing calls for sacrifice?[88]

Fri, a temporary holdover in the changing of Administrations, initiated MOPPS to deal with the administrative problem he faced of having too many developing energy technologies and their bureaucratic advocates under one budget. "MOPPS," he says, "was the result of a breakdown in the ERDA planning system."[89] The agency's beginning budget had been more than ample for getting started, and it had doubled within three years. But it could not be expected to keep up with the ballooning financial requirements of fusion, photovoltaics, the breeder, and other large R&D efforts, as well as costly demonstration programs for fossil, solar, and synthetics. It was time for a winnowing.

The winnow list, actually called "the winner's list," was prepared for Fri during November and December of 1976 by ERDA planning chief Roger LeGassie on the basis of modeling runs done for him by the same Brookhaven team he had used in the earlier ERDA studies. The list of winning programs was to be part of the required ERDA 77-1 report. But it caused a commotion among competing program managers. An emergency meeting of top ERDA brass shortly before the end of the year was, in the characterization of one attendee, "traumatic." Administrator Seamans, who would leave ERDA the following month, "was attempting to wrap up the decisions that could be made responsibly before the Carter Administration took office."[90] He questioned the analytical basis for the controversial winner's list, and had it pigeonholed. The ultimate ERDA 77-1 report was not what LeGassie intended.

Fri, still facing the budget problem and now a serious management problem as well, framed the MOPPS analysis to include the agitated program managers as direct participants. Their role, one of market evaluators rather than technology salesmen, required a reorientation. Fri asked them to examine competing technologies in relation to one another and to the market, focusing on likely dates of introduction, states of technical readiness, rates of penetration, and environmental impacts.

Carter was just assuming office the month MOPPS got started. The president had made clear his intention of merging the current energy bureaucracy into a new Department of Energy, so Fri knew MOPPS would be ERDA's swan song. With personal futures and organizational questions very much up in the air, it made sense to try to provide the new administration with a well-considered, focused research program based on a careful assessment of market needs. That was the purpose of MOPPS.

Fri, whose time was being taken up by a White House nuclear study, placed Philip C. White, ERDA's assistant administrator for fossil energy and a former executive of Standard Oil of Indiana, in charge of MOPPS. LeGassie served as vice chairman and chief liaison with the White House task force working on the energy plan.[91] Martin R. Adams, who reported to White within the ERDA organization, headed the analysis committee that ran the study on a day-to-day basis. Several working groups had responsibilities for various categories of energy demand, supply, and the environment. They were staffed by a large contingent of outside consultants and analysts from many parts of government, including about 100 people from ERDA. Contract support totaled $2 to $3 million. The time horizon was the year 2000, with intermediate points designated. A parallel study on "inexhaustible energy resources" addressed longer term considerations.

MOPPS treated energy policy as a given, exogenous to the analysis and in the domain of the National Energy Plan (NEP) whose drafting was proceeding simultaneously. Fri was critical of the plan's short time horizon. He wrote, "The NEP is a limited plan. By looking only as far as 1985, it does not solve the energy problem—nor, for that matter, does it expose its true magnitude."[92] Fri was especially concerned about the environmental implications of the use of shale and ever greater amounts of coal.

MOPPS employed an interactive gaming procedure within a model of the energy market that had participants playing the role of decisionmakers. Rather than "project future energy requirements and market shares based on extrapolation of historic energy consumption trends," MOPPS focused on the "underlying activities which generate demands for 'energy services'."[93] It stated energy demand in terms of work units such as vehicle miles of intercity travel and square feet of residential space, and tried "to examine a wide variety of technological substitution and conservation potentials," including competition among fuels.[94] Yet economist Charles J. Hitch, president of Resources for the Future at the time, expressed concern that the analysis might have neglected the full range of opportunities for replacing liquid fuels by gas.[95] MOPPS regarded expansions in the supply of gas as feasible and economic. Gas, it suggested, would have no trouble keeping up with projected demand whereas oil would. Hitch wondered whether this result would follow if all ways of substituting gas for oil had been included in the analysis.

MOPPS's central conclusion was that the United States had "an awesome transitional problem—especially in liquid fuels."[96] But its pronouncement on gas supplies was the subject for which it became best known and is most likely to be remembered. This recognition was neither sought nor relished. Gas, once again, was a political troublemaker.

When Fri initiated MOPPS, he had not anticipated the need for a curve of natural gas availability. The purpose of the study was to assign priorities to technologies, not to estimate resources. The gas curve was developed to provide a baseline against which to gauge the future requirements for nonconventional fuels. The first curve was put together with the help of outside consultants in a task force led by ERDA chemical engineer Christian W. Knudsen.[97] The task force's estimates indicated more low-cost gas becoming available than was commonly believed—considerably more, in fact, than the analysis committee was ready to accept. When the task force stuck to its guns, the committee commissioned an "unofficial" group to take a different approach. Its figures turned out to be "much more in line with common sense and industry estimates," says Adams.[98]

It had not been the practice of the U. S. Geological Survey to provide availability estimates that incorporated economics. This was not the way geologists tended to think about reserves. By introducing economics, Knudsen's projections cast doubt on the long-term gas shortage proclaimed by the administration and the media.[99] The programs of ERDA and the NEP no longer seemed so vital.

On April 7, preliminary MOPPS results were presented at a meeting of ERDA senior officials. Knudsen's projections, indicating that low-cost gas was still to be found in abundance, shocked most ERDA officials in attendance: It would mean a significant setback in ERDA's plans for all kinds of research and development projects for substitute technologies. As Knudsen later testified before the Senate Energy Committee, his low cost estimates "put many of the technologies that ERDA is developing in some question about their urgency and priority."[100]

The gas curves (both sets) quickly came to public attention. With their figures on marginal cost versus reserves misread[101] as price versus production estimates, they became a political football. Proponents of gas deregulation (such as the editors of the *Wall Street Journal*) cited the estimates to support their claim that higher gas prices would indeed bring forth significantly more supply. Opponents of price hikes cited the very same estimates to question the need for any price rises at all.

White, told by an industry-wise aide whose opinion he respected that Knudsen's estimates "would be 'laughed out of court' and would jeopardize the credibility of the whole MOPPS effort,"[102] returned Knudsen unceremoniously on April 26, 1977 to the regular ERDA job from which he had been on temporary assignment. The affair was widely publicized and blown up in congressional hearings.[103] An attempt to substitute the second set of gas estimates for the first caused a public brouhaha, complicating ERDA's predicament. Fri asked a group of geologists and resource experts with conservative views on gas availability to derive still another set of estimates. The result was the most conservative curve yet. All three curves were used for the final report.

The *Wall Street Journal* did a total of five editorials on the subject over the period from April 1977 to April 1978. Its piece called "ERDAgate" alluded to the Watergate coverup.

The innocent scientists and technicians in MOPPS had no idea what vested interests their simple calculations threatened. Even more to the point, bearing this unwanted message to the White House would be a black mark against the ERDA bureaucracy. The Federal Energy Administration would be in a position to gobble up all of the best spots when the two were merged into the new Department of Energy. Given these realities, there was only one answer to the no-energy-crisis crisis. The ERDA brass recalled the MOPPS study, and threw out all the charts that had been so innocently put together over the months. By April 6, it had a "revised" MOPPS study, with the charts looking much like those from the FEA.[104]

Fri disagreed. He believed the revised study still showed very substantial gas reserves, but at higher incremental costs than indicated in the original estimates. "Thus, we actually began to depart from the FEA (PIES) line even more, since our analysis showed that gas was available but at a relatively high price."[105]

The *Wall Street Journal's* editors did not relent. The "revised study," they wrote in the last of their editorials, showed a lot of natural gas would be available except "the administration was forbidding anyone from paying the prices necessary to recover it . . . If there was a supply curve—any supply curve—there was no energy crisis, no need for a $10 billion Department of Energy, and no need for its bureaucratic inhabitants."[106]

Given unavoidable uncertainties in resource estimates, the differences between the MOPPS curves were not that significant, especially at the margin. Why, then, asked political scientists Aaron Wildavsky and Ellen Tenenbaum, did they generate so much passion and hostility? Their answer:

> The disagreements at bottom are not about estimates but about ideologies and institutions. Where values and political aims are polarized and where certain numbers appear to strengthen or weaken one's position, small differences between the estimates are magnified. Even when the technical experts agree, the political process converts their agreement into disagreement.[107]

MOPPS was never officially published, although draft copies of the final report did get circulated. Schlesinger later acknowledged the superiority of the MOPPS gas estimates over those used for the NEP; but that was not the purpose of the study. In its attempt to work out a priority ranking of technologies by involving program managers, the project is reported to have recommended all but two of the technologies on LeGassie's original winner's list. One of the two not recommended was the breeder, included on the original list, says a former member of LeGassie's staff, primarily because it was a favorite of the Republican Administration.

Many have commented on the ingenuity of the MOPPS design and the enduring value of its analysis. Those involved were proud of what they accomplished, such as further development and understanding of the concept of energy services and end-use technologies that originated with the Brookhaven model.[108] But the MOPPS report was shelved in the transition to the new department. No one there appeared very eager to unshelve it.

There was one last mishap. The Department of Energy asked the Government Printing Office to produce a limited supply of MOPPS Volume I, the integrated summary report, for review. Somehow that request resulted in the report's being placed into a formal distribution mechanism and it was sent to depository librarians. "An administrative goof," says White. To correct the mistake, the printing office sent out the following note:

ATTENTION DEPOSITORY LIBRARIANS: The Department of Energy has advised this office that the publication *Market Oriented Program Planning Study (MOPPS), Integrated Summary Vol. 1, Final Report, December 1977,* should be removed from your shelves and destroyed ... We are advised the document contains erroneous information and is being revised.[109]

It was the Department of Energy looking very much like Big Brother.

ERDA had matured considerably in the years between its first plan and the completion of MOPPS. It had learned much about itself and its mission. But to be useful to the new administration, the knowledge and insights it acquired would have to find a place in the reorganized energy bureaucracy.

6 PROMOTING THE PROGRAM

As the Democrats entered the White House in January 1977, a second energy policy report by the Ford Foundation was nearing publication. It was to play a key role in the framing of Carter's nuclear energy policy. Meanwhile, the hard and soft path terminology of reformist Amory Lovins was being added to the vocabulary of the energy debate, and internationalist Carroll Wilson was trying to galvanize industrialized countries into action with an assessment that warned of impending energy shortfalls. President Carter used a still more unsettling analysis by the CIA to dramatize the need for his National Energy Plan. The controversy raged on, yet there was a perceptible shift in attitudes about the role of nuclear power, the availability of oil, and the future demand for energy.

FORD FOUNDATION FOCUSES ON NUCLEAR

Having blazed a trail in energy conservation with the Energy Policy Project, the Ford Foundation chose nuclear policy as its next target area. It saw the nuclear debate as an unyielding confrontation between optimists envisioning unlimited low-cost energy and pessimists fearful of worldwide nuclear calamity.

The public debate on nuclear power issues was poorly structured and undisciplined. The various actors were talking past each other to the crowd; irresponsible statements were going unchallenged; and implicit value judgments were unacknowledged.[1]

With states like California getting ready to conduct contentious referendums on whether to terminate nuclear development,[2] there was a tendency to think in highly polarized terms—"for or against nuclear power."[3] It was easy to forget that there were actually several levels of decisions facing policymakers, and many separate issues bearing on these decisions: reactor safety, proliferation, the waste problem, uranium availability, the costs of coal, rising oil prices, the need for energy, and implications for economic growth.

The Foundation initially planned to set up a neutral ground of impartial arbiters to whom partisans of both sides would come to present their views. After listening to the arguments, the referees would attempt to moderate extreme positions, narrow and define the differences, and provide a considered framework for intelligent discussion. Once again, the Foundation at first looked to Resources for the Future (RFF) as a possible site for the study but dismissed the idea. RFF simply did not appear to be the right place for what was intended. Before long, the Foundation also dismissed the idea of establishing itself as mediator. The cold reality was that both sets of antagonists were emotional, suspicious, and unlikely to trust the Foundation's motives. A confrontation of mindsets was apt to generate more heat than light.

Instead, it was decided to undertake a policy-directed study by uncommitted people not identified with either side. With the 1976 election coming up, the study could aim to help shape nuclear policy in the new administration. At that point, few people thought Jimmy Carter would be the next president.

In the summer of 1975, Ford Foundation president McGeorge Bundy chaired a meeting to explore possibilities for the new study. One of the attendees was nuclear physicist Hans Bethe, a pronuclear moderate who tried to provide a counterbalance to antinuclear forces at the meeting.[4] Bundy discussed the possibilities further that summer with nuclear physicist Wolfgang K. H. Panofsky and others at an Aspen workshop on arms control. Spurgeon M. Keeny, Jr., a former colleague of Bundy, was a participant in the Aspen group. He had served as assistant director of the Arms Control and Disarmament Agency until 1973. With twenty-five years of government service under his belt, Keeny was at the time in a "holding pattern" at the MITRE Corporation. Bethe and Panofsky, with whom he had recently been involved in a Reactor Safety Study of the American Physical Society,[5] both held Keeny in high esteem.

In the following months, Bundy invited Keeny to direct the Foundation's new effort and asked him to submit a plan of study. The MITRE Corporation agreed to handle project administration and the Foundation awarded it a sizable grant to do the job. By January of 1976, the Ford-MITRE study of nuclear energy policy was underway.

Chastened by its experience with the Energy Policy Project (EPP), The Ford Foundation determined that this next study would not be misconstrued as one man's crusade or a platform for advocacy. It would be a genuine group effort, operating very differently from the advisory board of the EPP. Its members would not merely express agreement or disagreement with the final report of the study. They would author the report themselves.

The blue-ribbon study group was set up to operate as a "free-wheeling think tank" with minimal staff.[6] Members were drawn from two main sources. About two-thirds of the twenty-one participants came from the academic community (half from Harvard, two from M.I.T., one from each of several other universities). A second and partially overlapping contingent were persons closely associated with Keeny in work on arms control, including many with whom he had long association and on whom he could rely. Others were added by Bundy—people he knew personally.[7]

To avoid a repetition of the Freeman/Tavoulareas experience, the Foundation stipulated that the study group "must be—and must be recognized as being—essentially open-minded on the general debate raging around nuclear power."[8] There were to be no hard positions going into the study. Only one participant (RFF economist Hans Landsberg) worked full time in energy policy research at the time.

But "open-minded" did not mean "of different opinion." Members of the study group were of unusually congenial temperament. According to Landsberg, they "clicked." The group dynamics were "perfect." One participant who had some complaints about how the study was run (but concedes all worked out well in the end) sensed a noticeable reluctance to disagree at meetings. He made a conscious decision not to rock the boat, since no one would have wanted him to.

Several members of the study group had been associated with each other the year before in an analysis of nuclear energy and national security directed by study group member Thomas Schelling and sponsored by the Committee for Economic Development (CED). Arms control and national security concerns were background for the Ford-MITRE study just as power-plant siting and environmental concerns had been for the EPP.

The CED analysis warned that in another ten years or so most nations would have the technical ability to construct nuclear explosives from reactor fuel. Plutonium, created in a conventional nuclear reactor as a by-product of generating steam for the production of electricity, becomes an available material for making bombs once it is extracted from the spent fuel of the reactor in the reprocessing procedure. Commercial reprocessing of plutonium, as envisaged, plus commercialization of the breeder reactor (which would produce more plutonium than it consumed) could lead to a "plutonium economy" that stocked and traded this hazardous, highly toxic fuel. "If the plutonium economy becomes a reality, a country interested in acquiring at least a limited nuclear weapons capability could easily disguise that interest by stressing the economic case for investing in nuclear reactors to meet its energy needs."[9]

Some in the Ford-MITRE group had thought through the policy imperatives of this proliferation danger during the course of the CED study. Several analyses had questioned the economics of reprocessing.[10] Such prior work, combined with the common arms control interests and good chemistry of the group, helped develop a consensus and get the Ford-MITRE project finished in the one year scheduled. The group's compatible outlook, a source of strength during the course of the project, was also a point of vulnerability after the fact.

Release of the Report

Despite the precautions taken, Ford-MITRE's recommendations were called preordained, just as EPP's had been. Both studies were accused of having started with conclusions and worked backwards. Criticism now came from each side of the aisle: from the Friends of the Earth, who praised the EPP yet accused Ford-MITRE of a "mania for massiveness" and an unwarranted assumption of continued growth in electricity demand,[11] and from nuclear advocates incensed at the proposal to terminate the $2 billion Clinch River prototype breeder project.[12] (The EPP had also urged its cancellation.)

In arguing that energy scarcities leading to higher energy prices need not force fundamental sociostructural changes in the industrialized world or retard economic progress,[13] Ford-MITRE provided the economic perspective and analytical basis deemed lacking in the EPP. Ford-MITRE assumed for its low case, demand increasing from 71 quads of primary energy in 1975 to 100 quads in the year 2000. This projection approximated the zero energy growth estimate of the EPP and the "most probable" figure of an Oak Ridge study,[14] both still considered heretical. For its base case, Ford-MITRE assumed energy consumption in the year 2000 of about 143 quads. This did not take into account the possibility of lifestyle changes. The analysis sought to err on the high side so that the fundamental conclusion that nuclear power was only of marginal economic importance would not depend on underestimation of future energy use. The sensitivity of the results were examined under a range of assumptions.

The Ford-MITRE report, although lucidly written, did not escape the ambiguity of interpretation that beset other policy studies dealing with controversial issues. In the eyes of the *Boston Globe* and the *New York Times*, the report had concluded that "nuclear power is not essential," and "America will not have to depend significantly on nuclear power in this century."[15] *The Christian Science Monitor* got the opposite message. In an editorial entitled, "We need nuclear energy," the *Monitor* wrote that while the report's "disaffection with the breeder reactor has received the headlines, its most significant conclusion is that the United States, nonetheless, needs nuclear power."[16]

Unlike the EPP, Ford-MITRE had no companion reports or media blitz. Keeny is said to have been a strong chairman who knew what he had to do to get the job done. He also seems to have understood what he did not have to do. He withstood a request to produce a fifty-page pamphlet anticipating important conclusions of the study and another to make an extensive statement to the press when the report was complete. Getting the book out was Keeny's number one priority.

Release of the book came on March 21, 1977, just weeks before President Carter was due to deliver his first major energy message. It was presented to the president in a personal White House briefing attended by most members of the study group as well as by former AEC chairman James R. Schlesinger, now Carter's energy coordinator. Five days after the briefing, Schlesinger, "once regarded as an apostle of nuclear energy," was reported in the headlines of the *New York Times* to be "Against Plutonium Fuel Use in Nuclear Reactors."[17]

There were also well received briefings for the Nuclear Regulatory Commission and Congressional aides on Capitol Hill. It helped that several participants in the study had advised Carter on nuclear policy during the campaign, and several were or soon would be taking on key policy positions with the new administration.[18] The interlocution between study and campaign helps explain why the report's conclusions meshed so well with Carter's nuclear policies.

The study group agreed early in the project that the book should be "on the street" by January 1977, in time to have maximum political effect. With Keeny reading galley proofs at the printing plant, that resolve was almost realized. Carter was so pleased with the way the book supported his views that he exuberantly signed the first copy at the White House on the day of its release and presented it to visiting Japanese Prime Minister Takeo Fukuda as a statement of his own nuclear policy postion. Fukuda must have been less than elated with the contents. His country was moving ahead with a nuclear reprocessing plant and looked with favor on the breeder, as did several oil-dependent West European nations. Many countries were well advanced on breeder programs and firmly committed to the technology. This would cause diplomatic problems for the United States in the coming months.[19]

Recommendations and Reactions

The report recognized that nuclear policy was more multifaceted than originally appreciated. It recommended two mutually consistent courses of action. "In any policy analysis," it pointed out, "the broad outlines of the results are determined by the assumptions."[20] That truth was in evidence throughout the Ford-MITRE report.

The report argued first for continued deployment of the light-water reactor. Second, it argued against reprocessing (as scheduled for the half-finished plant in Barnwell, South Carolina)[21] and against early commercialization of the breeder (as was to be demonstrated at Clinch River). The first part of the conclusion, which upset nuclear opponents, was based on the growth assumed in electricity demand, the environmental implications of utilizing coal, and the economics projected for nuclear power compared to coal. The second part, which angered breeder proponents, was based on arms-control concerns about the dangers of a plutonium economy as well as on optimistic projections for coal production, uranium availability, and energy conservation. The dual conclusion depended on assuming some, but not too much, growth in electricity demand; some, but not too much, economic advantage to nuclear over coal; and a relative abundance of uranium. Higher projections for electricity, larger advantages for nuclear, or a less optimistic outlook for uranium would have tilted the economics more in favor of the breeder.

The report did favor keeping the breeder option alive under a strategy that would encourage development of new proliferation-resistant concepts. With this qualified support for the breeder, plus endorsement for conventional light-water reactors, study group members were puzzled by the uproar their report caused within the nuclear community. They wondered if only the reviews were being read, and if the spate of publicity given to Barnwell and Clinch River was drowning out the rest of their message.

By the mid-1970s, proliferation was an issue of mounting concern. India's explosion of a "peaceful" nuclear device, using plutonium from the spent fuel of a research reactor supplied by the Canadians, had shocked the world. It was frightening for those who understood the technology of nuclear bombs to know that plutonium reprocessing plants were already (or soon to be) operating in the United Kingdom, France, West Germany, India, Japan, and the Soviet Union.[22] The dangers were highlighted in the study done by Willrich and Taylor for the EPP. Published in 1974, this landmark work clearly documented how a terrorist group might divert civilian nuclear materials for the manufacture of deadly explosives.[23]

Proliferation was one of three major issues bearing on the merits of the civilian nuclear energy program. The others were reactor safety and waste disposal. On safety, Ford-MITRE judged the methodology of the Rasmussen Reactor Safety Study flawed and the uncertainty of its risk assessments understated (a conclusion given greater currency later on by the Lewis critique). Still, it found the safety record of reactors to date excellent, the risks socially acceptable, and the doubts inevitable. It recommended that action be taken to ensure that extremely serious reactor accidents not increase in likelihood with greater numbers of reactors, and it argued for stricter criteria in the siting of new reactors. It stopped short of making safety a predominant issue.

On waste disposal, critics charged the report underestimated the seriousness of the problem by emphasizing the technical side, which was easier to deal with than the institutional and political aspects. On proliferation, the report declared the increase in the number of countries having access to nuclear materials and technology to be the most serious risk associated with nuclear power. To help lower this risk, it recommended: (1) deferral of plutonium reprocessing; (2) deemphasis of the breeder program and delay of commercialization; (3) lowered priority for nuclear power in R&D; (4) an end to active promotion of nuclear power at home and abroad.[24] These recommendations attacked the very premises and logic upon which plans for the long-term growth of the nuclear industry were based. To defer the breeder now, the industry protested, would represent a severe dislocation.

Some believed that the goal to achieve a breeder technology was the reason historically for developing nuclear power. "The unique promise of nuclear energy," wrote one senior economist, is to use the breeder to provide the world with a virtually inexhaustible source of energy. "It would be a supreme irony if at the very time that the world is desperately seeking technologies for tapping inexhaustible energy resources, the United States were to abandon its forward movement in the one technology which is closest to achieving this objective."[26] The writer questioned whether firm conclusions could be drawn from the uranium resource estimates then available. He faulted the Ford-MITRE study for minimizing the costs and problems of expanded coal production and for failing to take account of the response of foreign governments.

The neglect of international implications was echoed by others, including a Ford-MITRE participant who believed the study group was cavalier in its treatment of the probable response of foreign nations. According to this participant, the group never argued seriously about the options of Germany, Japan, and the developing countries and it later came to haunt the study.

Added a law professor in a published review of the report:

There is little analysis of how long the diffusion of plutonium technology can be delayed or of how effectively U.S. international economic leverage and political example can slow that diffusion ... The deeper interactions between U.S. and foreign energy policies are generally ignored—the book says little about the extent to which large U.S. oil imports might support high oil prices and economically encourage the spread of nuclear power and the sale of conventional weapons to cover balances of payments. Little is said about the politics of the international nuclear industry. Nor are the international institutional arrangements for nuclear power treated adequately.[27]

The report nevertheless struck a resonant chord and was taken very seriously. One critique provides an interesting example of countermodeling, a procedure for challenging the results of a modeling exercise by the use either of a different model or the same model with changed assumptions.[28] Alan Manne had run his model for Ford-MITRE under assumptions provided by the study group to assess the economic impact of delaying the breeder program. The probreeder Electric Power Research Institute (EPRI) charged that the modeling results supported the preconceived conclusions of the study by assuming that coal would be available at a low price almost indefinitely. Manne's former Ph.D. student, Richard Richels, an EPRI staff member, reran the model under less-optimistic assumptions about the future availability of uranium, oil, and coal. Richels, in collaboration with EPRI's Rene Males, showed that under the changed assumptions, a delay in commercializing the breeder could incur much more significant costs than the Ford-MITRE study indicated. Males and Richels argued that "the potential insurance value of the breeder could be very large indeed."[29]

Among the more controversial assumptions of the Ford-Mitre study were its uranium supply projections (initially drafted by

Hans Landsberg), which were optimistic compared to estimates of most geologists.[30] Economists obtain higher estimates of potential mineral supplies than do geologists (conservative by nature) by assuming higher prices stimulate new discoveries. Says environmental scientist John Holdren, "the economist's view is more reasonable early in the development history of a given mineral (as I would argue is the case for uranium), while the geologist's view is more realistic late in the development history (as I think is the case for oil)."[31] Holdren tended to agree with the Ford-MITRE estimates. Ironically, a declining demand for uranium resulting from the sharp cutback in nuclear plans eventually turned the problem of uranium supply into a nonissue.

Aftermath

On April 7, 1977, barely two weeks after publication of the Ford-MITRE report, President Carter announced a plan to restructure the breeder program and postpone reprocessing domestically and for export. The nuclear industry took this as confirmation of the president's lack of sympathy for its cause, and for the agreeable conclusions of a report on the breeder that ERDA had arranged for the White House at Schlesinger's request.[32]

The president's action on reprocessing and the breeder appeared to directly reflect the Ford-MITRE recommendations. Yet study group member Panofsky points out that Carter emphasized nonproliferation as his reason, whereas Ford-MITRE had put economic considerations first. Nonproliferation objectives were "decidedly secondary" with us, asserts Panofsky.[33]

Carter's decision caused a commotion at home and abroad. It was the start of a busy two years for Harvard government professor Joseph Nye, one of the youngest members of the Ford-MITRE group. Now in the State Department, it would fall to him to elaborate and explain the administration's nonproliferation policies. Carter's action was partly a reaffirmation of steps taken earlier by President Ford. Ford's initiative to suspend reprocessing deliberately avoided the breeder issue. It was a reaction to congressional concerns over the events in India and nuclear policy statements by candidate Carter. A special White House task force headed by Robert Fri found the economic benefit of reprocessing to be marginal and its commercial deferral not costly. This made easier Ford's decision to defer reprocessing on nonproliferation grounds.[34]

Fri, acting administrator of ERDA after Carter's election, met weekly with Nye, Keeny, and later, Philip Farley, all former members of the Ford-MITRE study group. It was, in effect, an informal transition coordinating committee for nuclear policy. Fri and Nye developed a close working relationship, and their frequent joint appearances before Congress became known on Capitol Hill as the "Fri-Nye show."[35]

Carter's strong position on reprocessing and the breeder created growing opposition from Congress, industry, and foreign countries. Says Nye, it became clear that

> some device was needed to introduce a longer-term thrust into international nuclear policy. Maintaining and refurbishing the international regime would require a general approach around which a broad group of nations could rally. The process of rethinking the conditions of the regime had to be shared beyond the United States alone.[36]

The result was a call by Carter in his April 7 message for an International Nuclear Fuel Cycle Evaluation (INFCE) open to all interested nations. The concept of INFCE evolved from the Reprocessing Evaluation Program of the Ford Administration. Carter broadened the idea to include other nations and other aspects of the fuel cycle besides reprocessing. "While officially INFCE was given a predominantly technical rationale," says Nye, "this was a means of attracting broad participation into what was really part of a political process of stabilizing the basis for the international regime."[37]

Forty countries and four international organizations attended the first INFCE conference in Washington, D. C., in October 1977. In addressing the attendees, President Carter sought to demonstrate the high cost of nuclear investment by comparing figures on the relative capital requirements of nuclear, conservation, and oil—figures obtained partly from Friend of the Earth and foe of the nuclear reactor, Amory Lovins.[38] Carter's message was lost on members of the audience whose countries did not have available the full range of energy options the comparison assumed.

Several authors of the Ford-MITRE report became involved in the INFCE effort, including Albert Carnesale who served on the Technical Coordinating Committee and Abram Chayes who headed it. The effort soon expanded to include sixty-six countries and five international organizations representing both suppliers and consumers, rich and poor, nuclear and non-nuclear, market and centrally planned economies, East and West, North and South. The participants worked two years to produce 20,000 pages of documents organized into nine volumes. "In the end, they showed little liking for America's purpose in launching the conference,"[39] and a great deal of enthusiasm for developing the breeder, despite their belief that supplies of uranium would be adequate through the end of the century. Participants estimated future demand for nuclear power at substantially higher values than those then being projected by the U.S. government. This was a bitter pill for Americans seeking a realistic international consensus.

For those who expected INFCE to be "a technical and analytical study and not a negotiation,"[40] the effort was a disappointment. For those who saw INFCE as staffed with nuclear partisans, its outcome was no surprise. "INFCE's assumptions were widely represented as its conclusions," declared Lovins, "ostensibly resulting from a careful assessment of alternatives which never actually took place."[41]

Yet, from the perspective of one who viewed the study as a means for gaining time and defusing an explosive political situation while drawing international attention to the dangers of plutonium, INFCE did serve a useful purpose. Nye, for example, felt the study had come out right along the lines he originally had in mind.[42] Said a State Department staffer afterward, "We are going into a period of quiet diplomacy."[43]

The ways of analysis are many and varied, the course uncertain, the results not always what one party or the other may have intended. There are ebbs and flows in the influence of analysis, as in the public mood and political agenda. Proliferation was a top priority national issue at the start of Carter's term when the Ford-MITRE study first made its appearance. By the end of his term—after INFCE—it was an issue non grata at the White House. Ford-MITRE helped usher proliferation into national focus. INFCE helped usher it out.

THE SOFT PATH OF AMORY LOVINS

One criticism of the Ford-MITRE study came from the irreverent Amory Lovins, whose provocative work and ideas we now pause to examine. Lovins believed the Ford-MITRE study, in cautioning against only plutonium fuel cycles, overlooked the proliferation threat posed by conventional (once-through, without reprocessing) cycles.

> So far, nonproliferation policy has gotten the wrong answer by persistently asking the wrong questions, creating a "nuclear armed crowd" by assuming its inevitability.[44] No one on the Ford-MITRE study asked the fundamental question of whether we should be building power plants. Not looking at whether you need nuclear power loses the opportunity to really avoid proliferation. The study started at the mid-point and made things worse by reinforcing the conventional wisdom that only incremental changes are viable. Every study that does not examine the fundamental assumptions, makes it harder to get them examined.[45]

Lovins argued there could be no adequate safeguards for *any* form of nuclear fission.[46] He believed that proliferation was not a product of certain processes and technologies of the nuclear industry so much as it was a product of the existence of the industry itself.

Lovins concluded that the supply-oriented strategy of nuclear development and homogeneous energy (the more the better) must give way to a demand-oriented strategy emphasizing energy efficiency and renewable forms whose quality and scale were appropriate to their end use. Lovins coined the terms "hard path" and "soft path" to distinguish between these two alternate routes to the era beyond oil and gas. He applied the same adjectives to a separate distinction between centralized large-scale (hard) technologies and decentralized small-scale (soft) technologies. The hard technologies—non-nuclear as well as nuclear—were those developed for producing and distributing bulk energy in premium forms such as electricity. Their use was a prominent feature of the hard path but was not synonymous with it.

Lovins insisted that the hard and soft *paths*, as opposed to the technologies, were mutually exclusive as coherent evolutionary patterns of socioeconomic development. In competing for the same resources, said Lovins, and in reflecting a different perception of the energy problem, these alternate routes presupposed wholly different tactics and institutions. Lovins believed the country already had more large central-station supply than it could use to economic advantage, given the characteristics of its energy use. Thus, cost-effective future additions to electrical capacity, if needed, must be of the smaller-scale, decentralized variety. To accentuate the point, he invoked "The Road Not Taken," Robert Frost's poetic metaphor on human choice. Frost, knowing that he could not venture forth simultaneously on both roads diverging before him in the wood, "took the one less traveled by." That, mused Frost, "made all the difference."[47] The soft path, argued Lovins, would make all the difference too—for the future of mankind.

Lovins's soft path analysis was much cited in energy policy discussions and controversies during the Carter years. It was sharply critical of all "establishment" studies that did not question the status quo. Was Ford-MITRE such a study? Yes, most definitely, said Lovins, who accused it of narrow, incremental analysis and of presupposing a growing demand for electricity[48] rather than examining the characteristics of this demand. Yes, also, said John Holdren, whose positions on most issues paralleled those of Lovins, but whose views on nuclear fission were less absolute[49] and whose reaction to Ford-MITRE was considerably more charitable. Holdren called the Ford-MITRE study a substantial step forward in "establishment analysis," and set the study alongside Lovins's work as "one of the new focal points of The Great Energy Debate."[50]

Delineation of the hard and soft paths identified three parties to the debate: first, nucleophiles and other hard-path proponents; second, champions of the soft path; and third, those in between—believers in a firm but pliant course. While receptive to further centralized systems development, the middle grounders were cautious about the extent to which large, costly technologies should be relied upon. They regarded the achievement of greater energy productivity and the perfection of renewable sources as worthy goals, but questioned their imminence and adequacy. Rejecting the two extreme positions, middle grounders argued for a pragmatic, flexible strategy.

The Ford-MITRE study diagnosed the long-range energy problem as one of higher costs rather than energy limitations. Its economic perspective suggested a marketplace solution tempered by consideration of social and environmental costs. Viewing the energy problem as fundamentally a contest to be argued out on Capitol Hill, on the other hand, implies the need for flexible political accommodation. Either way would avoid strict adherence to either the hard or soft path.

Lovins, arguing for removal of market imperfections and institutional barriers favoring conventional technologies, endorsed the principle of letting the market be the judge. He regarded the soft path as economically superior to continued large central-station development. For him, the soft path was both socially desirable and economically practical to pursue.

Many took issue with Lovins, some quite heatedly. They saw him using energy as a tool to change society. Lovins countered that it was the hard path advocates who were challenging the status quo. He viewed himself as a social conservative seeking to protect society from the changes in lifestyle imposed by the hard path. Adversaries received his verbal jabs as assaults on their beliefs and threats to their careers, which partly explains the intensity of their often *ad hominem* counteroffensives.

Lovins was indefatigable. He would defend his thesis repeatedly with technical calculations and copious documentation whose accuracy and appropriateness were constantly under attack. One of the arguments was with Nobel Laureate Hans Bethe and concerned a solar energy calculation. The exchange, placed in the Senate record, received wide publicity.[51] A criticism of Lovins of under 400 words and 3 references would receive a spirited rebuttal of over 1200 words and 19 references.[52] The aggressive volleys back and forth form a lively part of the story of energy policy analysis in the 1970s. We now look more closely at the ideas of this tenacious crusader as background for understanding how he came to receive such wide notice.

A Return to Fundamentals

Lovins was an analyst-phenomenon in the energy debates of the 1970s. His command of the energy literature and facility with figures were impressive. "A walking encyclopedia of alternative energy sources, a fountainhead of statistics on the economic follies of hard technologies," this "environmentalist's trump card" was described colorfully by one fan as consistently overwhelming his opponents "by speaking at a higher, multidimensional level of analysis" with a gift of "statistical fluency."[53]

Lovins traveled extensively. An American by birth, a European by residence, he had extensive contacts on both sides of the Atlantic. A hero to one group, another's nemesis, his name rarely evoked neutral reactions. It was not uncommon for adversaries to pay him grudging and even admiring respect. One referred to him as "superstar of the energy circuit,"[54] another called him "the *enfant terrible* of the energy left"[55] and worried about his "brilliant back-of-envelopism" while allowing that he was the "most articulate writer on energy in the world today."[56] A reviewer who interpreted Lovins as wanting to abandon centralized energy sources in favor of soft technologies wrote that "for all its flaws, the Lovins critique is easily the most comprehensive and technically sophisticated attempt to put together an energy program compatible with environmental values."[57]

A member of the baby boom that followed World War II, Lovins grew up in the shadow of Hiroshima and Nagasaki and went to school during the days of the Vietnam conflict. He began his professional career at the age of seventeen, consulting in experimental physics to help support himself at Harvard. A mountain climber, he was intellectually restless and of iconoclastic bent. Feeling constricted at Harvard and running out of money, he left for a less expensive research program at Oxford. Lovins considered himself a generalist/synthesizer and wanted to cross-pollinate disciplines. He was influenced by the position of the Limits to Growth School and the ideas of Hannes Alfven, David R. Brower, Herman Daly, C. S. Holling, Thomas Jefferson, Henry Kendall, Aldo Leopold, Lao Tse, and adversary Alvin Weinberg, whom Lovins credits with first having stirred his concerns about the hazards of fission.[58]

Though politically conservative, Lovins was viewed as a rebel with three major causes. He was against the development and use of nuclear energy, which he considered uneconomic and proliferative. He was opposed (as a Jeffersonian and free-marketeer) to Hamiltonian autarchy, technocratic rule, and governance by an elite. And he despaired of the sociopolitical costs imposed by the increasing scale, vulnerability, and remoteness of large energy producing systems such as central electric.

The three causes came together to form a compelling logic for Lovins.

> If I seem to be presenting advocacy as well as analysis, it is not because I began with a preconceived attachment to a particular ideology about energy or technology, such as the "small is beautiful" philosophy that some have tried to read into my results. It is instead because the results of the analysis so impressed me ... As one brought up in the high technology tradition, I was surprised that the analysis made a much stronger case for much more unconventional conclusions than I had expected.[59]

Lovins considered large centralized modes of producing and distributing energy wasteful at the margin—vulnerable monuments to indiscriminate development. He was not opposed to massive technologies in the same absolute way he was against nuclear energy, but he was against their further expansion. High levels of energy use were to him no more an indication of economic well-being than high levels of traffic congestion. Both, he claimed, were signs of society's inefficient investment. We do not yet understand how to use our power wisely, he contended, and we are not giving adequate priority to keeping our systems resilient and flexible. The principle of "the more energy the better" was backward, charged Lovins. Goals to provide energy in ever increasing amounts confused ends with means.

Lovins did not stop with energy. He detected in energy problems symptoms of problems that beset other aspects of society, such as threats to the environment, proliferation, unemployment, and alienation. Lovins found in the measures available to deal with energy problems insights that could help with these additional problems as well.

The soft path prescription was to focus on the heterogeneous thermodynamic structure of energy requirements rather than on the homogeneous concept of primary energy supply. In terms of the "spaghetti chart" that traces energy flows from origin to end use, the soft path approach began with the right side of the chart instead of the left. To minimize direct economic costs, Lovins favored technical fixes—current cost-effective technologies that could raise energy productivity without significantly altering lifestyles. And he was for supplying energy in a form, scale, and quality (entropic level) that matched the tasks it was asked to perform. By such means, he argued, end-use efficiencies could be doubled by around the turn of the century, tripled or quadrupled within thirty to forty years after that, and there would still be room for further improvement.[60] This, he concluded, would reduce the need for big supply technologies.

These heroic visions were not forecasts of what would happen, but estimates of what could happen. They were certainly not projections of what would eventuate if things continued as they were. Lovins (quoting Niels Bohr who said, "It is difficult to make predictions, especially about the future") had little use for forecasting models he considered elaborate extrapolations.

> Such models have trouble adapting to a world in which, for example, real electricity prices are rapidly rising rather than slowly falling as they used to ... Extrapolations have fixed structure and no limits, whereas real societies and their objectives evolve structurally over decades and react to limits. Extrapolations have constants, but reality only has slow variables ... Extrapolations assume essentially a surprise-free future even when written by and for people who spend their working lives coping with surprises such as those of late 1973. Formal energy models can function only if stripped of surprises, but then they can say nothing useful about a world in which discontinuities and singularities matter more than the fragments of secular trend in between. Worst, extrapolations are remote from real policy questions.[61]

Lovins's way of dealing with the future was an extension of the scenario method of the EPP. He used the term "scenario" in the specific sense adopted by the film industry to mean a description sufficiently vivid for the audience to be able to imagine itself participating in the events described. Lovins credited the EPP with having incorporated fundamental attitudes toward energy into its scenarios. To these he added attitudes about centrism, autarchy, vulnerability, and technocracy—the root issues, he charged, that establishment studies almost always neglect. By working backward from a desired future rather than forward into a gap—thus "turning divergences into convergences"—and by being explicitly normative, Lovins felt he was able to obtain insights unavailable in traditional modeling exercises.[62]

Lovins's Legacy

So much for Lovins's philosophy and methodology. How did he come to attract so much attention? There is a story that can be told. It starts with David Brower, dedicated conservationist, former president of the Sierra Club, and founder in 1969 of Friends of the Earth. Two others who play roles are Carroll L. Wilson and William P. Bundy. They will be introduced shortly.

Brower, in one of the first book projects of his organization, published a photo-essay on the Eryri mountains of Wales that Lovins had put together in 1970/71 with a photographer companion and fellow climbing enthusiast. Lovins was still at Oxford at the time but was tiring of the laboratory. His friendship with Brower, an articulate mentor on the fragility of the natural environment, blossomed. When the university raised an eyebrow at his desire to devote himself to energy and resource policy, Lovins resigned to become British representative for Friends of the Earth. He began working on land-use planning, resource issues generally, and finally, energy policy.

M.I.T. professor Carroll Wilson first met Lovins in Stockholm in June 1972 while Wilson was serving as an advisor to the United Nations Conference on the Human Environment. Lovins came to Wilson with "page after page of calculations" relating to a study of climate that Wilson had led the summer before.[63] After showing the calculations to two colleagues on the study, who were impressed, Wilson invited the "extremely perceptive" young man to the first meeting of his Workshop on Alternative Energy Strategies (WAES), the project that we discuss next. Lovins served as a consultant to WAES during 1974/75. His controversial ideas, according to a WAES colleague, mixed careful analysis with militant advocacy. As the advocacy gradually took on a more reasoned tone, according to this observer, it began to gain legitimacy for Lovins's views.

Those on the project with strong industrial and pronuclear sympathies, like the French, found Lovins's views repugnant. Wilson was having a hard time holding the complicated multinational study together. He knew he would have to take an even-handed approach to nuclear energy, and feared the project would compromise Lovins's freedom of action. The two parted on good terms about halfway through the project.

One afternoon in 1975, Lovins had a discussion at WAES with someone from Shell Oil. The interchange led Lovins to begin formulating his ideas in a paper that eventually became the seminal article on the hard and soft paths.[64]

In considering where to submit the paper for publication, Lovins thought of *Foreign Affairs*, the respected journal of the Council on Foreign Relations. An article by Wilson warning against America's growing oil dependence on the Mideast—arguing for expansion of coal and nuclear power and for a strategic petroleum reserve—had appeared there shortly before the Arab embargo.[65] Lovins, although on a different tack, felt the journal would be "a good way to get to the establishment—the diplomatic, banking, and foreign policy audience." So, in the spring of 1976, he sent the eighth draft of his paper to editor William P. Bundy.

Bundy (brother of McGeorge) had no prior contact with Lovins but got a favorable report on him from his good friend Wilson, who was a long-standing member of the Board of Directors of the Council on Foreign Relations. Wilson urged Bundy to take a close look.

After some delay, Lovins received a thoughtful response saying that the piece had great strengths and weaknesses, and would require work. Four revisions later, with extensive help from Bundy and an able assistant editor, the article was ready for publication. "The editors drew out more substance and clarity than I knew was there," admits Lovins gratefully. With thirty-two pages and thirty-six footnotes, it was the longest, most annotated paper *Foreign Affairs* had run up to that time.[66]

The article appeared in the October 1976 issue. The same issue had a piece by a leading Israeli politician with a map that became a diplomatic faux pas and received much attention from the press. Lovins's article also received attention. The journal received more orders for reprints of it than for any article since a "Mr. X" piece on the Soviet Union, written anonymously by George Kennan in 1947.

Also in October 1976, Lovins participated in an Oak Ridge symposium on "Future Strategies for Energy Development," for which he wrote a lengthy contribution containing much of the technical backup for the article in *Foreign Affairs*.[67] Shortly afterward, while a guest of the International Institute for Applied

Systems Analysis in Austria, he revised his Oak Ridge work and brought it together with his article as a book.[68] The article and book were to become his manifesto on the soft path.

The establishment clearly had mixed feelings about Lovins. On the one hand, organizations wanting to avoid the impression that they were sanctioning his views would deny him entrance to a building or use of an auditorium to present an invited talk. On the other hand:

- On October 18, 1977, Lovins had a private meeting with Jimmy Carter attended by Presidential Science Adviser Frank Press, Secretary of Energy James Schlesinger, Domestic Policy Chief Stuart Eizenstat, and Eizenstat's White House assistant on energy issues Kitty Schirmer, after which Schlesinger had one of his staff members go to Europe to follow up Lovins's comment about coal-fluidized-bed developments abroad.

- That same year, the Edison Electric Institute, principal association of investor-owned electric utilities, devoted an entire issue of its "Electric Perspectives" to critiques of Lovins's position.[69] The Senate's Small Business and Interior Committees had hearings on the same subject, later published in two bulging volumes[70] and subsequently edited into a more readable compendium of selected pieces.[71]

- In 1978, the Institute of Gas Technology filmed a movie of Lovins as the basis for a panel discussion at its annual meeting, while being careful to point out that its having organized the symposium did not "constitute an endorsement of the soft path view."[72] In the spring of that year, Lovins presented three Regents' Lectures at the University of California at Berkeley. In 1979, a prominent New York brokerage firm invited him to address its Research Conference on Electric Utilities to explain why he was predicting a limited future for central electrification.[73]

Lovins did not work in isolation. Disciples and fellow advocates in many countries were operating in the decentralized, individual style of the soft path. In order to provide coaching materials for this international network of coworkers, Lovins developed a "Do-It-Yourself Soft Energy Path Study Kit." Here are some extracts.

Put most of your analytical effort into demand, not supply ... Start writing early ... Be modular and recursive ... Don't get hung up on a number ... Don't be overprecise ... Keep assumptions explicit and clearly in mind ... Keep your calculations scrutable, documented, and transparent ... There is no such thing as accuracy in 2025: just do the best you can ... Keep thermodynamic structure of end-use foremost in your mind ... Keep an eye out for analogies, precedents, and vivid examples ... Beware of multiplying alternatives ... Test sensitivities ... Imbed local case-studies to lend concreteness ... Anticipate major criticisms in your text (but don't be defensive) ... Remember that numbers aren't everything ... Keep networking and pass on your experience.[74]

Not bad advice for any analyst.

What in the final reckoning will be the significance of this defender of individual initiative, this prophet of a softer way? At the very least, he will have made the case that consumers merit a wider range of options and suppliers need to pay more attention to how energy gets used. If government subsidies, institutional barriers, and market imperfections discriminate against small, decentralized technologies, and if correcting these inequities favor soft-path technologies, then they might indeed gradually gain ascendancy over hard-path alternatives. Lovins claims this has already been happening, with renewables contributing more new energy than any other source and efficiency improvements contributing many times more than all supply sources combined.

If, on the other hand, the doubters are right, then the fork in the road could be illusory. Repudiation of further large-scale energy system development could be a mistake that serves to narrow rather than expand consumer choice. Since the soft path specifically rejects coercion and centralized planning, the Chinese Cultural Revolution is not a fair analogy. But it does suggest a peril to which detractors have alluded.[75] The fear is that adoption of a soft path could, in the name of greater humanity, turn out to be socially oppressive.

Yet, if history be our gauge, there is not much cause for worry. The records show that policy analysis, no matter how persuasively argued, is more often the material for intellectual debate than social reform. Policy analysts, at their best, are more likely to be shakers of cherished assumptions than leaders into the promised land.

THE WAES OF CARROLL WILSON

On December 8, 1946, as the civilian Atomic Energy Commission (AEC), which he chaired, was about to take over the Manhattan District with Carroll Louis Wilson as acting administrative officer, David E. Lilienthal made the following entry in his diary:

> Carroll is proving the wisdom of our choice more every day, by his sweetness of character, his firmness, his skill with other people. That was the most important decision since the Commission membership itself was selected ... "Carroll, are there any more at home like you?"[76]

Wilson had just been appointed to the post the month before. He was to serve as first general manager of the AEC from 1947 to 1951, the critical starting period for the civilian nuclear enterprise. Years later, after the accident at TMI, in trying to puzzle out what went wrong with the nuclear program, he recalled that no one had spent much time looking at the total cycle from fuel enrichment through reactor operation to reprocessing of spent fuel and disposal of radioactive wastes. No one, reminisced Wilson, had much interest in the back end of the cycle. "There was no awareness that the whole system must function or none of it might be acceptable."[77]

This man, whom Lilienthal was so glad to have hired, had a lifelong association with M.I.T., spanning the years prior to and following his AEC career and a stint with industry. He also had a continuing relationship with M.I.T.'s distinguished leader, Vannevar Bush, whom he assisted after receiving his B.S. degree from the Institute in 1932 and then again during the war years when Bush directed the Office of Scientific Research and Development. In 1959, Wilson returned to M.I.T. to accept a faculty position with the Sloan School of Management. Its dean then was Howard W. Johnson, later president and chairman of M.I.T.

Wilson did not fit the mold of the traditional faculty member. One former student refers to him as a "nonprofessorial professor" who "dealt with people rather than concepts." He was a very effective seminar leader, according to this admiring student, who especially enjoyed Wilson's course on Technology and Social Issues.

Wilson's missionary and ecumenical bent was clear from the start of his teaching tenure at M.I.T. During the 1960s, he developed a Fellows in Africa program for students interested in obtaining high-level working experience in developing countries. In the summer of 1970, he led an interdisciplinary Study of Critical Global Environmental Problems, which he initiated because he "saw no likelihood that anybody in the U.N. or U.S. Government was going to do homework for the U.N. Conference on the Human Environment." The following summer he organized a Study of Man's Impact on Climate in which he brought together thirty-five atmospheric scientists from fifteen countries.[78]

Wilson was a Sloan School colleague of Jay W. Forrester, originator of the systems dynamics modeling methodology and propounder of the world dynamics conception that led to the highly publicized *Limits to Growth* study.[79] Wilson invited his controversial colleague to the meeting of the Club of Rome at which Forrester first presented his ideas on world dynamics.[80] Wilson was attracted to Forrester's views on the interdependence of industrialization, population, food production, environmental pollution, and resource exhaustion, although he did not fully subscribe to the formulation adopted in the world dynamics modeling approach, an approach that has been criticized as simplistic and biased against industrialization.

When Wilson was invited by the Council on Foreign Relations to give a set of three Elihu Root lectures in the Spring of 1973, he chose to present the first on his study of climate, the second on a proposal for gaining energy independence from growing reliance on Mideast oil, and the third on a suggested mechanism for performing global assessments. The second lecture became the aforementioned article in the July 1973 issue of *Foreign Affairs*, thanks, says Wilson, to a "first-class job" of editing by his friend William Bundy.[81] With references to the moon program and the Manhattan project, its plan for what Wilson wishes he had called "self reliance" instead of "energy independence" foreshadowed Nixon's Project Independence later that year.

If Wilson's article inspired Project Independence, his goal of limiting oil imports to 5 million barrels a day by 1985 was not reflected in the project's hopeless target of 0 oil imports by 1980. Wilson's second goal was to keep oil prices below $6 a barrel (twice the price at the time of his writing). The futility of that quest was soon apparent.

The third Root lecture provided the design for the WAES study to which we now turn. Wilson organized the study in 1974 as an outgrowth of a suggestion made by former AEC commissioner William T. Golden at a dinner after the lecture. Asked Golden, referring to Wilson's proposed scheme for making global assessments, "Why not do it on energy? If you need a 'grubstake', I'll help."

Wilson's talent for bringing together influential people in common explorations, and for raising funds to support such endeavors, was vital to the feasibility of WAES and set its style. Wilson's intent in launching this "experiment in international collaboration" was not to develop another institution but rather to conduct a private and informal study outside of normal channels that drew on the resources of existing organizations without disturbing their operations. Central to the design was the surety in advance that the project would dissolve upon its scheduled date of completion. The small headquarters staff at M.I.T. would disperse back to former responsibilities or on to new ones. There would be no organizational "residue," no "politics of structure," no promotions, jockeying for position, or acting out career ambitions. "The only credits," stated Wilson, "would be in doing a good job."[82]

Wilson selected members for his study from government, industry, finance, and academia in the United States and abroad, people who were in a position to promote change in their countries. The thirty-some participants came from fifteen mostly industrialized nations that used about 80 percent of the energy in the World Outside Communist Areas.

Participants were generally at the level of director of a center or chairman of a company. Since their time and capacity to work was limited, they were invited to choose technical associates, typically from their own organizations, who would devote substantial effort to WAES under their direction. It was a clever stratagem to obtain valuable analytical support for the project while at the same time keeping senior participants involved and interested.

The stated aim of WAES, "global energy assessment to the year 2000," was sufficiently ambitious and diffuse that no one had

a very well defined idea of what the product would be. The Ford
EPP was just concluding. Its preliminary report had been distributed, and its final report would be released when WAES held its
initial meeting in October 1974. The EPP provided a beginning
focus. WAES adopted its methodology for estimating future energy demands under alternate scenarios. Says Wilson, "it was abso-
lutely vital in the first year to be able to start out with something
we could get our teeth into."[83]

The Message

Given the influence of the EPP and the conservation predilections
Wilson had expressed in his 1973 *Foreign Affairs* article, one
might have expected a strong conservation conclusion to have
emerged from WAES, supported by authoritative country-by-
country analyses on the feasibility of moderating energy consumption. But the WAES demand studies, relegated to a technical appendix, did not receive emphasis in the main report.[84] To
WAES consultant Steven C. Carhart, who had been a staff member of the EPP,[85] WAES missed the role of energy productivity in
its treatment of conservation. "It came out with a coal/nuclear
solution instead." Coal and nuclear were also among Wilson's
predilections.

WAES did not ignore conservation. Indeed, Wilson's statements to the press (like his article) advocated enforcement of conservation with "wartime urgency," a phrase that *Time* magazine
pointed out was "strikingly reminiscent of Jimmy Carter's call for
the moral equivalent of war."[86] It was a call for curtailment in
energy use more than for greater productivity. To staff member
Paul Basile, WAES "quite extraordinarily" reached a "consensus
among oil companies and big industrial consumers that conservation was a front-line strategy that would succeed."[87] The final report affirmed that "much energy can be saved" and this "may
well be the very best of the energy choices available." The report
also acknowledged that conservation's "opportunities and challenges deserve much more careful analysis, much more detailed
probing, than we have been able to do."[88]

Given the nature of the WAES membership and style of its
director, the report's restraint in emphasizing conservation is un-

derstandable. Many participants from industry and foreign countries were antagonistic to the anti-industrial tone of the EPP and its rousing appeal to curtail energy use. Their major concern was that their countries have energy supplies adequate for future growth. Wilson was sensitive to this concern. Unlike Freeman, according to Carhart, he was looking for agreement. Consensus and conciliation were his watchwords, not confrontation.

The corollary is that WAES tended to overestimate future energy demand, as did most others at the time. Its assumptions about economic growth were optimistic and its oil prices were too low in most scenario runs. When WAES analysts took the difference between projected energy demand and the amounts they felt could be satisfied by non-oil sources, and when they compared this difference with their estimates of available oil, they were left with significant and growing disparities beginning in the 1980s. Associate David Sternlight, chief economist of Atlantic Richfield, labeled these disparities "notional gaps." These gaps, he felt, served to signal a tension or impending problem, even though the supply/demand imbalances they denoted would never actually arise in practice. (The higher prices, government constraint, and slower economic growth induced would restore balances through dampened consumption, accelerated production, intensified fuel switching, and the like.)[89]

Some wondered why WAES took this gap approach rather than insisting on price projections that equated supply and demand, as in traditional economic analysis. One reason it did not confront the price question directly had to do with the representation on the project of large oil companies who were seen as having significant price-setting influence. A collaborative effort focused on future oil prices might look to the antitrust authorities like collusion. That was the apprehension. Prices were a sensitive area.

WAES ran five scenarios from 1985 to 2000 under different growth and price assumptions. One of the five, shaded in the scenario chart of the final report, was not published in the same detail as the others. It was the so-called low-growth (2.8 percent a year), high-price (rising to $17.25 a barrel by 1985) scenario labeled D-3, the one, it turns out, that came closest to the economic conditions that in fact ensued.[90] It was the only scenario in which projected supplies covered demands through the end of the century. The fact that it was not given equal status with the other runs was said to have been a concession to the anxious Japanese who feared their government would interpret inclusion of the scenario as official sanction of its assumptions and expression of a willingness to pay the higher oil price.

The $17.25 price, measured in 1975 dollars, converts to about $30 a barrel in early 1980 prices. That was remarkably close to actual prices at the time. Had WAES presented the balanced D-3 scenario as a price projection, it would have been in the select company of those who anticipated the sharp post-1977 oil price rises. "If we were doing it again," says a chagrined Sternlight, "I would insist that D-3 be published in full."

What was the accomplishment of this Workshop on Alternative Energy Strategies? It was the first global, comprehensive analysis of energy performed by an international group. By its own appraisal, its supply studies, especially its analysis of the range of future world oil production, were probably its chief contribution.[91] "WAES was a misnomer," points out Amory Lovins, who participated in selection of the name. The report did not really explore alternative strategies. "Its most important conclusion was that business-as-usual does not work."

WAES warned that the industrialized world was headed for a serious energy shortfall and needed to engineer a quick transition away from petroleum. It did not develop the role of prices and the market in avoiding the shortages it was predicting. As Wilson put it,

> My conclusion at the end of this Workshop is that world oil will run short sooner than most people realize. Unless appropriate remedies are applied soon, the demand for petroleum in the non-Communist world will probably overtake supplies around 1985 to 1995. That is the maximum time we have: thirteen years, give or take five.[92]

What remedies did the study propose? It was difficult for it to reach an explicit consensus. Wilson was intent on avoiding the dissents that he felt "smothered" A Time to Choose and spoiled the reports of the Committee for Economic Development. He was determined to announce that "all Workshop members agree on the general analysis and the main findings of the report"—and so he did.[93]

Conservation got muted recognition. Nuclear energy, even more controversial and most apt to splinter the group, was treated with similar reservation. Each country provided "maximum likely" and "minimum likely" nuclear forecasts based on its own best subjective judgment. The nuclear chapter was the hardest to write, attests Wilson, who wrote it. Handling its subject was ironically, or perhaps appropriately, the severest test of the diplomatic skills of this first general manager of the AEC.

Coal was closest to being the transition fuel of choice, although not acclaimed by all. Japanese associate Kenichi Matsui (who complained that the WAES meetings tended to be dominated by the "Anglo-Saxon" countries) came from a country considered pronuclear and cool to coal. We interviewed him in May 1979, after Three Mile Island. Wilson was already well along on a successor study to WAES called WOCOL, a WAES-type analysis of the needs and prospects for world coal.[94] Looking back, Matsui said he felt Wilson had the objective (a successful one of which Matsui now approved) of promoting the use of coal worldwide. With continuing pressure from the International Energy Agency, Japan was beginning to accept the idea of importing coal to burn in place of oil.[95]

Disappointments

Media coverage of the WAES report in May 1977, a few months after Carter took office, emphasized looming shortages and shocked many readers. Journalists made no distinction between long-term imbalances that could be corrected nondisruptively through gradual market adaptation and overnight shortages that would cause personal hardship. Said Sternlight, "the press got the impression from the summary charts that WAES was forecasting doom, disaster, and major gaps. There was no discussion of alternate scenarios or mention of prices."[96] It was a major disappointment to WAES members who regarded their study as "a call for action, not a cry of despair."[97]

A second disappointment was the failure to reach the highest levels of government. One key legislative aide felt that WAES was quoted much less than he would have expected, although Frank Potter, in another part of Congress, declared the study to have been extremely useful to the subcommittee staff he directed for Congressman Dingell.

Perhaps the greatest disappointment was the inability to get to Carter personally. Whether it was because of the sensational character of the reporting, a fear that it might provoke precipitous policy action, an impression that WAES was trying to build headlines, or its focus on shortages and gaps can only be a matter for speculation. The direct route to Carter would have been through the president's science adviser, Frank Press, a former faculty colleague of Wilson at M.I.T. who later became president of the National Academy of Sciences. Press, when asked, said he did not recall "why the WAES report was not brought to the President in person," but noted that the approval of many White House advisors would have been required.[98]

In any case, Carter did not invoke WAES as support for his energy plan. He did use, weeks before, another analysis that also prophesied oil shortages, this one by the CIA. It is the next study we review. The president's highly publicized use of the CIA analysis was generally received as a political maneuver, possibly poisoning the air at just the time the similarly sounding WAES study was being released. WAES may have had a different reception at the White House had it come a month earlier.[99]

SURPRISE SPOTLIGHT ON THE CIA

If there was ever any question of the importance of timing to the attention a report receives, the CIA's April 1977 study on the "International Energy Situation: Outlook to 1985" removes all doubt. The written report was delivered to Carter at precisely the time he was preparing to present his National Energy Plan to the nation. The classified CIA analysis seemed to offer just the sense of urgency that advisers had been telling the president he needed to create the proper climate for congressional action.[100]

In the absence of greatly increased energy conservation, projected world demand for oil will approach productive capacity by the early 1980s and substantially exceed capacity by 1985. In these circumstances, prices will rise sharply ... By 1980 ... the USSR will be losing its status as a net oil exporter to the West. We estimate that in

1985 the USSR and Eastern Europe will be net importers of 3.5 to 4.5 million b/d.[101]

Carter publicized the CIA findings—calling them "deeply disturbing"—at a press conference on April 15, three days before his first major energy address to the nation.[102] Analysts in the CIA's Office of Economic Research, caught off-guard, had to scramble over a hectic weekend to get the report declassified for general distribution to a large new audience. They had prepared their report for "busy noneconomists," not professionals who would severely criticize a weak demand analysis for which the CIA claimed no originality. On the supply side, where the analysts felt their work was sophisticated and detailed, they were unable to present much supporting evidence because of the nature of their sources. The result was suspicion, disbelief, and a political boomerang.

The president was was not alone in finding the CIA's conclusions deeply disturbing. The projection of continuing price rises was bad enough news for a nation still trying to adapt to the first quadrupling of oil prices. The spectre of competition developing with the Soviet Union (the world's largest oil producer) for the coveted petroleum resources of the Middle East was even more dire. This was not part of the generally accepted prognosis. The allusions to the Communist countries' being transformed within eight years from exporters of about 1 million barrels of oil per day (net) to importers of 3.5 to 4.5 million barrels put the CIA's predictions well outside the range of conventional economic prophecy.[103]

Other studies had tended to assume the Communist Bloc would not impinge upon the Western oil market in any significant way—either positively or negatively—for the foreseeable future. The WAES report, for example, (although not yet published) said:

A large uncertainty in our work has been the possible future role of the U.S.S.R. and China as energy exporters or importers. Although the U.S.S.R. and China are major world energy producers and consumers, their trade with the rest of the world has to date been relatively small. We assume that their future role in world energy trade will continue to be small.[104]

We have seen that the energy studies were often accused of tailoring their conclusions. No study was attacked on these grounds more savagely than the CIA analysis. Yet CIA analysts had no idea of the role their report was to play.[105] They had been following Soviet oil production for some time and had only recently begun to notice a combination of factors that made them question future production levels and their own projections of continued increases into the 1990s. They briefed Carter on these disquietudes before his inauguration, then provided him with a classified statement of their findings in April 1977. His subsequent public announcement stunned them. They had not prepared the report for general consumption. Media pressure and the desire of Agency Director Admiral Stansfield Turner to make publicly available as much CIA research as feasible led to the report's quick declassification and wide distribution. A companion analysis and a series of follow-up reports followed.[106]

Many energy experts, while expressing confidence in the CIA's Office of Economic Research, rejected its estimate of a Soviet import requirement of 3.5 to 4.5 million barrels a day by 1985. There was sharp disagreement within the administration as well.[107] Energy Secretary Schlesinger and others called the CIA's projections overly pessimistic.[108] Critics felt the Soviets could and would avoid becoming net importers of oil in the 1980s if only by virtue of their special ability to force oil conservation and compel diversion into other energy sources. The CIA, they contended, had ignored possibilities for expanded use of natural gas as well as increased utilization of coal and nuclear power. They questioned the CIA's low estimates of Soviet oil reserves and argued that greater efficiencies in use and production could be expected with growing tightness in the world oil market. Furthermore, the Soviet Union needed hard currency. Russian specialist Marshall I. Goldman observed that "petroleum exports in 1976 accounting for about 50 percent of the Soviet Union's hard-currency earnings" made it difficult "to see where the Soviet Union will find the wherewithal to import, particularly if the real price of petroleum continues to climb."[109]

The CIA was not unaware of the currency problem.[110] It agreed that the Soviets could not "afford to lose that kind of hard currency in the international oil market" and would "do virtually anything to prevent" it. Retreating from its controversial estimate, the CIA in later publications agreed that the assumption that "any shortfall of oil automatically would be imported" was "unwarranted." It regretted its allusion to an import requirement of 3.5 to 4.5 million barrels a day.

> It lends itself to misinterpretation much too easily. It was meant to express the Agency's belief that *unless* the Soviet Union alters its energy consumption pattern, significantly increases its conservation practices, or greatly reduces economic growth and the oil consumption traditionally correlated with that growth, there will be a shortage of 3.5 to 4.5 million b/d of oil by 1985. No one ... believes the Soviet Union will import oil at that magnitude.[111]

"We had intended these projections to be a measure of the potential gap between demand for and supply of oil," says CIA economist Maurice C. Ernst, "but did not make this clear in the original study." This led to "widespread misunderstanding and criticism."[112] The CIA, like WAES, derived its supply and demand estimates independently, then produced the mistaken impression that the difference between these projections was an impending shortage.

Later in 1977, Moscow gave notice to its customers in Eastern Europe that they should be looking more to the international market for oil supplies.[113] This warning, though made off and on since 1966, may have sounded more emphatic this time and may have appeared to be confirmation of the CIA prognosis. When, in 1979, the CIA testified before Congress that Soviet production was "currently stagnant or declining everywhere except in the western Siberian fields," one reporter concluded this was "fulfillment of CIA projections made two years ago." The reporter had misread the CIA forecast. The estimated importation of 700,000 barrels of oil a day by 1982 referred not to Moscow but to the USSR and Eastern Europe combined, "with East European imports offsetting reduced Soviet exports."[114] In addition, the forecast would hold only if the USSR and Eastern Europe were unwilling "to accept a substantial slowdown in economic growth." The CIA was now convinced that "the USSR could not afford to become a net oil importer."[115]

Economic forecasters are sometimes charged with hubris. But they have a humility too. They rarely allow for the fact that their forecasts may themselves affect the conditions on which the forecasts are based, as if they do not believe they will be heeded. Although the CIA studies did not have much influence in the U.S., despite the large amount of attention they received, they may have had a decided impact on Soviet policy.

> In rapid order, the Soviets signed contracts for a drill bit plant, a factory to produce secondary recovery chemicals, and a $220 million gas-lift process to replace water injection in the West Siberian fields. Most of these projects had been pending for years, some from 1973, but only after the CIA report came out was the money allocated.[116]

The lesson is that policy analysts can never be sure of who their audience will be. "Soviet officials always pay more attention to Western criticism than to their own," asserts Goldman, "even if our criticism merely repeats what has been said there. As a result, the 1977 [CIA] report virtually guaranteed that its predictions would turn out wrong."[117] Goldman called these predictions "self-defeating prophecy." Countered the CIA: the exploration technology that the Russians added had too long a lead time to affect the agency's forecasts; the size of their reserves was the real constraint.

In May 1981, Soviet oil production was going strong and the Defense Intelligence Agency accorded it a bright future.[118] But the CIA did not relent. Although it softened its forecast and postponed the predicted decline in Soviet oil production, it held fast to its belief that there would be a significant falloff. It was, said the CIA, just a matter of time.[119]

7 MICROCOSMUS POLITICUS

The study of the Committee on Nuclear and Alternative Energy Systems (CONAES) of the National Academy of Sciences extended over almost half the decade of the seventies and involved many of the elite who gained expertise and influence in energy matters as the nation responded to the energy crisis. CONAES provided its own miniature version of the energy debate. The controversy it evoked reflected deep-seated economic, political, and value-based differences that resembled those deadlocking energy policy for much of the decade. The energy crisis jolted consensus and challenged accepted truths. Simple extrapolations about energy growth no longer applied, and pat assumptions about the price of energy and the availability of fossil fuels and nuclear power had to be rethought. With consideration of CONAES in this chapter, we continue to examine the role of policy analysis in this reevaluation.

25 December 1979

Dear Mr. Secretary:

I have the honor to transmit a report entitled *Energy in Transition, 1985-2010* prepared by the Committee on Nuclear and Alternative Energy Systems (CONAES) of the National Research Council ... Mr. Secretary, the National Research Council is pleased, proud, and considerably relieved, to make this report available.[1]

185

With candor and a sigh of relief, Philip Handler, chairman of the National Research Council (NRC) and president of its overseeing National Academy of Sciences (NAS), officially presented to newly installed Energy Secretary Charles W. Duncan, Jr., on Christmas Day 1979, the long-awaited CONAES report. No less festive a date would have been appropriate for the presentation. The CONAES study had finally come to a close over four years after it was first requested by Robert C. Seamans when he took over at ERDA, a forerunner of Duncan's Department of Energy.

The CONAES report might well have been titled *Energy Study in Gestation, 1975-1980*. In the four-plus years of its labors, CONAES had outlasted both Seamans and his successor at ERDA, as well as James Schlesinger, Duncan's predecessor in the Department of Energy. During its toil, CONAES had witnessed a changeover in the White House from the Gerald Ford Republicans to the Jimmy Carter Democrats. It had seen ERDA merged into a new Department of Energy and had watched wistfully as a host of other energy studies with missions similar to its own had come and gone. It had experienced the incident at Three Mile Island, the Iranian revolution, an end-of-the-decade upsurge in oil prices, and the mixture of confusion, anxiety, and public outrage aroused by this unrelenting series of very unsettling events.

CONAES got started at a time when the American public was only beginning to awaken to its energy problems. By the time the study came to a close, world events had shaken all sense of complacency. What had first appeared in 1974 as a sudden and suspicious rise in oil prices, accompanied by exasperatingly long gasoline lines, was by 1980 being viewed as a dangerous resource and financial problem for much of the developed and developing world. The possibility of war was discussed openly and seriously.[2]

Released at the close of the decade, CONAES illustrated much of the energy analysis, political debate, and mobilization of experts that gave the seventies its special character.

ANALYSIS BY COMMITTEE

CONAES formed a microcosm of the intellectual disagreements that raged in the United States over energy during the 1970s. It was beset with controversy in both its execution and reception.

Opinion was divided on whether it constituted a "weighty contribution to the energy debate," or was "a culture-bound artifact of
America's technological society at the start of the 1980s."[3] One
approving congressman, contrasting it with previously published
studies "received with skepticism in policy-making circles," referred to the CONAES report as a "rational, thoroughly balanced, and scholarly document that will be an essential resource
to members of Congress" as they consider critical energy issues.[4]
Detractors, on the other hand, unimpressed by the report's copious analysis, scoffed that the CONAES elephant had labored
mightily only to produce a mouse. From the kitchens of the National Research Council/National Academy of Sciences, according
to one unfriendly reviewer, had emerged a "two pound waffle."

It is tempting to take the letters of an acronym whose original
significance has become obscure or obsolete and attach new meanings retrospectively. CONAES lends itself to such sport; for example, "Committee on Negotiating an Emergent Stalemate."
This cynical rendition points out the parallel between the strug-
glings of experts in CONAES and the analogous exertions by energy policymakers in Washington. U.S. energy policymaking in
the seventies has often been viewed as an exercise in futility.
CONAES has too.[5] We believe these assessments to be overly
harsh and hasty.

That there was enormous frustration in the seventies in both
energy policymaking and analysis cannot be denied. But frustration does not imply futility. "Energy is much too large a layer of
the cake to have a policy about," says economist Kenneth E.
Boulding, who served on the CONAES overview committee. "Policy should be confined to smaller issues . . . slices of the cake rather than layers."[6] A similar point could be made with respect to
energy policy analysis. CONAES chose not the slice and not the
layer, but the whole cake.

ERDA Administrator Robert Seamans set off the chain of
events that was to become CONAES in a letter to Handler dated
April 1, 1975, thereby setting a precedent for the commemorative
date choosing observed by Handler with his Christmas Day letter
four years and nine months later. Seamans was following up on
an offer of help by Handler. What he asked for was a review of
"the overall Breeder Program and the Clinch River Project" to
determine whether it would be possible to assess the suitability of

breeder reactors for commercial operations. A judge's ruling had required that such an assessment be made by 1984.[7]

Much of the ERDA budget was devoted to work on the breeder and Clinch River. This raised the issue of R&D priorities. To Jack M. Hollander, study director of CONAES while on leave from the Lawrence Berkeley Laboratory at the University of California, that issue was central to the CONAES effort.

> In the United States the supply of energy had traditionally been an activity of the private sector. The case of nuclear energy was special. The legacy of government monopoly development of nuclear energy, following its use for military purposes in World War II, led to a government R&D programme on nuclear energy but very little on other energy forms. The combined factors of rising energy prices and the oil embargo led to the conversion of the Atomic Energy Commission (AEC) to ERDA in January 1975 and to increased emphasis in federal R&D on non-nuclear energy ... Since ERDA was given the responsibility of R&D in all energy sources, it was natural for ERDA to question whether there was a proper balance between the breeder R&D programme and the other programmes for which it was now responsible.[8]

There was growing public resistance to the uranium-thrifty, plutonium-producing breeder reactor. Seamans was seeking an impartial assessment of this controversial technology from a respected source of scientific and engineering expertise, one that he knew well from long-standing personal association.[9] He asked for "a detailed and objective analysis of the risks and benefits associated with alternative conventional and breeder reactors as sources of power."[10] But the Academy had no desire to pass judgment on the U.S. breeder program based on a narrow technical study in a "contextual vacuum."[11] According to Hollander:

> The position of the National Academy of Sciences was that an effective evaluation of the role of nuclear energy could be made only in ... the context of available alternative energy sources and of the possibilities for reducing future energy growth without detrimental economic and social consequences. This view was accepted by ERDA.[12]

Another view comes from deputy study director John O. Berga, then a member of the Academy staff. "I did not believe ERDA was looking for 'balance' in its programs." Nor did Berga feel that possibilities for reducing energy growth entered into the Academy's considerations.[13]

Seamans acquiesced to the broadened mandate without ever completely embracing it. He instructed that most effort be devoted to the details of reactor safety, proliferation, and waste disposal, but this emphasis was not reflected in the study's work statement.[14] Instead, the Academy set out to "assess the appropriate roles of nuclear and alternative energy systems in the nation's energy future,"[15] looking at a wide range of social, economic, and political factors as well as purely technical considerations. The project was now conceived to be "nothing less than a detailed analysis of all aspects of the nation's energy situation."[16] Its purpose had become diffused.

The fact that the CONAES study as executed was very different from the study intended by its sponsor was hardly its main distinction. That characteristic, as we have seen, is not uncommon in U.S. policy studies.[17] Nor was it the newsworthy fact that the $4.1 million project far exceeded its original schedule and budget. The fundamental feature of CONAES that distinguished and separated it from other energy studies of the seventies was the extraordinarily large number and wide variety of people it brought together and the highly discordant views expressed at its meetings. It has been said that the way to get a study done on time is to form a group of like-minded people who respect each other and operate well together, then set them to work with a clearly defined objective.[18] CONAES, involving so many people of disparate motivations (referred to as "desperate" in a slip by a CONAES official), violated that maxim on all counts.

Handler considered CONAES the most complex task ever attempted by the National Research Council. Critics regarded it as having strayed too far afield in too politically sensitive an area, given the National Research Council's technical expertise but political innocence. But Handler took the project's expanded charter very seriously. He was determined to achieve balance and impartiality in the selection of members to ensure credibility for the study's findings. This resolve is reflected in the composition of the overview committee appointed "after wide consultation with appropriate individuals and organizations . . . 5 engineers, 3 physicists, 1 geophysicist, 2 economists, 1 sociologist, 1 banker, 1 physician-radiobiologist, 1 biological ecologist, and 1 'public interest' lawyer." Of the sixteen members at the start of the effort, Handler speculated that "about one-third were negative, perhaps 3 were positive, and the others were genuinely open-minded concerning nuclear energy."[19]

Not everyone viewed the leanings of the members this way. Antinuclear activists charged that the newly formed committee was a stacked group with heavily pronuclear membership.[20] CONAES redoubled its efforts to achieve representativeness. An opportunity opened up through the organizational strategy that evolved.

Many overview committee members were inadequately informed technically about the subject of the study when it began. Most had been chosen for their prominence or position rather than their energy expertise. The project leadership felt its first job was to provide the energy novices on the committee with the specialized information and analysis they needed to understand the issues and form judgments. This task was to occupy much of the first year of the study.[21]

The energy apprentices on the overview committee did not necessarily consider *themselves* the neophytes. Dudley Duncan, an outspoken social scientist, eventually resigned[22] because he felt the committee, dominated by "technocrats," was not listening to him. (It was, although not all members liked what they heard.)[23] Duncan believed his engineering colleagues needed an education at least as much as he did. He perceived an "asymmetry" in the committee's use of tacit assumptions concerning the social component of the study:

> On the one hand, if a social scientist (particularly one other than an economist) offers a careful and tentative statement based on some kind of systematic evidence and formal analysis ... it is hooted down, unless it agrees with people's preconceptions. On the other hand, if an engineer pronounces on the future of the birth rate or the determinants of public opinion, the pronouncement is entertained as a piece of serious discourse. People don't even seem to realize when they are talking ... about social science questions.[24]

The approach taken to inform the overview committee was to divide the subject area into four fields of interest, each to be covered by a panel with its own chairman and a membership almost as large as that of the overview committee itself. The Supply and

Delivery Panel (Supply) and the Demand and Conservation Panel (Demand) were to investigate the two basic components of energy—its production and consumption. The Risk and Impact Panel (Risk) was to examine the consequences of energy use for individuals, society, and the environment. The Synthesis Panel was to integrate the work of the other three panels—a job the dimensions of which were the most difficult to define.

The intention was for panel results to be fed to the overview committee for its edification and review. But panels felt the need themselves for specialized data and analysis. Energy supply, for example, included many different sources: oil, gas, coal, solar, geothermal, as well as nuclear. Each source had its own methods of production and coterie of experts. Demand for energy, similarly, occurred in several different sectors of the economy: residential, industrial, transportation, and buildings were singled out. Each had its own special considerations.

To meet the needs of the panels for specialized information, CONAES created a second tier of research activity. It empowered the panels to assemble resource groups to delve into specified fields of interest. About two dozen such groups thus came into being. Each one initiated its own independent research. In certain cases, resource groups commissioned the preparation of outside studies, introducing still another level of investigation and another task and time requirement to the schedule of work.

The organizational strategy drew into the study a variety of extremely able people of diverse backgrounds and ideological positions: among them, engineers W. Kenneth Davis and Floyd L. Culler (Supply), physicist John H. Gibbons (Demand), public-interest lobbyist David Cohen, political scientist Charles O. Jones, economist Tjalling Koopmans, cultural anthropologist Laura Nader, and many, many more—in all, over 300 very busy people working without pay.

The time spent by individual participants was significant. Jones was a member of the Synthesis Panel and the decisionmaking resource group. Although outside of the central deliberative process, he estimates he devoted as much as fifty days to the study in the first year of his participation—20 percent of his professional working time.[25] Most of this time was taken up with meetings—and he did not attend all of the ones to which he was invited. Had he done so, he gauges conservatively, it would have added another 40 percent to the number of days given over to CONAES. He complains that despite all of the meetings, there was no closure on a plan of study for the panel on which he served. Though goals were frequently discussed, they "were never clearly established. Rather, talk on that subject would trail off in the late afternoon as people disappeared one by one to catch planes."[26]

This was not an isolated experience. Many members were unclear on objectives and on how their efforts were to relate to the rest of the study. Communication among panels and resource groups was poor. The Demand Panel and its five resource groups were able to work in a reasonably integrated way. They adopted an input-output framework with common assumptions on price, population, and the rate of economic growth in the absence of war or major interruption. But price was not as meaningful for the Supply Panel. Its eleven resource groups, operating semi-independently, focused instead on ability to produce. The result was constant interpanel tension. The Supply Panel would ask the Demand Panel how much electricity or liquid fuels would be demanded at some future date. The Demand Panel would reply, "We don't know unless you tell us the price."

CONAES was beset with problems of timing and coordination from the start. It took almost a year for the Demand and Conservation Panel to be formed and longer still for some of the other groups. The Synthesis Panel met contemporaneously with the panels whose analyses it was supposed to be integrating. This compounded the difficulty posed by the Supply Panel's inability to produce price-supply curves to cross with the price-demand curves being prepared by the Demand Panel.

CONAES did not establish mutually consistent, plausible assumptions for its panels. It did not even agree about the year on which to base dollar valuations. Different panels used different years. They could not possibly serve the overview committee as originally intended.

It is tempting to conclude that with a tighter organizational plan and a better defined set of marching orders, the problems of CONAES could have been avoided. Certainly having a clear research design spelled out in advance and understood by all would have helped. But that simple diagnosis tends to overlook the difficulties inherent in any attempt to integrate the work and ideas of strong-minded intellects from varied disciplines and of different persuasions. Critics maintain that since there was no common perception of the problem, CONAES was trying to integrate the unintegrable—forcing closure rather than illuminating differences. CONAES was accused of making "hidden value judgments"[27] and neglecting "possible changes in social and political behavior that could bring about an early transition to reliance on renewable energy sources."[28] In fact, the overview committee did consider the possibility of social change but dismissed it as not likely enough "to provide a satisfactory basis for prudent planning."[29]

Some charged CONAES with inertia, top-down thinking, and placing excessive constraints on participants. A study that accepts the existing institutional framework as a point of departure will tend to favor nondisruptive courses of action. Most policy studies do. That upsets those who advocate fundamental revision and question the status quo.[30]

SYNTHESIS AND SCENARIOS

The principal device planned for integrating the decentralized work of CONAES was scenarios, a set of estimated energy demands under a range of assumptions about energy prices and technological responses. The Synthesis Panel, asked to produce these scenarios, never met with the overview committee "to get its charge," attests Jones.[31] Responds Hollander, "The committee clearly instructed the Synthesis Panel very early in the study ... that its primary charge was to produce a set of supply-demand scenarios as a major integrating element."[32] Those instructions did not register. Says panel chairman Lester Lave, it was "not clear what they wanted."

The panel viewed its mission rather broadly as one of:

1. Coordinating and integrating the work of the Supply, Demand, and Risk Panels;
2. Modeling the future supply and demand for energy, and assessing their impact on the economy;
3. Investigating the processes by which energy decisions are made;

4. Investigating the implications of various energy levels for
    consumption, location, occupation, and production patterns.[33]

To pursue these objectives, panel chairman Lave formed a
Modeling Resource Group, a Decisionmaking Resource Group,
and a Consumption, Location, and Occupational Patterns Group
headed respectively by Tjalling Koopmans, David Cohen, and
Laura Nader. Members of the Koopmans group tended to be traditionalist in bent and those of the Nader group reformist in
bent. The resulting dissonance was not what CONAES intended
when setting up the Synthesis Panel.

The overview committee saw the scope of the Synthesis Panel's
purview in considerably more restrictive terms than did the panel. Panel members were not aware of the desire to limit their mission to the production of integrated scenarios. Had they been,
some said, they would not have agreed to serve. Scenarios to
them were means, not ends.

Use of the term "scenario" in the energy field does not refer to
a script in the Hollywood sense, although it shares with the script
the ability to convey a mood and explore a chain of events and
dependencies. The military services have used the scenario approach extensively as a training and exploratory device.[34] The Energy Policy Project was the first to adopt it as a means for
making energy projections.

A scenario is an experiment of the mind, not a prediction. It is
a reflective exercise of the imagination, a projection, a "what if"
statement that suggests the expected results of more or less plausible assumptions about future events according to some self-consistent model.[35] The model could be a mathematical
representation of the energy system, a set of calculations of key
energy relationships, or a descriptive portrayal of what depends
on what and on how things might reasonably be expected to
change under a specified alteration of base conditions. In this portrayal, there should be an internal consistency between the supply and demand sides of the energy picture.

As it turned out, the Demand Panel, the Supply Panel, the
modeling group, and the Nader group all developed their own scenarios for different purposes and under different assumptions.
The final integration was devised by Hollander and a staff member based on the scenarios of the Demand Panel.[36] The integrated

scenarios had projected energy consumption levels in the year 2010 ranging from just under the amount of energy then being consumed to over twice that level, including a high-side scenario[37] requiring aggressive technological development. Scenarios were essentially "surprise free," a simplification most analyses adopt to circumvent otherwise unmanageable uncertainties.[38]

Through scenario analysis, an energy study can provide a medium for airing disagreements and presenting different outlooks. CONAES became a forum for exploring and challenging opposing views. With so much effort going into developing alternative scenarios, it could be called—with justification—the Committee on Negotiating Alternative Energy Scenarios. These scenarios laid the battleground on which barricades were erected and positions taken. CONAES gave participants a "view from a side of the barricades unfamiliar to them ... As a nation we can transcend our divisions only by recognizing that serious people manned both sides of those barricades."[39] CONAES, with all its problems, provided many thoughtful people with the opportunity to hear and consider arguments on both sides of a large number of the central issues of the times.

COMPLETING THE STUDY

CONAES meetings, afflicted with problems of organization and communication on the one hand, and with sharply conflicting points of view among participants on the other, often seemed to go nowhere. The difficulties were mutually compounding and morale suffered. CONAES foundered seriously on more than one occasion, and it became progressively harder to bring it to completion.

The Academy committee formed to review the CONAES report rejected the first draft. The reviewers "included many individuals with little prior knowledge of energy issues." They "raised so many questions and criticisms ... that the writing and circulation of new drafts went on and on with only marginal improvements being gained."[40] The result was further loss of momentum.

When the study ran on beyond its scheduled termination date, Hollander returned to his post at the Lawrence Berkeley Laboratory. When funds ran out the first time, ERDA provided replenishment; when they ran out a second time, the Department of Energy declined the encore and the National Academy of Sciences was forced to dip into its own reserves. "A 'malaise' . . . developed in CONAES, with some members fearing that the report would never come out."[41] Many sidewalk observers shared this pessimism.

It is a tribute to those who never gave up that the project was finally brought to a responsible completion. Handler's honest expression of relief in his transmittal letter was shared especially by Harvey Brooks, persevering cochairman of the study, who felt under the gun to the last.

Brooks, a member of the National Academy of Sciences and the National Academy of Engineering, both overseeing bodies of the National Research Council, was a Harvard-based applied physicist long associated with public policy concerns emanating from science and technology. Handler succeeded in persuading him to chair CONAES only after offering to provide a cochairman, Edward L. Ginzton, head of a major electronics firm in California, who had recent experience managing a large national study effort.[42] The apprehension Brooks felt about presiding over such an ambitious endeavor seems prescient in retrospect.

Brooks deserves primary credit for getting the study finished. According to his critics, he also contributed to slowing it down. Some believe that if he had been less insistent on having the committee "find answers" and reach consensus, its deliberations would have proceeded more expeditiously. By their reasoning, limited goals such as improving the definition of the problem, posing alternatives, and outlining areas of disagreement would have been easier to attain than definite conclusions and positions.

Brooks was seen as overly dominant by those participants who either felt uncomfortably remote from the center of decisionmaking or favored an open forum for discussion. Brooks believes that if he erred it was on the side of permitting too much freedom for discussion and debate. He was criticized for not taking votes in committee. He found the idea of doing so inappropriate to what he regarded as essentially intellectual issues. Intellectual acceptance (if not agreement) was for him a fundamentally different process than political compromise. Critics charged it was naive to aim for intellectual acceptance given the fundamental disparity of opinion. They would have been satisfied with illumination of the

opposing positions. The overview committee did agree finally to
settle for a consensus on what the differences were, according to
Brooks. Development of this strategy was for him the biggest accomplishment of the study.

The final CONAES report was a combination of attempts both
to outline disagreements and take positions. It has been faulted
by those who would have preferred greater decisiveness and crisp-
er conclusions, as well as by those who would have given more
emphasis to the uncertainties and differing points of view. Committee member Kenneth E. Boulding, for example, wished the report had "stressed more the divergence of the feeling on the
committee instead of trying to concentrate on areas of agreement." He considered those divergences "a very important part of
the picture."[43]

Positions taken in the report were duly qualified. An appendix
registered over 200 dissents and amplifications by committee
members. About 20 percent of this appendix was written by
Brooks, another 25 percent by committee member John P. Holdren whose persistent questioning of nuclear energy and steadfast
attention to environmental concerns set the tone for many committee meetings.[44] According to Holdren, the report is beset with

> conditionals, qualifiers, and admissions of ignorance; yet at times . . .
> seems to overlook the complexities it elsewhere reveals, portraying
> choices as simpler and less ambiguous than those we really face. Some
> early news accounts homed in precisely on these oversimplifications
> and transformed them into generalizations more sweeping and pre-
> scriptions more clear-cut than any the report contains.[45]

Newspaper coverage of the CONAES report was bewildering to
many readers,[46] a result of the time pressures and space limitations—deadlines and column-inches—under which journalists
work. The tendency of reporters to select and oversimplify is especially precarious in covering a subject as technically and politically complex as energy.

MEDIA RESPONSE

The first newspaper account of the completed CONAES study
was on the front page of the Sunday edition of *The New York*

Times on January 13, 1980. It carried the headline: "Science Study Urges Conservation as First Priority for Energy Policy" ("Major 5-Year Assessment for Government Concludes that Nation's Economic Growth Need Not Suffer").[47] Not until the fourth paragraph does author Anthony J. Parisi mention nuclear reactors. Parisi, a business reporter for the *Times* who had earlier written a feature on energy efficiency as a way out of the energy bind,[48] portrayed CONAES as rather negative on nuclear. He wrote that breeders "may not be necessary" and "the need for even conventional nuclear reactors may decline after 1990."

Parisi's account of the study's conclusions was misleading by "omission rather than commission"[49] through "selective quotation," according to Brooks.[50] CONAES had indeed attributed critical significance to conservation. The covering letter for the report declared that "as energy prices rise, the nation will face important losses in economic growth if we do not significantly increase the economy's energy efficiency. Reducing the growth of energy demand should be accorded the highest priority in national energy policy."[51]

Yet the same covering letter assigns second highest priority to the development of a domestic synthetic fuels industry and stresses the desirability of a balanced combination of coal and nuclear fission as the "only large-scale intermediate-term options for electricity generation." It also notes the need to keep the breeder option open. These points did not receive equal weight with conservation in the *Times* article.[52] Nuclear advocates were distressed at how the article swept aside the technology they consider essential to the nation's energy and economic well-being.

Thomas O'Toole, staff writer for the *Washington Post*, had another reason for being upset with the *Times* article. Parisi, known to be sympathetic to an antinuclear, conservationist point of view, had obtained his copy of the CONAES report in advance from private sources. O'Toole's copy, on the other hand, was one of the official volumes distributed to the working press by the National Research Council's Office of Information. Official copies carried a customary several-day embargo on publication, a gentleman's agreement that allows reporters time to read a report and consult with others in its review. The embargo was to run until Tuesday, two days after the unexpected appearance of the

Times scoop. Parisi had an exclusive, in effect. This may explain why the *Times* put the article on the front page. Some believed CONAES did not warrant such notice.

With publication of the piece in the Sunday *Times*, the embargo was no longer in effect so far as the *Post* was concerned. O'Toole's editor wanted a review of CONAES, and he wanted it fast. Since the review came a day earlier than scheduled, there was not ready space for it—but some was made available.

On Monday morning, the day after the *Times* account, the *Post* had its own front-page article on CONAES under O'Toole's byline. With the title, "Energy Analysis Stresses Use of Coal, Atom Power,"[53] the article did not seem to refer to the same study. The contrast with the *Times* coverage could not have been sharper. The *Post* account hardly acknowledged CONAES's conservation recommendation, making only a fleeting reference to it in the tenth and eleventh paragraphs. That was not the news so far as O'Toole was concerned. Several other energy studies had recently promoted energy efficiency,[54] and besides, the conservation theme had already been played by the *Times*. O'Toole had no desire to echo Parisi.

To O'Toole, what was different about the CONAES study was its conclusion that the only way for the United States to meet electricity demand for the next thirty years was to burn coal and build nuclear power plants. The other recent studies had tended to come down harder on coal and nuclear. By O'Toole's reading of the CONAES report, nuclear represented America's best option for generating electricity. The study, by his interpretation, had recommended that the nation continue to develop the fast breeder nuclear reactor. Now it was opponents of nuclear energy who were upset at the unbalanced reporting.

Put "end-to-end," said Brooks, the *Times* and *Post* articles combined might have been regarded as reasonably responsible reporting. Taken individually, they were one-sided accounts. Be that as it may, CONAES had fulfilled the aspiration of every national energy study—coverage on the front pages of both the *New York Times* and the *Washington Post*. According to one close observer of the Washington scene, "nothing has political reality in Washington unless it appears in the *New York Times, Washington Post* or *Wall Street Journal*. That is how the day's agenda gets set in D.C."[55]

What were the factors in the attention CONAES received? The prestige of the National Academy of Sciences? Maybe so: some news reports did refer to CONAES as a committee of scientists, despite the banker, lawyer, and economists in its membership. The study's $4.1 million cost? One wag remarked that unless a study spent a few million dollars, no one would pay any attention to it. The anxious mood of the nation at the time? Not unlikely: Iran had taken fifty American hostages, Carter had frozen Iranian assets held in the United States, Russia had invaded Afghanistan, the gold market was setting a dizzying pace, and oil prices were again on the escalator up.

Or was the reason that this was the first time a large outside effort had been officially sponsored by a government agency? Parisi mistakenly referred to CONAES as a "major Government work."[56] Was it that the study was so long in coming? Representative Morris Udall, chairman of the House Subcommittee on Energy and Environment, had called publicly for an explanation of the delay[57] as did Ralph Nader's Critical Mass Energy Project."[58] Was it the study's exhaustiveness and elaborateness? The fact that so many prestigious people were involved? Or was it just plain curiosity and suspense?

Whatever the explanation, the prerelease, front-page stories in the *Times* and *Post* gave CONAES great exposure. The upshot was an overflowing auditorium at the National Academy of Sciences on Monday, January 14 for the press conference to announce the CONAES report. In the days that followed, the Committee on Science and Technology of the U.S. House of Representatives held hearings on CONAES, requests poured into the Academy and *Post* for prepublication copies of the still undistributed 753 page report,[59] and a spate of articles and editorials appeared. They included a *Times* editorial that presented a supply orientation very different in outlook from that of the earlier news article.[60]

The same was true of an editorial that appeared in the *Post*. It also did a "reverse twist" of its earlier news article,[61] testifying to the autonomy of the editorial office within a news organization. The *Post*'s editorial showed propensities closer to Parisi's than O'Toole's. CONAES's "most important conclusion," wrote its author, was that "managing energy demand rather than supply was the key to energy policy." Further, if the transition is managed gradually and sensibly, economic output can be doubled without increased energy use, or, put technically, the country's energy/ GNP ratio can be halved.[62] The editorial imputed this finding incorrectly to the economic models employed by CONAES. Models often get credit for things not their due.[63]

The *Post*'s editorial writer found the paper's news coverage of CONAES "way off base," and the CONAES report's message "hard to delve out." "In fact, it did not have a single message. People could pick up whatever they wished."[64] This point was echoed by an editorial in the *Wall Street Journal*.[65] Nor was there disagreement from fellow journalist O'Toole. "Everyone read the report as they wanted," said O'Toole, "including myself."[66]

Letters flooded the offices of the *Post* in the days following publication of its mutually contradictory editorial and news article. Readers chose to react to the review with which they disagreed. Nuclear advocates wrote indignant letters to the editorial staff, while nuclear opponents and solar enthusiasts wrote equally irate letters to O'Toole. "It was a Battle Royal. The situation was totally polarized. There was no clear voice in the middle."[67]

POLARIZATION AND THE ABSENT MIDDLE

The CONAES experience, when observed over successive stages retrospectively, was like reflections through a set of mirrors. The media response, review of the report, working sessions and debates, creation of the research groups, appointment of the overview committee, and background for initiation of the study, repeated each other in a sense. The polarized newspaper accounts[68] echoed the one-sided positions taken in CONAES discussions, which were rooted in the conflicting views of participants, which in turn revealed deep-seated divisions among experts at large.

In part, CONAES was a confrontation between those whose image of reality was rooted in the past or in the world as they saw it—traditionalists—and those convinced of the need for fundamental change—reformists. The fact that there were representatives of both camps on the overview committee complicated its functioning and drove Brooks on occasion to fits of despair. The primary center of the first camp was the Supply Panel, with its allegiance to continued energy production and technological development. The influential center of the second camp was the Demand Panel, although its brand of reformism was not as orthodox as that of the Nader group. A distinguished nuclear engineer not on CONAES once said to Laura Nader, "You are asking me to deny the last twenty-five years of my life." That protest captures the tension that existed between many traditionalists and reformists involved in CONAES.

Not everyone, by any means, fell into one camp or the other. Nor did all participants leave the study with the same attitudes they had at the start.[69] Partly because of the existence of the outlying Nader group, the Demand Panel was accepted as moderate and professional, even though the respective projections of energy consumption by the two bodies were not very different. The Demand Panel based its engineering projections on plausible assumptions about energy price. The Nader group, on the other hand, started with the goal of minimizing the use of energy, which seemed going about things backwards to many of the engineers. Eventually, Laura Nader, despite her feelings about technical experts who work more with numbers than people added numerical estimates to her group's qualitative analysis of lifestyles and attitudes.

The Supply Panel eventually came to accept the possibility of lowered demand projections for two principal reasons. First was its own perception of the difficulty the economy would have in obtaining increasing amounts of energy. Second were the mutually supporting arguments of the Demand Panel and the Modeling Resource Group that the economy need not necessarily suffer from reduced energy growth or higher energy prices. The conclusion that the economy could continue to prosper at substantially lower levels of energy use than had previously been projected was regarded by many as the most significant result of CONAES. It was interesting to observe how different people developed or assimilated this important idea.[70]

In the summer of 1976, *Science* magazine asked Jack Gibbons, chairman of the Demand Panel, for an article on low-energy futures. Gibbons agreed, with the approval of the overview committee. He intended for his article to come out after the final CONAES report, whose distribution was then expected in December 1977. When the CONAES report was delayed, Gibbons was allowed to proceed. His account of the work of the Demand Panel appeared in *Science* twenty months before the final study report was released.[71]

The *Science* article, with projections considerably below the general thinking of the times, was widely quoted. The fact that the final CONAES report gave conservation first priority was somewhat surprising given the hard-liners on the overview committee. The report also mentioned a "middle ground" involving a "balanced mix between supply expansion and demand growth reduction."[72] Some called it "consensus by exhaustion."[73]

More than anything else, CONAES was an educational exercise at a crucial period in the nation's preoccupation with its energy problems. The study provided an illuminating if frustrating experience for large numbers of intelligent people, many new to one aspect or another of its multidimensional subject. Economists met with engineers, social scientists with physical scientists, and gradually they began to understand their different perspectives and methods of analysis. The project groomed experts, many of whom continued to work on energy problems, some in key government posts. The process of learning and response to a policy crisis may not always be efficient in a democratic society, but it taps the best minds and it works—and the society is the better for it.

8 END OF THE SEVENTIES

In the second half of 1979, shortly before appearance of the CONAES report at year-end, three research teams—one at the Harvard Business School and two at Resources for the Future—published important studies on energy in quick succession. The analyses reinforced each other in some conclusions, disagreed in others, and demonstrated how expert thinking and attitudes about energy had matured during the decade. The relatively friendly reception to the studies suggests that the public was by now convinced that the energy problem was genuine and needed treatment. Weary of the protracted energy debates, policymakers were searching for agreement. The three end-of-decade studies are the final ones in our review. The last of them, by another panel of the Ford Foundation, capped the efforts of energy policy analysts during the seventies with a timely call for realism. Suspicion and denial had given way to pragmatism and a hard look. The nation seemed ready to confront realities.

Policy analysts are not reticent about declaring the merits and appropriateness of the actions they recommend. The following avowals, reading like a lead-in to a set of analysts' recommendations, would therefore hardly be noteworthy were it not for the fact that they were made in not one, but three separate studies released within months of each other in the latter half of 1979.

> Now is the time for the United States to come to terms with the realities of the energy problem, not with romanticisms, but with pragmatism and reason.[1] Some hopeful signs of a national awakening and a developing sense of purpose are now becoming visible.[2] Our study group has tried to identify some of the fundamental realities that, in our view, define the energy problem.[3]

The country had experienced a double dose of energy shock in the period immediately preceding publication of these studies. Another round of sharply rising oil prices had followed the upset in Iran. Then came the alarming nuclear accident at Three Mile Island. A politically injured President Carter, having taken pains to defuse the charged energy problem during his last State of the Union address, had just returned from rethinking his positions at Camp David and was now promoting a revamped set of measures to help restore confidence in his administration. Energy was once more in the news, and the public's attention was being drawn to what little policy ostensibly had been agreed to since the first oil price hikes nearly six years before. The allusions in the quotes above to a need to "come to terms with the realities of the energy problem" and to a "national awakening" capture the mood at the time.

OPPORTUNITY FOR STATESMANSHIP

The situation afforded the occasion for statesmanship. Each of the three 1979 studies, as we shall refer to them, rose to the challenge with sweeping analyses and appeals for reason. So did the long-gestating CONAES effort which foaled at the very end of the year.

Though providing reinforcing signals, the three studies had basically different missions and thrusts. The first, Stobaugh-Yergin (after editors Robert Stobaugh and Daniel Yergin), was explicit in suggesting policy initiatives. It was an effort spread over seven years by researchers at the Harvard Business School, each responsible for surveying work and findings in a different energy area. Stobaugh and Yergin, seeking to integrate the individual re-

sults of their associates and extract policy conclusions of pragmatic value, took as the "cornerstone" of their thinking "that
conservation and solar energy should be given a fair chance in the
market system to compete with imported oil and the other traditional sources."[4] Eschewing the regulatory approach of the Ford
Energy Policy Project, they urged that reliance be placed on
financial incentives and the pursuit of profit. Theirs was a marketplace solution. It favored federal intervention only to correct
for market failures and imperfections. It sought to relieve petroleum dependence without resort to a tariff on imported oil.

The second study, RFF-Mellon, shied away from suggesting
such a plan or solution. Principally funded by the Andrew W.
Mellon Foundation, the study consisted of a small group of senior
analysts at Resources for the Future (RFF). The RFF analysts
made a decision early in their project not to aim for recommendations. Instead, they strove for synthesis and education—clarifying
and summarizing fundamental facts about energy to help illuminate the national decisions that needed to be made.

The third study, Ford-RFF, also involving Resources for the
Future, was the largest of the three in number of participants.
Funded by the Ford Foundation, it operated in the style of the
Ford-MITRE panel. Experts met monthly to hear briefings, make
presentations, and debate the issues. They then returned home to
write and rewrite drafts reflecting collective judgments. Ford-
RFF, unlike CONAES, made no attempt to represent a full spectrum of views or disciplines. About half the nineteen participants
were economists. They developed a framework for understanding
energy problems that posited a number of energy fundamentals
as the basis for making a broad set of policy recommendations.

A NATIONAL BEST SELLER

If anyone had been prescient or rash enough to predict at the
start of 1979 that an energy report with seventy-two pages of
footnotes published in July would go to seven printings and
106,000 copies in hardcover, then be succeeded within a year by a
revised and updated paperback edition gracing the windows of
the best bookstores, that person's judgment would certainly have
been called into question. By then, the nation already had en-

dured five years of energy studies and energy education. Its appetite for further edification seemed about spent. The hubbub over the energy problem was dying down, and the administration was trying its best to defuse what had been a politically explosive set of issues. Surely there could be little left to say, and few left to want to listen. That was the opinion of more than one publisher at the time.

Yet such a book was published and, much to the surprise of its editors, became a best seller. Out of stock for five weeks following publication, it was referred to enthusiastically as "a model of what university research and monograph writing on a major question of policy should be."[5] The book's title was *Energy Future*; it's subtitle, *Report of the Energy Project at the Harvard Business School*. The publisher who decided to take it on (Random House) worked closely with editors Stobaugh and Yergin in helping to make it eminently readable. The promotional campaign for the book was aggressive, but even more propitious for sales was the fact that the book hit the stores at the very time gasoline lines were forming again. "It was unbelievable luck," allows Random House editor Grant Ujifusa, ignoring for the moment the other characteristics of the book that contributed to its popular appeal.

To critic Paul Joskow, the success of *Energy Future* could be explained as follows:

> First, the book came out when the country was reminded once again of our vulnerability to energy supply interruptions and price increases resulting from our dependence on imported oil. Second, the book makes a serious effort to explain the origins of our energy problem, our difficulties in implementing an effective energy policy, and the options that are open to us, in a way that is easily understandable to the informed layman. Third, it has at least an aura of being very "reasonable." Villains are of course identified—energy suppliers, economists, engineers, etc.—but not in the conspiratorial context that has become popular. Finally, and most importantly, the book claims to provide a solution to our energy problems that costs very little (or actually saves money), that minimizes controversies over income redistribution, that causes no additional environmental damage, and that yields substantial benefits in terms of reduced dependence on foreign oil relatively quickly. Given the controversies and frustrations of the past six years, such a program must attract serious attention.[6]

Appearing at the time of the Camp David retreat and President Carter's subsequent policy turnabout, *Energy Future* was referred to frequently by Congress and the media in the controversy over the administration's crash program to develop synthetic fuels. To the *Wall Street Journal*, the book was perhaps "the most important contribution yet made to the energy debate."[7] Agreed the *New York Times Book Review*, "it is the best single examination of America's energy problem in print."[8]

The country was emotionally ready for some peace between its warring energy factions. Here, it appeared, was the hard-headed Harvard Business School taking a soft-path line that brought cheers from Amory Lovins and the Friends of the Earth at the same time that it evoked favorable review from respected members of the business community like Robert O. Anderson, chairman of the Atlantic Richfield Company, and C. C. Pocock, managing director of Royal Dutch Shell. Yergin had written convincingly in a chapter titled "Conservation: The Key Energy Source" of the impressive gains several companies had made with their energy conservation programs. Soft-path enthusiasts were quick to interpret this as sound management justification for their position. With publication of *Energy Future*, it must have seemed like rapprochement was near.

But appearances were somewhat deceiving. The Stobaugh-Yergin estimate that as much as 20 percent of America's energy needs could be met by renewable energy sources in the year 2000 hardly jibed with the calculations of the energy industry. The authors were criticized for the same kind of reckless optimism, superficial reasoning, and neglect of institutional factors in making their estimate that they had attributed to the advocates of coal and nuclear technology. This damaged their credibility with the selfsame traditionalist camp with which many readers had identified them. In responding to our study questionnaire, traditionalists gave considerably higher marks on quality to the other two 1979 studies than to the Stobaugh-Yergin effort.

Be that as it may, Stobaugh's discerning historical analysis of the oil crisis in *Energy Future* was a masterful explanation of the reasons for the price rises and the jeopardy into which the United States had fallen. The interpretation, both interesting and understandable, was a model of good communication. Had we asked respondents in our questionnaire to rate the fourteen studies for success in conveying their message to a lay audience, Stobaugh-Yergin is likely to have come out first on the list.

The study covered world oil, natural gas, coal, nuclear, conservation, solar, and, in an appendix, economic models. Besides Stobaugh and Yergin, its authors were I. C. Bupp (nuclear and natural gas), Mel Horwitch (coal), Sergio Koreisha (models), Modesto A. Maidique (solar), and Frank Schuller (natural gas).

RFF IN THE POLICY LIMELIGHT

Resources for the Future must have felt more than an ordinary amount of pride and accomplishment upon publication of two reports in August and September of 1979 in which it played key roles. The Ford Foundation, benefactor in RFF's establishment many years before, had finally chosen it to administer a Foundation-sponsored energy analysis after having passed it by on two previous occasions (Ford Energy Policy Project and Ford-MITRE). RFF economist Hans Landsberg led the Ford-RFF effort with the help of a small staff. Landsberg's longtime colleague, Sam H. Schurr, was already under way with four other RFF associates on the RFF-Mellon study, wholly an RFF effort. The final reports of the two projects, released within a month of each other, enjoyed the same auspicious timing.

The first part of the RFF-Mellon report—its overview—was distributed a month before July 1979 publication of *Energy Future*, receiving far less attention. When the full RFF-Mellon report came out two months later, press coverage in the *New York Times* consisted simply of a news item on page 35. But the reviews that did appear were good ones. An enthusiastic commentator wrote, "Probably the best energy study yet to come along . . . a veritable bible of information on energy."[9]

RFF-Mellon took the position that all interests could be satisfied, but not to the full extent each might desire. The economy could advance without high rates of energy growth, but holding growth to zero was not necessary. Environmental conditions could be improved with technologies currently under development, and a reconciliation with environmental values was possible. If zealots on both sides of the debate would back down from their rigid stances, reasonable compromise could be reached.

In contrast to Stobaugh-Yergin, RFF-Mellon refrained from advocating a particular program, and chose, in customary RFF fashion, not to be identified with any one recommended course. It did say that both coal and nuclear options were needed and that steps should be taken to make these energy sources more acceptable. It thought solar offered great potential but was just too expensive to be competitive for most purposes during the current century without a breakthrough in the availability of cheaper materials or a concerted program of government subsidies. Some soft-path advocates considered RFF-Mellon pronuclear and antisolar, but most readers accepted it as basically objective. The study provided a wealth of useful, well-documented information and analysis—the goal it had set for itself.

Working with Schurr on the study were Joel Darmstadter (energy consumption and conservation), Harry Perry (mineral and synthetic fuels), William Ramsay (energy supply and risks), and Milton Russell (economics and public policy). Specific research contracts let as part of the project supplemented information available from elsewhere and supported the staff's work. Darmstadters's illuminating examination of energy conservation was one of several parts of the research singled out for special commendation by other investigators.

THIRD ENTRY BY THE FORD FOUNDATION

McGeorge Bundy, stepping down from his eventful thirteen-year presidency of the Ford Foundation, was like a proud father as he wrote the foreword to the Ford-RFF report:

> This is the third major independent study of energy problems that we have commissioned in the last eight years, and like its two predecessors, it appears at a timely moment. The central message delivered by the authors of the first study, *A Time to Choose*, was that this country could and should get along with less energy than historic patterns of growth suggested. That message was right and timely in 1974. The central message of the group that did the second study, *Nuclear Power Issues and Choices*, was that over the next generation we could and should handle nuclear power in ways that would reduce the dangers of nuclear proliferation, without giving up light water reactors for the present and other nuclear options for the future. That

message was right and timely in 1976/1977. The central message of
the present report is that energy—expensive today—is likely to be
more expensive tomorrow and that society as a whole will gain from a
resolute effort to make the price that the user pays for energy, and
for saving energy, reflect its true value. And I myself think that the
message will prove right and timely in 1979.[10]

Bundy had a good sense of what was policy relevant. Though
he says he did not "predict the conclusions of any of the three
studies before they were undertaken,"[11] each was skillfully tuned
to political priorities of the period. In mounting its third analysis,
the Foundation sought to avoid earlier mistakes. Since it considered the Ford-MITRE format a big improvement over that of the
Energy Policy Project, it adopted the same format for Ford-RFF.
It wanted to make sure that its new study would not be criticized
for neglect of market forces—as had the EPP—or for slighting international considerations—as had Ford-MITRE. It gave both
themes major emphasis in the Ford-RFF study.

The Ford-RFF team was commissioned to examine the prospects for coal, but soon decided (as had CONAES with the breeder issue) that the subject could not be treated meaningfully in a
contextual vacuum. Setting out to develop a comprehensive and
consistent economic framework, study members sought to identify costs and tradeoffs for comparing coal to the alternatives.
They ended up addressing a wide range of issues bearing on the
energy problem.

Ford-MITRE and Ford-RFF bracketed the CONAES study in
time. They had a big advantage over the protracted study, with
whom there was a bit of friendly competition. The Ford Foundation panels were compact, compatible groups. They were not chosen to represent all disciplinary backgrounds or points of view. In
Ford-RFF, for example, economists dominated the group. They
took as a given that "energy is an economic good" and that
"firms and individuals can, given the proper motivation, find
ways of satisfying their needs with less" of this good.[12]

Contributing to Ford-RFF's dispatch was the fact that many
members had participated in earlier energy studies, sometimes together. Kenneth J. Arrow, Richard L. Garwin, Hans H. Landsberg, George W. Rathjens, Larry E. Ruff, John C. Sawhill, and
Thomas C. Schelling had all been associated with the Ford-MI-
TRE effort. Several had been connected with other major energy

studies as well. Robert W. Fri had been directly involved with the ERDA and MOPPS analyses and with the White House task force on nuclear that was contemporaneous with Ford-MITRE. William W. Hogan had participated in both the Project Independence and CONAES efforts, Robert Stobaugh was directing the project at the Harvard Business School, and Harry Perry was involved in the RFF-Mellon research. Other members included economist Francis M. Bator, law professor Kenneth W. Dam, World Bank director Edward R. Fried, MITRE chief scientist S. William Gouse, nuclear physicist Theodore B. Taylor, conservationist Grant P. Thompson, public health professor James L. Whittenberger, and geographer M. Gordon Wolman. All told, the group had considerable experience in energy and related fields and brought several years of thinking about energy problems with them to study meetings.

The Ford-RFF study adopted a twenty-year horizon. Its final report, *Energy: The Next Twenty Years*, came out in September 1979, about the same time as the RFF-Mellon report and just two months after *Energy Future*. The book sold well, as such studies go, but still less than *Energy Future* by a factor of about ten.[13] The report began with seven assumptions phrased as fundamental realities. The intent was to put to rest the major myths that underlay scapegoating and misunderstanding.

1. The world is not running out of energy. It is incorrect and misleading to define the long-run energy problem in terms of a gap, shortfall, or shortage, as though there were some natural definition of energy needs and some physical supply limits preventing these needs from being met.
2. Middle East oil holds great risks, but is so valuable that the world will remain dependent on it for a long time.
3. Higher energy costs cannot be avoided, but can be contained by letting prices rise to reflect them.
4. Environmental effects of energy use are serious and hard to manage.
5. Conservation is an essential "source" of energy in large quantities.
6. Serious shocks and surprises are certain to occur.
7. Sound R&D policy is essential, but there is no simple "technical fix."

These realities led the authors to propose six general principles and objectives for U.S. energy policy:

1. Use market forces to price and allocate energy efficiently, because doing so will simplify most other aspects of energy policy.
2. Develop ready contingency programs for a sudden interruption in oil imports, because such a cutoff is almost certain to happen somewhere, sometime, with potentially disastrous consequences.
3. Recognize that economic and political factors in world oil markets make the cost of imported oil greater than measured by its price; any explicit measures to reduce imports can be economical if the per-barrel cost of achieving the reduction is not too high.
4. Encourage energy production—including its equivalent, energy conservation—anywhere in the world, because world energy and economic systems are so interdependent that developments in one part of the globe affect everyone.
5. Change some of the procedures and programs now used to deal with environmental conflicts ... the costs of mismanagement are too high to be ignored.
6. Use government research and development policies to define and develop a wide range of information and options, but not to push premature adoption of one or another technology.[14]

The Ford-RFF team fashioned these guidelines into a set of specific recommendations that emphasized reliance on market forces over government intervention and bureaucratic controls.

We believe our recommendations, taken together, provide a sound framework for facing an energy future that will be difficult but manageable. The fundamental changes we anticipate in the world energy situation over the next twenty years—such as a relative decline in oil and gas use, rising energy costs and prices, and increased efficiency in energy use—will occur whether the United States handles its energy policy wisely or foolishly.

The difference that enlightened policy could make, ventured the authors, would be to make the inevitable changes "relatively smooth and easy" rather than costly and disruptive. It was a fitting theme for an anchor analysis of the energy decade.

SIMILARITIES AND AGREEMENT

The 1979 studies had noticeable affinities, we observe, despite differences in thrust and goals. They were similar in timing, of course, and in the broad scope of their analyses. They were also similar in drawing attention to the seriousness of U.S. vulnerability to disruptions in the flow of oil from the Mideast. Each asserted that energy would continue to cost more. Each attached importance to energy conservation. And each favored the removal of controls from petroleum prices.

All agreed on the need to give consumers better signals on the true cost of oil and gas. RFF-Mellon had said, "Effective energy policy is best served by a pricing system that tells each consumer accurately and directly what it costs our economy to use more energy."[15] Added Ford-RFF, "It is a dangerous misconception to believe that government can somehow provide dependable, clean, and plentiful energy cheaply,"[16] a point underscored by Stobaugh and Yergin who called it "a dangerous delusion for a country that imports almost half of its oil and is the largest buyer of OPEC oil." Denouncing the irrationality of the sytem of price controls by which the nation was effectively subsidizing the importation of foreign oil ("a great disservice" to American consumers), Stobaugh-Yergin nevertheless noted that "it would be a mistake to regard decontrol as a miracle solution."[17]

The Harvard study seemed to have its own "miracle solution" in its enthusiasm for conservation and solar energy. At least that was the charge of M.I.T. economists Robert Pindyck (who referred to the Harvard prescription as an "easy panacea")[18] and Paul Joskow (who preferred "free lunch" and "economic counterpart of alchemy").[19] Retorted Stobaugh and Yergin, "If there is one thing that we did do, we pointed out that ... 'No easy remedy will solve the energy crisis'."[20]

Stobaugh and Yergin had denounced some of the economic analysis applied to energy in the 1970s as overly optimistic or irrelevant. Their book contained an appendix titled "Limits to Models," which found fault with three specific econometric analyses of the energy markets, including one at M.I.T. with which Pindyck and Joskow had been associated. Thus, the brickbats flew in both directions. Stobaugh and Yergin criticized the M.I.T. economists. The M.I.T. economists criticized them. Finding the policy conclusions of the Stobaugh-Yergin study inconsistent and unconvincing, they compared the study unfavorably with the other two more analytically oriented efforts.

Although all three 1979 studies agreed on the desirability of price decontrol, that was by no means the view within the population at large. Yet a political force was developing. Conservationists and environmentalists were beginning to see decontrol as an effective means for curtailing energy growth. Their backing strengthened the long-standing position of market economists and industry. Both the *Washington Post* and *New York Times* had by now joined the *Wall Street Journal* in calling for oil price deregulation. President Carter, previously resistant, was initiating steps to put decontrol into effect. The 1979 studies reinforced this political conversion without provoking the outbursts triggered by earlier decontrol recommendations.

The 1979 studies shared a conviction that large gains were possible from increased energy efficiency, and that reduced energy growth need not impair the economy. Thus bolstering the case for energy conservation, the studies were affirming positions that had gradually been gaining currency inside and outside the analytical community.

Former Project Independence analyst William W. Hogan, citing flexibility in the economy, wrote an extensive plausibility argument for lowered demand expectations as part of the Ford-RFF report. His views on the subject had developed during close involvement in the CONAES modeling runs and a follow-up analysis of energy/economy interrelationships which he directed for the Energy Modeling Forum.[21.] Hogan remembers having had his "eyes opened" by an exploratory exercise in which he participated as a member of a CONAES task force asked to formulate an illustrative low-demand scenario.

The CONAES Demand Panel, as noted earlier, had made a strong case for the feasibility of curtailing energy growth. Stobaugh and Yergin cited this work in presenting their optimistic outlook on the country's conservation potential.[22] It was looking as if a modicum of agreement was forming at last. "The failure to achieve a national consensus," wrote the Harvard team, has "proved to be one of the greatest obstacles to a rational energy policy."[23] Was the long-awaited consensus finally at hand?

The RFF-Mellon analysts believed that national agreement on goals, though difficult, was desperately needed for progress in energy policy. The inability to reach such agreement had been the bane of energy policymakers throughout the 1970s. The immobilizing debates waged in CONAES committee meetings symbolized the disagreements. CONAES was now coming to completion. The three 1979 studies were being well received. Were these signals that the battle was finally coming to an end?

The Antagonists

Who was it that was doing the fighting? On the surface, it seemed to be a struggle between traditionalists and reformists; but the situation was not that simple. There were, in fact, many variations of these polar stereotypes, as described in Appendix A.

RFF-Mellon analysts described the debate as one between two conflicting perceptions: a traditional penchant for economic and technological progress versus a more recent (but recurrent) view of the world as finite in resources and limited in potential to support growth. Holders of the first position asserted that continued expansion in energy use was needed to broaden the available range of goods and services so as to keep consumers satisfied. Adherents to the second position felt an imperative to redistribute resources and restrict demand. Ecologist Paul Ehrlich referred to the two factions as "Cornucopians" and "Neo-Malthusians,"[24] recalling an earlier reference by Harrison Brown to the "great cornucopian-conservationist debate" between technological optimists and conservationists.[25] The distinction was not the same as the one between the hard and soft paths of Amory Lovins, although individual allegiances fell along similar lines.

Stobaugh and Yergin pictured the conflict more as a tug-of-war over energy and money. Who would benefit and who would lose from the post-1973 energy shortages and petroleum price increases? The size of the "windfall" from revaluation of U.S. oil and gas reserves was, they suggested with some theatrical flair, an awesome $2 trillion ($25,000 per family) as of the end of 1979.[26]

This windfall created a distributional dispute between consumers and producers. It also pitted blocs within groups against each other. At odds were various regions of the country, multinational and domestic companies, small and large refiners, and different socioeconomic strata. Economists called the policy issue a trade-off between "equity" and "efficiency,"[27] but having a name for a problem did not resolve it.

Central to the maneuvering and wedged in the middle was government. We have seen in the reaction to the Ford Energy Policy Project the deep-seated antipathies between those calling for more regulation and those firmly committed to a competitive free-market. Many in the former group believed the public was the victim of a plot by the oil companies. Those in the latter group countered that the energy problem would take care of itself once free market forces were restored.[28] At issue was the appropriate degree of government intervention.

Political scientists Aaron Wildavsky and Ellen Tenenbaum, in discussing dissension over oil and gas estimates, attached special importance to the element of mistrust in energy politics.[29] Their dichotomy between those with a "quantity" perspective and those with a "price" perspective was an attempt to summarize the differences that have characterized energy politics since early in the century. Quantity-oriented preservationists and consumerists have historically looked to government to regulate the amount of energy produced or available, either to preserve resources or satisfy consumer wants. Price-oriented industrialists have preferred to rely on the marketplace. They consider the price of energy to be the best determiner of who gets how much and what is produced—at least during periods of rising demand. When the market grows soft, business has been known to appeal to government too—to control against overproduction and falling prices.

In the theory of the markets, quantity and price are two ways of looking at the same thing. In the political world, they have significantly different meanings. A quantity perspective, exemplified by net energy analysis,[30] is leery of the fairness and long-term efficiency of the unbridled market. A price perspective has faith that self-interest can be employed for the general good and doubts that any government can know better than people themselves about their own desires and needs. Wildavsky and

Tenenbaum view the history of energy politics as a constant struggle between members of these two ideological perspectives. They believe the fundamental distrust engendered by this struggle explains the disappointing record of energy policy in the seventies.

The quantity and price perspectives could be used to classify some of the energy studies we reviewed. EPP, WAES, and the CIA analysis were quantity-oriented—especially the latter two with their references to gaps between supply and demand. PIES, MOPPS, and Ford-MITRE were price-oriented, as were RFF-Mellon and Ford-RFF.

One may or may not agree with the usefulness of the quantity/ price dichotomy, but the underlying polarizations are very real. We have seen them repeatedly in the conservation controversy, the widening nuclear rift, and the debate over natural gas deregulation. They are so ingrained that full consensus in energy matters seems unrealistic. Periodic setbacks appear inevitable as changing conditions exacerbate the intrinsic conflicts.

Path to Reconciliation

Yet conciliatory shifts in policy positions by interested parties do occur, and government can make constructive adjustments in its policy approach. Chapter 2 notes how the support of the pipeline companies for price controls waned when gas supplies fell short. Congressmen from consuming states came to realize that what was good for producers could also be good for their constituents and thereby found grounds for favoring deregulation. Resource conservationists also grew more accepting of higher energy prices, and a positive outlook on both energy conservation and oil price decontrol gradually took hold.

Even the generally pessimistic Wildavsky and Tenenbaum had some suggestions for satisfying the parties in conflict,[31] although they did not go nearly as far as the RFF-Mellon analysts. The RFF-Mellon team put its hopes in improved knowledge of the facts, formulation of a shared world view, and acceptance of a decision process that all contending parties could respect.[32]

Our own analysis of elite attitudes supported the possibility of reconciliation in that the dissimilar viewpoints expressed by the elite were not entirely in opposition. Experts often were talking past each other rather than being in direct conflict. We concluded that there was room for accommodation if a convincing case could be made for serving mutual interests or the national welfare.

The 1979 studies began to suggest the outlines of what such accommodation might be. The debate had run its course. The country, convinced now of the validity and severity of the problem it faced, was looking for harmony.

Analysis may or may not have a part to play in effecting a reconciliation. We have some misgivings, but are generally more sanguine on this point than Wildavsky and Tenenbaum. We know that policy studies can lag rather than lead the tide of public opinion, and that they are prone to become targets and instruments of disagreement rather than mediators. We realize that the conflicting results of analysis can add to the confusion and hurt the cause of negotiation more than help it. But analysis can also have a salutary effect. It can be an antidote to scapegoating, it can reveal the futility of denial, and it can expose the roots of mistrust.

The influence analysis has depends partly on the nature and phase of the political dispute—the environment for the analysis. Analysis can affect as well as be affected by this environment. Thus, the question of the role of analysis is considerably more complicated than it may first appear. Analysis has many parts to play and many possible effects. Its consequence in a given context is itself a subject for analysis.

We return to this theme in subsequent chapters. For now, we note in passing that policy studies rarely relate their results to the interests of the parties in contention, and often do not acknowledge who these parties are or even that they exist. There is a certain gentlemen's agreement—an enforced naivete—in the way policy analysis tends to be conducted. We refer to this professional detachment from political reality as the "blinders syndrome." It seems to us counterproductive and ill-advised.

9 ANALYTICAL ISSUES ACROSS STUDIES

Many of the energy studies dealt with the same issues, but often in different ways. In comparing the approaches and assumptions of a few selected studies, we observe how they reflected the change in attitudes that took place over the course of the decade. A significant revision occurred, for example, in the outlook for energy demand and in the economic impact attributed to energy scarcity. The change is traced through the conclusions of successive energy analyses. Agreement in these analyses was generally difficult to achieve on issues of both energy supply and energy demand for much of the period. Most intractable were estimates of "externalities"—the environmental, safety, social, and international issues not easily quantified or translated into dollars. Too often analysis merged with advocacy. There was no easy or acceptable way to resolve value conflicts.

We now compare how several of the energy studies treated certain analytical issues central to the energy debate. The comparison is summarized in four tables dealing respectively with energy demands (Table 9–1), supplies (Table 9–2), model formulation (Table 9–3), and externalities (Table 9–5). Since these tables are intended primarily for the specialist, the vocabulary used may be unfamiliar to readers without training in economics. For those not acquainted with terms such as price elasticity of demand, partial equilibrium, present value, and externalities, we offer brief definitions of these concepts in the text where they first appear.

ENERGY DEMANDS

In most analyses, it is assumed that the GNP growth rate is one of the primary determinants of energy demands. If it is assumed to be the *only* determinant, demands can basically be extrapolated. This approach worked reasonably well during the 1950-1970 era when there were no abrupt discontinuities in energy prices.

For the post-1973 world, however, demand trend extrapolation has proved distinctly unsatisfactory. Increasingly, it has become apparent that energy models must allow for *price-induced* conservation. This can be analyzed either from an engineering or economic viewpoint. Engineers tend to relate conservation to specific design changes such as insulation in homes and other structures, industrial use of heat exchangers and cogeneration, and diesels instead of internal combustion engines in automobiles. These changes become cost-effective with sufficient rise in the price of energy. To summarize conservation possibilities, economists employ the concept of price elasticity of demand, here denoted by the symbol e. *Price elasticity of demand for energy* is the percentage reduction in energy demands associated with a 1 percent increase in the price of energy relative to other commodities. For estimating price elasticity, it is supposed that GNP and other factors influencing energy demand remain unaffected by energy price increases.

Each rise of 1 percent in energy prices leads to a reduction of e percent in energy demands—given sufficient time for adjustments in equipment, structures, and lifestyles. If we suppose that energy prices and the GNP growth rate are independent of each other—that is, if we neglect energy-economy interactions—a simple demand forecasting equation would be written:

$$[\% \text{ change in energy demand}] = [\% \text{ change in GNP}]$$
$$- [\text{price elasticity } e] \times [\% \text{ change in energy price}] \quad (9\text{-}1)$$

As an example, suppose there is a consensus that U.S. labor force and productivity will continue to expand so as to sustain a long-term annual GNP growth rate of 3 percent. Suppose also that future energy prices will rise at the annual rate of 4 percent (in dollars of constant purchasing power) with the depletion of domestic and international petroleum resources. According to Equation (9–1), the implied annual growth rate of energy demands would then vary anywhere from 3 percent (for $e = 0$) down to 1 percent (for $e = 0.50$). Projected over a thirty-year period, this difference between 3 percent and 1 percent would compound to an 80 percent difference in energy demand forecasts.

There are wide variations in the numerical value of e employed in the selected analyses shown in Table 9–1. Within the CONAES Modeling Resource Group (MRG) alone, the variety of econometric approaches adopted produced price elasticities ranging from virtually 0 to 0.50. Some price elasticities were estimated from time series data collected on the United States as a whole, others from regional and international cross-sections of data, and others from *pooled* combinations of both.[1]

We should not be surprised, therefore, to see a wide range of demand estimates emerging from these studies. Some of the variation is due to differences in assumptions about GNP growth, mandatory conservation, or energy prices, but most of it seems attributable to differences in the price elasticity of demand for energy.[2] In the Ford-MITRE and MRG studies, the price elasticities were evaluated explicitly. For the rest, they were implicit, and one can only surmise qualitatively whether they were high or low. In Figure 9–1, note that the three alternative EPP scenarios of the Ford Energy Policy Project (historical growth, technical fix, and zero energy growth) cover a sufficiently wide range so that they bracket *all* the subsequent estimates of the demand for primary energy in the year 2000 by the ERDA, Synfuels, Ford-MITRE, WAES, and MRG studies. Also shown in Figure 9–1 are the FEA and CIA medium-term projections for 1985.[3]

To buttress its conclusions, the EPP employed the innovative Hudson-Jorgenson (H-J) econometric model, which attempted to capture consumer responsiveness to higher energy prices. EPP's conclusions had originally been derived based on technical feasibility, mandatory conservation, and public regulation.[4]

Table 9–1.   Energy Demands

	EPP	FEA	Synfuels	ERDA
Price responsive-ness of consumers	EPP conservation estimates were based primarily on technical feasibility. Appendix F did, however, provide econometric estimates of price responsiveness based on U.S. time series.	Econometric estimates based on pooled regional cross-section and U.S. time series.	Fixed future demands for "end-uses"; interfuel substitution based on market share price elasticity assumptions.	Fixed future demands for "energy services"; interfuel substitution based on engineering process analysis.
Implied price elasticities of demand for primary energy, η	High	Medium	Low	Low
Non-price aspects of energy conservation	Need for educating consumers and managers; mandatory fuel economy standards; shift away from auto and air to truck and rail transport; cogeneration of electricity and process heat, etc. (ch. 2).	Model employed to evaluate a wide variety of mandatory conservation and price control proposals.		Heavy emphasis on the need for developing technologies leading to improved efficiencies in end use.
Energy-economy interactions; employment and labor productivity	"Suprisingly small" impact of energy conservation on economic growth; "economic impact of gradually reducing the long term growth rate of energy consumption is fundamentally different from the impact of a sudden and unexpected interruption in energy supplies" (p. 71).			None explicit, but described as under study through linkage with the Hudson-Jorgenson model (ERDA-76, p. 97).

Ford-MITRE	MRG	WAES	CIA
Alternative assumptions on own- and cross-price elasticities.	Methodology differed among six U.S. models. One (Nordhaus) employed a pooled cross-section and time series for 6 industralized nations.	Methodology differed from country to country.	Informal reductions of energy demands to allow for: "higher prices and conserva-tion measures" (p. 4).
.35 (base case); alternatives: .25 and .50.	Low to .50; see MRG Appendices C and D.	Low	Low
Included list of reasons for *not* counting on significant "life-style changes" other than those that are price-induced (p. 56), but allowed for income elasticity of 0.9 through year 2000.	10% non-price induced reduction in aggregate energy consumption projected by 2010 (p. 12).		
None explicit. This follows from view that "for the long run, we can say with confidence that there is no direct relationship between energy cost and the number of jobs" (p. 48).	Clear distinction drawn between cases "where the curtailment of supply is abrupt or gradual" (p. 44). "The negative feedback from a curtailment of energy use. . .is moderate, in fact small, unless *both* the curtailment is large—say by 50%—*and* the price elasticity of demand is small in absolute terms—say substantially below one half" (p. 114).		

Figure 9–1.   EPP Scenarios: Energy Use in 1985 and 2000

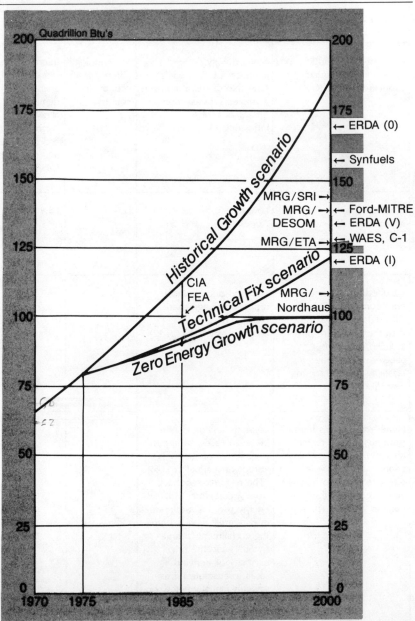

Source: Energy Policy Project of the Ford Foundation, *A Time to Chose: America's Energy Future.* (Cambridge, Mass.: Ballinger Publishing Company, 1974), p. 13; superimposed with other projections.

Our calculations assume that in the future, users in the industrial sector will be more aware of energy costs (and therefore more responsive to using energy in an economically efficient way), and that the market imperfections that inhibit investments to save energy in the residential, commercial, and transportation sectors can be removed by specific government actions ... We believe that if the nation adopts energy conservation as a goal, and adopts the set of policies that we shall analyze in detail below, we can achieve the level of savings in the *Technical Fix* scenario, and by so doing alleviate concerns about supply, environment, and foreign policy without appreciable energy price increases.[5]

Although EPP acknowledged the potential usefulness of market mechanisms, it argued that "market imperfections" created a need for new regulatory institutions. Political conservatives rejected its proposals for expanding the role of government out of hand. By way of contrast, the ideologically middle-of-the-road Ford-MITRE and MRG studies showed little inclination to advocate nonprice routes to energy conservation, except to stipulate a 10 percent nonprice-induced conservation by the year 2000 from, for example, a shift in consumption patterns away from manufactured goods toward services. Neither Ford-MITRE nor MRG counted upon widespread adoption of an energy conserving ethic.

ENERGY-ECONOMIC INTERACTIONS

Equation (9–1) is often described as a *partial equilibrium* model, for the price of energy and the rate of GNP growth enter as independent variables. In a partial equilibrium model, the energy sector is assumed to have only minor impact upon the long-run rate of GNP growth. By contrast, a general equilibrium model allows for indirect impacts of the energy sector on GNP—hence requires more data inputs.

With energy-GNP interactions, energy prices and GNP growth are not independent of each other. Rising prices and energy scarcities could lead to a slowdown in aggregate growth, together with general inflation, balance-of-payments deficits, a drop in labor productivity, and a rise in unemployment. One general equilibrium analysis of the 1973/74 oil crisis, quantifying these effects, concluded that:

The oil price increase added nearly 1.8 percentage points to the U.S. inflation rate in 1974 and 1975; by 1975, the energy crisis had raised the unemployment rate by 1.7 percentage points; real GNP was reduced by 3 percent in 1974.[6]

Unlike this macroeconomic study, none of our selected analyses were designed to analyze the short-term employment, inflation, and balance-of-payments effects of energy shortages. Except for the CIA study, all were intended to provide insights into the more distant future under the assumption of gradual and foreseeable energy price increases. Only the H-J model allowed for feedbacks from the energy sector to the balance of the U.S. economy, concluding that:

1. Energy conservation along the lines of the *Technical Fix* or *Zero Energy Growth* scenarios is possible within the existing structure of the economy;
2. The result of moderating energy use in terms of higher inflation and reduced real incomes and output is significant but not catastrophic.[7]

The other studies assumed that general equilibrium effects were negligible. That is, GNP growth would be virtually unaffected by energy scarcities. (See the quotes on the bottom of Table 9–1 excerpted from the EPP, Ford-MITRE and MRG studies.) The typical view was that expressed by the MRG:

> The negative feedback from a curtailment of energy use below the Base-Case path is moderate, in fact small, unless *both* the curtailment is large—say by 50 percent—*and* the price elasticity of demand is small in absolute terms—say substantially below one half.[8]

Except for some rough side calculations and approximations,[9] the MRG did *not* undertake a detailed analysis of the assumption that energy demands and GNP growth could be decoupled. This was by design. According to the group's minutes of 3 June 1976, early in its deliberations:

> Koopmans turned the discussion toward a consideration of what priorities should be given to the various questions that the MRG is capable of answering. He suggested that highest priority be given to estimating the benefits from investing in alternative R&D packages. The next highest priority would be accorded to assessing the environ-

mental aspects of the alternative R&D decisions and resulting technologies. Middle-level priority would then be assigned to these questions: (1) estimating the effects of other important policies; (2) examining the extent to which energy and GNP can be "decoupled"; and (3) assessing the capital requirements to meeting the energy demand in the future.

This technical limitation was not fully understood outside the MRG. Our interviews revealed that many CONAES participants were under the impression that the MRG performed an explicit analysis of energy-economy "decoupling," an analysis that could not have been formally undertaken since the H-J model was virtually the only tool available in the United States for this purpose at the time (1976), and it was not among the models represented in the MRG.

After 1976, the Energy Modeling Forum conducted intermodel comparisons of energy-economy feedback effects using six models,[10] including both the H-J model by itself and the equilibrium framework created by linking the H-J model to the Brookhaven model,—"the first time that a macroeconomic growth model had been linked to an interindustry sectoral model and subsequently linked to an energy technology oriented resource allocation model."[11] By mid-1977, Charles J. Hitch, then president of Resources for the Future, was able to organize a symposium to compare several alternative approaches to modeling energy-economy interactions. By that point, a considerable degree of professional consensus had emerged.[12] Each of the general equilibrium studies reached conclusions consistent with the MRG findings—that the feedback effects are small unless the energy curtailment is large and the price elasticity is substantially below 0.5. What had been relative heresy in 1974 at the time of the EPP became a new orthodoxy three years later.

ENERGY SUPPLIES

During the period from 1974 to 1977, as the outlook for conventional energy supplies grew increasingly gloomy, it became essential to examine whether new technologies would permit a transition away from oil and gas. Table 9–2 indicates how the selected studies dealt with the question of supplies. In 1974, EPP

Table 9-2.   Energy Supplies

	EPP	FEA	Synfuels
Price responsiveness of producers Proven reserves and ultimate resources	"The amount of oil and gas in the ground will not be a limiting factor in supplying historical energy growth for the rest of this century. A more likely prospect is that oil and gas production will clash with other social values, and those constraints will slow the pace at which resources can be discovered and brought to market" (pp. 26–27).	1985 domestic oil production would be 44% higher if domestic prices were to rise from $1.2 to $1.9/MBTU. Natural gas deregulation could make a 38% difference in domestic supplies by 1985 (PI, pp. 5, 6).	Region-by-region estimates of long-run supply curves for crude oil, natural gas, etc. Marginal cost, excluding lease bonus payments, plotted as a function of cumulative committed production (II, p. 23).
Capacity limits, conventional technologies		Capacity limits projected for existing energy facilities and also for nuclear power plants through 1990. Because of these limits, nuclear energy production was, in effect, an exogenous input to the model rather than an output.	
Dates and rates of introduction of new supply technologies (typically post-1990) Abbreviations: MBD: million barrels daily GW: gigawatt (typical size for a nuclear reactor)			With a federally subsidized "information program" in place by 1985, the decision tree provided for corporate decisions leading to synfuels capacity of up to 4 MBD by 1995 (I, p. 46).

ERDA	Ford-MITRE	MRG	WAES	CIA
See Figure III.1. No cost-quantity estimates cited.	Total of domestic plus imported oil and gas available to U.S.: 3000 quads at $2/MBTU; 5.5 million tons of uranium available at prices up to $100/pound (p. 69).	Total domestic gas and gas resources available to U.S.: 1720 quads at $2/MBTU. 7.8 million tons of uranium available at prices up to $120/pound (p. 13).	Two alternative estimates cited for ultimately recoverable oil resources in WOCA: 1.3 and 1.6 trillion barrels (p. 126). No cost-quantity relations cited.	
	"Delayed nuclear" case based on projection of 100 GW of nuclear capacity for 1985 (p. 61).		Oil discovery rates, together with lower bounds on reserve/production ratios, implied bounds on world oil supplies—even in absence of OPEC government-imposed limits on production (pp. 127–130).	Entire analysis hinged upon the agency's ability to estimate OPEC's oil production capacity through 1985, and to a lesser extent, Soviet production.
Scenario V projected the following unconventional supplies for year 2000: coal-based synfuels of 7 MBD, shale oil of 4 MBD, 300 GW of conventional uranium-fueled light water reactors plus 75 GW each of two advanced reactor types—one gas cooled and the other a plutonium-fueled fast breeder. See Apendix B, ERDA-48. No rationale provided for this technology mix. In retrospect, these official goals appear far too optimistic.	Shale oil limited to 2 MBD in 2000; FBRs beginning either in 2000 or in 2020 for a "delayed breeder or alternative" (p. 69).	FBR and other advanced electrical generating technologies limited to 6 GW of demonstration and prototype capacity in year 2000; ceilings on annual growth rates during subsequent years were specified as: 40% during first 5 years, 30% during second 5 years, etc. These growth rates were based upon analogies with LRW program as viewed during early 1970's.	Shale oil limited to 2 MBD for U.S. in year 2000; additional unconventional fossil fuels production from tar sands and heavy oils in Canada and Venezuela (pp. 217–221); "unlikely that power from FBRs could provide more than 5% of nuclear energy by year 2000" (p. 210).	

Table 9–2.   Continued

	EPP	FEA	Synfuels
Oil import projections	"Most current energy supply planning is based on the premise that the United States cannot count on large supplies of imported energy, particularly oil. For the near to intermediate term, however, the limits on increases in imported supplies are largely political and, therefore, subject to change" (p. 31). No details provided on feasibility of importing 32 quads of oil and 5 quads of gas in year 2000 for "high import" case (p. 28).	Oil imports serve as "swing fuel" in PIES model to close the gap between domestic supplies and demands. Imports could vary between 7 and 26 quads in 1985— depending primarily on whether U.S. domestic prices are $1.9 or $1.2/MBTU (PI, p. 7). These estimates underwent continual revision subsequently in the process of evaluating alternative policy measures.	Oil import prices viewed as one of the principal uncertainties confronting synfuels development. Depending on cartel behavior and world energy demand, international oil prices could range from $2 to $3 to $3.6/MBTU in year 2000 (II, p. 26).

had not viewed the availability of oil and gas resources as a twentieth-century problem. By 1977, however, both the WAES and CIA reports were expressing grave concern about the outlook for world oil production over the next two decades. The CIA had estimated production capabilities based on information from international oil companies and equipment suppliers, and added up the capacities of individual fields, pipelines, and port facilities in a procedure sometimes called "bean counting." It did not rely upon long-term geological estimates of petroleum resources.

The 1973/74 trauma increased the credibility of those expressing concerns about the finiteness of oil and gas resources. One prophet in the wilderness was M. King Hubbert, whose views during the 1960s were largely ignored. By 1977, Hubbert's geological estimate of total recoverable world oil resources (past and present) of about 2 trillion barrels was quoted widely and was included in President Carter's National Energy Plan.[13]

Except for the PIES effort of the FEA, none of the selected studies performed an independent professional analysis of oil and gas resource estimates. Later studies referred to U.S. Geological Survey (USGS) Circular 725, a study performed for the FEA during 1974/75 that resulted in a substantial downward revision in

ERDA	Ford-MITRE	MRG	WAES	CIA
"All scenarios, except V, are unacceptable individually: they show increasing imports. That is, only the successful development and implementation of a large number of technologies in a combination of approaches can make importing fuel a matter of choice" (p. S-5).		Total imported oil and gas resources available to U.S.: 1000 quads at $2/MBTU (p. 26). "Base case" imports vary between 14.2 and 23.3 quads in 1990 (p. 52).	For year 2000, "unconstrained" scenario C-1, oil imports would be 22 quads. Altogether, there would be a 27% prospective excess of oil demand over supply in WOCA (p. 238).	See Figure III.2.

resource estimates more nearly in line with Hubbert's views than previous USGS figures.

The estimates in this report of undiscovered recoverable oil and gas resources for the United States were made: (1) by carefully reviewing a large amount of geological and geophysical information gathered on more than 100 different provinces by over 70 specialists within the Survey; (2) by applying a variety of resource appraisal techniques to each potential petroleum province; and (3) through group appraisals and the application of subjective probability procedures.[14]

"Subjective probability procedures," inevitably viewed with suspicion, nonetheless remained indispensable for energy policy analysis. More objective were the data originally cited by Hubbert—historical comparisons between discovery and production rates, averaged over five-year periods to dampen short-term fluctuations. As put by the Exxon Corporation in 1978:

Until the 1970s oil discoveries substantially exceeded production, and known reserves were increasing. But for the last few years production has exceeded discoveries. This relationship is not expected to be reversed for any long period in the future. Thus, on balance, discovered reserves are being drawn down ... It appears that the limited availability of discovered reserves could cause the oil production growth rate to taper off, and possibly reach a plateau before the end of this century.[15]

Given the shaky nature of the evidence available, geologists were skeptical of quantitative attempts to gauge the responsiveness of oil and gas producers to changing prices. The following comment was all that was to be found on this topic in USGS 725:

> The higher price-cost ratios existing in 1975, if they should continue or increase even higher, would likely increase estimates of both undiscovered recoverable resources and reserves significantly—some economists think perhaps by half again.[16]

Despite the reluctance of geologists, economists persisted in estimating price-responsive supply functions. As seen at the top of Table 9–2, such functions were employed in the FEA, Synfuels, Ford-MITRE, and MRG analyses. Little empirical justification was provided other than the expert judgments of individual consultants (ICF for the FEA's supply functions and Stanford Research International for the oil and gas supply functions of the Synfuels study). These mostly undocumented estimates were crucial in the debate over price controls and the long-term energy outlook.

Unlike the hundred-year history of the oil and gas industry, there had been only thirty-five years to accumulate evidence on uranium. It is no surprise that geologists differed with economists even more about nuclear fuel than about petroleum. ERDA's official uranium resource estimates were viewed with considerable skepticism in both the Ford-MITRE and CONAES studies, and became inextricably linked to the controversy over breeder reactor development.

> While we recognize the concern and the uncertainties, our review suggests that the official statistics err substantially on the low side. A history of low prices for uranium supplied to a single customer—the U.S. government—has discouraged exploration, expecially for lower-grade ores that are more expensive to mine. Assuming the incentive of higher prices, more intensive and extensive exploration should produce enough uranium, both in the United States and abroad, to supply the domestic and foreign requirements of light-water reactors (LWRs) well into the next century.[17]

The CONAES MRG polled individuals on their subjective probabilities of uranium resource availability as part of a decision analysis of advanced nuclear reactor development.[18] Economists gave significantly higher estimates than did geologists, and no meaningful probability consensus emerged. It was difficult to establish standards of professional expertise in so controversial an area.[19]

MODELING ASPECTS

We now discuss some issues of central importance in the formulation of models, and compare how selected studies dealt with these issues.

Aggregation and Computation

Level of aggregation is a key decision that must be made in the design and implementation of a model. Each individual policy analysis may require a somewhat different division into time periods, geographical regions, and physical commodity specifications. In models designed by an interagency committee, there is a virtually irresistible temptation to include the greatest possible level of detail that might be of any conceivable interest to any of the participants. But disaggregation has a price. It can make it enormously time consuming to obtain reliable data inputs and to perform numerical computations.

Table 9–3 shows that the United States was viewed as a nationwide aggregate in all but two of the selected analyses. The FEA had to be able to respond to Congressional inquiries on the regional impact of alternative energy policies. Decontrol of crude oil prices might hurt Massachusetts but help Texas. To ease the difficulties of data collection and computation, the FEA designed PIES so the United States would be subdivided into just nine demand regions instead of fifty individual states. Massachusetts was included within the New England region, and Texas was combined with other west-south-central states. This made the project manageable, but it also meant that state-by-state questions would be beyond the scope of the study. In retrospect, it was a wise decision. No one model can be disaggregated in enough detail to be capable of answering all energy policy questions.

Unlike the FEA, Ford-MITRE did not try to do everything with a single model but employed a less formal, eclectic approach.

Table 9–3.   Model formulation

	EPP	*FEA*	*Synfuels*	*ERDA*
Methodology, time horizon and intervals	Technical feasibility and econometric projections for 1985 and 2000.	Interregional, multi-product, price-quantity market equilibrium projected through iterations between a linear programming model of energy supplies and an econometrically estimated model of price-responsive demands in 1985 (PI). Solutions computed separately for 1985 and 1990 (NEO).	Interregional, intertemporal multi-product price-quantity equilibrium analyzed through a recursive network algorithm. Covered 52 years from 1973 through 2025 subdivided into 17 time intervals of unequal length.	Reference energy system employed as an accounting framework for interfuel substitution in 1985 and 2000. Linear programming for cost minimization.
Geographical regions	US totals.	US subdivided into 9 demand regions.	US subdivided into 8 demand regions.	US totals.
Treatment of uncertainties			Probabilistic decision analysis based on an aggregated version of network model. Sequential resolution of uncertainties between 1985 and 1995. Probabilities based on "the best collective judgment of the Task Force" (II, p. I1).	

Ford-MITRE	MRG	WAES	CIA
Intertemporal price-quantity equilibrium computed through ETA, a linear programming model of energy supplies and a price-responsive model of energy demands. Reduced gradient optimization algorithm. Covered 75 years from 1970 through 2045, subdivided into 16 time periods of equal length.	Comparative analysis of five individual models—including the three employed by FEA, Synfuels, and Ford-MITRE. In addition, included two dynamic linear programming models: DESOM (based on ERDA's reference energy system) and Nordhaus' simulation model (incorporating econometrically estimated price-responsive demand curves). Results reported for three representative time periods: 1990, 2000, and 2010.	Country-by-country projections for 1985 and 2000 under common assumptions with respect to international oil prices and economic growth. To close the projected gaps between supplies and demands, WAES employed both a "highly constrained linear programming routine" (p. 255) and also informal expert judgments.	Year-by-year projections through 1985.
US totals.	US totals.	WOCA (world outside Communist areas) subdivided into: North America, Western Europe, Japan, OPEC countries and others.	"Free world" subdivided into: US, OECD Europe, Japan, Canada, other developed countries, OPEC countries, and others.
	Two probabilistic analyses of R&D on advanced nuclear reactors. One (based on ETA) dealt with sequential resolution of uncertainties on breeder development. Other (based on Nordhaus) compared breeders with advanced converter reactors. Probabilities determined by averaging responses to poll conducted within MRG.		

It addressed nuclear reactor safety through a detailed critique of the Rasmussen Report, and assessed the health hazards of coal-fired electricity units by comparing two studies that had each arrived at quite different conclusions.[20] Geographical disaggregation was essential to the analysis of health effects and nuclear reactor safety but (because of the low transportation costs of uranium) not to the issue of uranium resource exhaustion where dynamics were crucial and one's time perspective had to extend into the distant future. From the top of Table 9–3, we see that Ford-MITRE's time horizon extended well into the twenty-first century.

To analyze the long-term effects of uranium resource exhaustion—and hence the economic rationale for plutonium-fueled reactors—Ford-MITRE employed the ETA model for energy technology assessment.

> The ETA model ... is simple, as energy models go, including just enough detail to deal with the questions we are interested in. It is designed for exploring broad alternatives, not for detailed planning ... The ETA model does not tell us about regional differences, does not distinguish among energy types in detail, and does not analyze demand in a disaggregated way ... As with any tool, one must keep its limitations in mind and use it with care.[21]

Level of aggregation can be a crucial choice for both builders and users of an analysis. A more technical and less controversial issue is the choice of solution procedure. The top of Table 9-3 shows that variants of mathematical programming were employed in all of the selected studies except for EPP and CIA. Mathematical programming can deal with a high degree of disaggregation, with literally thousands of unknowns linked to each other through thousands of constraints. This can result in heavy computational costs and, more significantly, problems for decisionmakers. The larger the system, the more difficult it becomes to question counterintuitive results, some of which may simply be the product of errors in data or computer programming.

Price Determination and Discount Rates

Despite superficial differences, the solution techniques employed in the selected studies—optimization, market equilibrium, decision analysis, and simulation—were fundamentally similar. Only in the CIA's trend projections did price and price-induced substitution not play a key role. In the optimization procedures of mathematical programming, prices emerged automatically as dual variables—the long-run costs of individual constraints within the system. In the competitive equilibrium, prices were what consumers were willing to pay and what producers were asking for the next increment of a given commodity.

Energy markets are competitive in many respects, but with the existence of OPEC and the international oil companies, they also have monopolistic and oligopolistic features. And there is price control by a variety of state and federal regulatory agencies.

In order to deal with the price regulation controversy, PIES included a simulation submodel to evaluate legislative proposals for pricing formulas other than those based upon a competitive market solution. It was used for countermodeling to clarify the differences between proponents and opponents of price decontrol, as in the House-Senate conference on the omnibus energy bill for 1975.[22]

In energy politics, fairness has proved to be a far more powerful criterion than technoeconomic efficiency. Yet none of the selected studies focused upon distributive economics—that is, "who get what?" The FEA, WAES, and CIA analyses were each designed to project oil and gas imports in considerable detail, but except for PIES, none were designed to deal with the far more divisive issue of who would gain and who would lose from price decontrol. Even in PIES, these issues were addressed primarily through side calculations. Distributive economics did not appear as a major theme in the FEA reports. Energy pricing provided an active theater for ideological warfare between populist politicians and oil and gas lobbyists. The analytic community did little to promote a legislative consensus.

In line with academic customs, economists have been reluctant to make interpersonal comparisons regarding gains and losses. Intertemporal comparisons, on the other hand, have been important in economic analyses of long-range investment decisions. These analyses use the concept of present value. The *present value* of a future stream of costs and benefits is the capital sum whose economic value today is equivalent to that future stream given some assumed discount rate. The higher the discount rate, the lower the present value of distant future costs and benefits.

Consider, for example, the following simplified benefit-cost calculation. Suppose that the investment costs of a dam are $100 million, and that the project produces net annual economic benefits of $5 million (in constant dollars, net of inflation) over an indefinitely long period of time. The present value of these benefits is that capital sum ($5/$r$ million), which when invested at an annual interest or discount rate of r percent yields an annual return of $5 million in perpetuity. Since the present value of benefits will exceed the $100 million investment cost at all rates lower than 5 percent, the project's advocates will tend to favor low discount rates, its opponents high discount rates.

Within the United States, controversy over discount rates dates back to the 1930's and the federal government's first major investments in the energy industry—power projects in the TVA region and the Pacific Northwest. It was clear that difficulties would arise if one employed different discount rates within public and private sectors. Were the U.S. Government to base its project appraisals on a 5 percent rate, and private utilities on a 10 percent rate, most power projects would be built in the public sector. A gain in economic productivity would result if the U.S. Government were prepared to lend to private utilities at any rate higher than the 5 percent opportunity cost of capital. Of course, the U.S. Government could not have access to indefinitely large quantities of investment capital at a 5 percent rate. Another complication was the fact that the cost of capital for private enterprises included federal income taxes, and these taxes provided income to the U.S. Government.

Among our selected studies, only Ford-MITRE and MRG addressed the discount rate issue explicitly. Both attempted to compare federal R&D costs for plutonium-fueled breeder reactors with economic benefits to be derived by electricity consumers during the twenty-first century when low-cost uranium resources might be exhausted. Environmentalists opposing the breeder argued for high discount rates, whereas breeder supporters within the nuclear power industry advocated low rates.

Economist Kenneth Arrow participated in both the Ford-MITRE and MRG studies. His suggestion concerning the discount rate was made too late to be implemented by Ford-MITRE but was adopted by the MRG:

The real pretax discount rate, most familiar to those in the private sector charged with making pricing and investment decisions, is taken to be the rate producers use to discount profits from investment projects. This is the rate that is equivalent to the cost of capital in the private sector. The other discount rate, the posttax rate, is taken to be the rate at which consumers discount the benefits they expect to derive from their future consumption ... The MRG has implicitly assumed that the rate at which consumers discount their private consumption possibilities is identical to the rate they would use as social decision-makers in discounting the social benefits from having alternative energy technologies. The corporate income tax creates a disparity between the pretax return earned by private corporations and the aftertax return earned by stockholders of such corporations. A wedge is driven between the pretax rate of return used in making investment decisions and the aftertax rate identified with the rate of discount on consumption ... Arrow suggested that both rates of discount be employed, but each in a different role. Since, R&D aside, most energy investments would be implemented in the private sector, Arrow suggested that the pretax rate of return be used to calculate capital costs of various technologies and, therefore, to determine which technologies would be used to meet assumed sets of energy demands. He further suggested that the aftertax rate of return be used to discount the streams of consumption benefits from adopting alternative energy technologies.[23]

To implement Arrow's suggestion, the MRG adopted for its base case a 13 percent pretax rate of return for pricing and investment decisions in the energy sector, and a 6 percent rate for discounting the benefits and costs of alternative technology options—both measured in constant dollars net of inflation. There was little time to undertake a detailed investigation of the returns to capital investment in the U.S. economy. Net of inflation, the 6 percent and 13 percent rates appear considerably higher than the returns actually realized by investors during the 1970s.

Decisions under Uncertainty

All of these studies included caveats about uncertainty and the need to hedge against it by conserving energy or expanding future supplies. There are two ways to deal analytically with uncertainty: by sensitivity analysis and by decision analysis. Sensitivity analysis is more open-ended than decision analysis. It leads one to explore a wide variety of possible contingencies. Decision analysis, by contrast, focuses on the operational question of what to do next in the face of uncertainty. Taken literally, a sensitivity analysis implies that all uncertainties can be resolved before any actual choices are to be made. First learn, then act. A decision analysis, on the other hand, is a systematic way to think about purchasing insurance against various types of contingencies. First act, then learn.

Consider the following illustrative example based upon the single uncertainty of whether the world's petroleum resources will ultimately turn out to be high (3 trillion barrels) or low (1.5 trillion barrels). These alternative states of the world are set in column headings of Table 9-4. The first two row headings refer to the alternative decisions of whether or not to commit an additional $50 billion to energy development as an insurance premium. The insurance will have been largely wasted if it turns out that the world's petroleum resources are high. But with low resources, the added expenditure would avoid future GNP losses due to energy shortages and rising costs.

The various economic consequences are summarized by the two entries within each parenthesis of Table 9-4. The first entry refers to the immediate costs of the development program: 0 in the first row and $50 billion in the second. The second entry refers to the future GNP loss. It is convenient to measure this loss relative to

Table 9-4.   Economic Losses.[a]

Decision	World Petroleum Resources		Expectation
	High[b]	Low[c]	
No Development Program	(0 + 0)	(0 + 400)	$400p$
$50 Billion Program	(50 + 0)	(50 + 200)	$50 + 200p$
Subjective Probability	$1 - p$	p	

[a] In billions of dollars, present value.
[b] 3 trillion barrels.
[c] 1.5 trillion barrels.

the case of high petroleum resources. Thus, with 3 trillion barrels, the second entry is 0 regardless of the development decision. Future GNP losses are positive only in the case of low petroleum resources. According to Table 9-4, these losses add up to $400 billion without a development program and $200 billion with such a program.

This type of statement is about as far as one can proceed with a sensitivity analysis. Unless we know which state of the world is going to occur, there is no analytic basis for deciding whether or not to proceed with the program. To compare these options quantitatively through a decision analysis, subjective probabilities must be assigned to each possible outcome. Each course of action is then evaluated through its expected (weighted average) outcome. If we are uncertain as to the numerical values of the probabilities, it is convenient to view them as unknowns and then explore the consequences of alternative values. If the probability p is assigned to the case of low resources and $(1-p)$ to the other state of the world, the expected value of the combined total losses is indicated by the rightmost column of Table 9–4.

Suppose that $p = 0.25$. Then the expected losses from both policies will be the same: $100 billion. Thus, it will pay to undertake the development program if and only if p (the likelihood of low resources) exceeds 0.25. For the program to be worthwhile, it does *not* require fifty-fifty odds that it will actually be needed. This reflects the asymmetry between two different types of error: the costs of an unnecessary energy development program versus the costs of future energy shortages.[24]

Decision analysis can allow for many types of uncertainty and for their sequential resolution over time using a tree diagram with a sequence of decision and chance nodes. Through this process, policymakers are led to define a set of feasible alternatives, uncertainties, and possible outcomes. The goal is to identify a robust optimal strategy—one that maximizes expected net benefits given a wide variety of risks.

The methodology is straightforward in principle, but can be difficult to implement. In a controversial policy area, it is not easy to obtain a consensus on probability distributions for chance nodes fifteen or more years in the future. It requires considerable care to define the set of feasible alternatives. If the problem is to be reduced to manageable and comprehensible dimensions, it becomes necessary to restrict the set of possible outcomes and feasible alternatives. "Tree pruning" must be done carefully. It is particularly important to avoid bias in the assessment of probabilities.

The decision analyses employed by the MRG and Synfuels studies were very similar. We shall concentrate on the latter.[25] The Synfuels study, representing the final report of the Synfuels Interagency Task Force formed in 1975 to examine the alternatives for early commercialization of synthetic fuels, recognized that synthetic fuels would not be economically competitive at 1975 oil import prices. It related the net benefits of a synfuels commercialization program to the following factors: the assumed strength of OPEC and the future international price of oil, U.S. domestic energy supplies and demands, the future cost of synthetic fuels, and the effectiveness of an R&D program in reducing costs.

Much of the Synfuels Task Force's analysis was based upon the SRI-Gulf model which, like the Brookhaven model, was an energy-flow network describing conversion efficiencies for individual processes, resource availabilities at alternative prices, and capacity constraints associated with specific technologies. The model contained considerable regional detail, reporting end-use data for eight U.S. demand regions and output data for twenty U.S. supply regions. Supplies and demands were balanced simultaneously at each node at each point in time over a fifty-two-year planning horizon.

Figure 9-2 represents part of this energy network. At each node, there are price-responsive functions to estimate the market shares captured by individual technologies. Because of the high degree of intertemporal, regional, and process detail, the model contained thousands of such nodes, requiring the solution of 100,000 nonlinear equations. This massive computational task used decomposition (divide-and-conquer) techniques.

The first step in identifying an optimal R&D strategy was to have the SRI-Gulf model calculate the economic benefits of alternative synfuels development programs. The optimal near-term strategy turned out to be sensitive to changes in input assumptions. Plausible scenarios were found to support options ranging from a maximum R&D effort to abandoning the entire program.

Figure 9–2.   Schematic Diagram of the Network Structure of the SRI-Gulf Model.

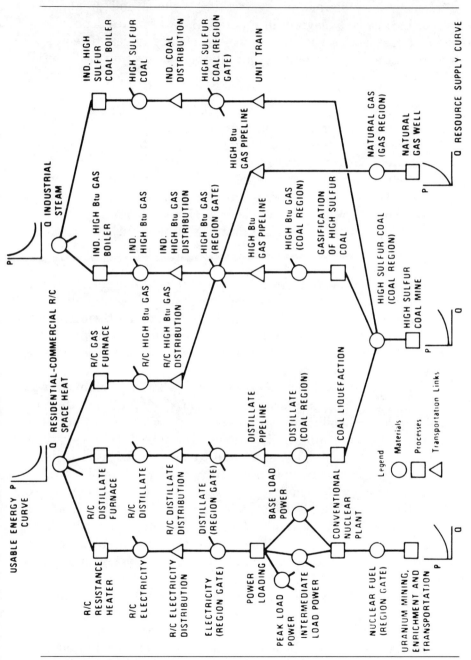

Source: Edward Cazalet et al. (SRI Decision Analysis Group), *A Western Regional Energy Development Study: Economics*, Volume I-SRI Energy Model Results, (Menlo Park, California: Stanford Research Institute, December 1975), p. 215.

The problem was then reformulated in a decision analysis framework for simplicity, aggregating regional and product details. On the tree diagram in Figure 9–3, two squares denote decision nodes and seven circles denote chance nodes. The initial decision was a choice among four alternative government-financed R&D programs:

1. No program—no official commercialization program, but continuation of research and development;
2. Information program—a program designed to produce approximately 350,000 barrels a day of synthetic fuels by 1985;
3. Medium program—a program designed to produce approximately 1,000,000 barrels a day of synthetic fuels by 1985;
4. Maximum program—a program designed to produce approximately 1,700,000 barrels a day of synthetic fuels by 1985.

These options are represented by four branches emanating from the 1975 (square) decision node. The three chance nodes (circles) denote improved information that will become available between 1975 and 1985. Based on this information, there is a private corporate decision in 1985 on subsequent expansion of the synthetic fuels supply, and further decision and chance nodes in sequence through 1995.

Each chance node is viewed as a point where uncertainties will be resolved. Uncertainties are represented numerically through subjective probability distributions incorporating prior hunches, convictions, and information about the likelihood of various states of the world. This is the most controversial aspect of the procedure. Critics say that probability theory cannot be applied to such types of ignorance as the state of the OPEC cartel two decades hence. The decision analyst, on the other hand, argues that policymakers can nonetheless scale their hunches about vague but relevant probabilities, then use these probabilities for making decisions. To act otherwise, says the analyst, would ignore important information.

After considerable interagency debate, the task force agreed to a nominal set of judgmental probabilities. The optimal choice turned out to be the first alternative, no program, since the expected discounted net benefits of the maximum, medium, and information programs were all negative. The synfuels decision

Figure 9-3.   Synthetic Fuels Decision Tree.

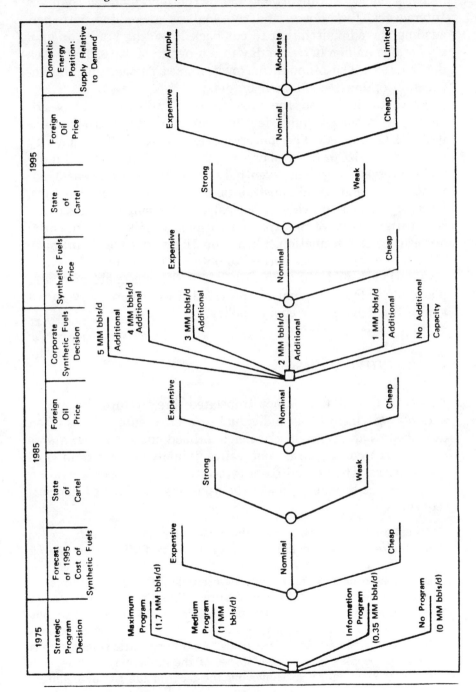

Source: Synfuels Interagency Task Force to the President's Energy Resources Council, Recommendations for a Synthetic Fuels Commercialization Program, Vol. II (Washington, D.C.: Government Printing Office, 1975), p. 54.

analysis played an important role in the report of the Synfuels Interagency Task Force, and the analysis is credited with persuading the administration to cut back President Ford's original goal of one million barrels a day to a program of 350,000 barrels a day by 1985. The proposal narrowly missed passage and the no-program option was adopted by default.

Synthetic fuels and the breeder were two of the many advanced technologies under development that competed for a limited R&D budget. The uncertainties facing each technology included the size of fuel resources, feasibility of alternative supply and conservation options, eventual commercial attractiveness, environmental and social constraints, and future energy demands. Decision analysis provided a logical tool for long-term R&D assessments because of its explicit treatment of risk and sequential acquisition of information. Used inductively, it could help make explicit the assumptions needed to obtain a given outcome. These assumptions could then be scrutinized for reasonableness and plausibility. Decision analysis provided a systematic way to organize the debate over energy policy.

EXTERNALITIES

A cost or benefit that is not translated directly into the profit-and-loss statement of an individual project or enterprise is often described as an externality or neighborhood effect.[26] *Externalities* are indirect consequences not conventionally measured in economic terms. They entail costs typically borne by people who do not enjoy the benefits or make the decisions that lead to the consequences.

> An external effect occurs when the costs or benefits of an activity are not limited to the individual or firm responsible for it but are shared by outsiders. The early discussion of externalities concentrated, for some reason or another, on bucolic examples. The bees of farmer A collect honey from the flowers of farmer B, with subsequent loss to farmer B; lumbering operations on a hillside lead to excessive runoffs and floods, damaging properties in the valley; cinders from locomotives set fire to adjacent wheat fields; or straying cattle trample crops on other people's land. These stories of the birds and the bees, formerly thought useful in conveying the facts of life to our children,

seem to have seized the imagination of nineteenth-century writers on externalities. When Milton Friedman lumps the whole problem of externalities under the innocuous phrase "neighborhood effects," he seems to be thinking in much the same way. When one looks, however, at the extensive pollution of air and water resources, interference in the ratio spectrum, and the struggles in the West for access to public lands and public waters, it is difficult to sweep the problems of externalities that are involved under the rug of "neighborhood effects".[27]

Domestic

Two common illustrations of an externality were the smokestack emission during the era prior to environmental controls, and the coal-fired electric power plant that emits sulfur compounds with an adverse effect upon human health. In the absence of environmental controls, the sulfur emissions constitute a cost to the community not included in the costs borne by the electric power plant. Environmental regulations take the form of technical standards such as "90 percent sulfur removal" or "best available control technology." Such standards tend to be arbitrary and inefficient. A more rational means of regulating emissions might seem to be a tax per ton of effluent. The community's health costs would then be internalized by the power producer. If the tax rate reflected health costs accurately, the plant's sulfur emissions would no longer constitute an externality.

Such ideas are not regarded with universal favor, and the concept of air pollution taxes has been attacked as a license to pollute. Those placing low (perhaps negative) value on additional economic output and high value on environmental goals are not interested in examining economic benefit-cost tradeoffs. Their view is the less pollution the better, regardless of economic costs. Increasing GNP, for them, is not an automatic good. For an environmentalist such as Amory Lovins, freedom from centralized technologies and bureaucracies may be far more important than economic growth per se.[28]

These value conflicts entered centrally into the debate over nuclear power. In the studies of reactor safety, even if there had been full agreement on the likelihood of a major core-melt accident, there would still not have been a consensus on the size of the insurance premium to assess against this contingency. To the antinuclear movement, preoccupied with low-probability, high-consequence events, there was no acceptable level of reactor safety. It is as though one were gambling in a casino where only the stakes and not the odds mattered. Nuclear power (and the breeder in particular) became a symbol of value conflicts. Passions ran high, and it appeared these technologies were no longer amenable to dispassionate analysis.

Because of the rising tide of sentiment against nuclear power, the MRG investigated a previously unthinkable possibility—a moratorium on the use of civilian nuclear power reactors within the United States. The MRG, reporting only on the economic implications of such a moratorium, did not attempt to quantify the health, environmental, and safety externalities represented in the top section of Table 9–5. Using standardized energy cost assumptions and GNP growth rates, three long-range models of the energy sector were run in parallel: ETA, Nordhaus, and DESOM.

> In these similations ... both ETA and Nordhaus between them project the costs of either a "nuclear moratorium" or limits on coal and shale oil to lie between $46 and $159 billion. These costs, translated into per capita (1976 population) terms, suggest lump-sum discounted costs of between $200 and $750. At a 6 percent rate of discount, this range of per capita lump-sum costs implies a range of annual losses in perpetuity of between $12 and $36, all in 1975 dollars. These estimated costs are considerably higher, however, if both a nuclear moratorium and coal and shale oil limits are in place. In that case, the ETA and Nordhaus lump-sum estimates climb to $358 and $457 billion, respectively, or on a per capita basis, to between $1,700 and $2,100. The cost estimates made by DESOM are even higher than those in ETA and Nordhaus in the cases of nuclear moratorium and/or restrictions in coal and shale oil production.[29]

Participants in the MRG effort were primarily economic modelers. A wider range of professional disciplines was represented in the Ford-MITRE study. Ford-MITRE accompanied its economic analysis with a review of the controversy over the health, environmental, and safety aspects of nuclear power. Treating each of these externalities separately, it stopped short of translating them into economic equivalents. As indicated at the top of Table 9–5, only the Synfuels study attempted such a translation.

Ford-MITRE is best known for its concern over nuclear weapons proliferation and its recommendations for slowing down development of reprocessing and plutonium recycling technologies. Despite impressions to the contrary, Ford-MITRE did not oppose the use of uranium-fueled nuclear power plants. It emphasized the uncertainties in calculating the health, environmental, and safety aspects, but concluded that "new coal-fired power plants ... will probably exact a considerably higher cost in life and health than new nuclear plants."[30]

International

Ford-MITRE was particularly concerned with the international impact of U.S. plutonium-fueled reactors. It feared that expanded use of plutonium would encourage the spread of nuclear weapons to a growing number of nations and to subnational terrorist groups. But it made the error of concentrating its economic analysis upon the United States in isolation. Japan and Western Europe vigorously resisted the Ford-MITRE analysis. They faced an altogether different picture with respect to uranium availability, coal costs, and the prospects for energy conservation, and had very different economic incentives for plutonium-fueled reactors than the United States.

In the other energy studies, international aspects entered primarily through the world price or availability of petroleum, as indicated in the middle of Table 9–5. None of these analyses made a formal attempt to estimate the influence of U.S. policy upon international oil prices. To have estimated such feedback effects would have been difficult. It would have compounded the uncertainties, and would have required analysis of the energy prospects of all other major oil-producing and oil-consuming nations. Despite the pioneering efforts of WAES, international energy analysis is still in an early stage of development.

Many of the studies we reviewed did not address the broad social and political issues. Some were confined largely to technoeconomic matters. In the early years of the energy policy debate, this omission did not appear serious. But as the debate grew more intense and more politicized, it became increasingly clear that the externalities must somehow be considered explicit-

Table 9-5.   Externalities

	EPP	FEA	Synfuels
Health, environ- ment and safety	"Translating a decision about pollution control to a cost vs. benefit formula is difficult, and often misleading, because in many respects the health and environmental benefits represent intangible values... As we have seen, there are only small margins of safety, if any, in the air pollution standards we now have... Despite a great deal of effort on the part of the AEC and the nuclear power industry, it is not possible to be sure that the public is being adequately protected against the risks of nuclear power... The impor- tance of *Technical Fix* is that it would buy the time to be sure we could 'test first and build later', before the breeder went commercial" (pp. 201–217).	"The major environmental issues associated with energy will focus on regional development questions" (NEO, p. 45).	Distinguishes between "compliance costs" (borne directly by private producers) and "externalities" (borne by public at large). For synfuels, the environmental costs would not exceed $1.00/B. (II, p. 130).
OPEC capabili- ties and interna- tional relations	"Under the Environmental Protection option of *Technical Fix*, the United States would import 5 million barrels of oil a day by 1985 and a fairly small amount of gas... The United States would be free to buy oil from a diverse group of relatively friendly nations under this option" (p. 172).	"The events of the past two years have indicated an ability by the oil producing cartel to maintain the high prices of oil established during the embargo, even in the face of substantial declines in world oil demand due to the high prices and reduced rates of world economic growth" (NEO, p. 14).	A 50-50 probability is assigned to the cartel being "strong" or "weak" in 1985 (I, p. 42). Depending also on foreign oil price uncertainties, the international oil price could range from $6 to $19/B (II, p. 112).
Nuclear weap- ons prolifer- ation	"We have raised the nuclear weapons proliferation problem here, under the High Nuclear option, because it seems to us that it is a most pressing problem for humankind and that the United States, rather than being the world's leading nuclear salesman, should take the lead in stopping the spread of the ingredients for making bombs—which includes power plants—before it is too late" (p. 171).		

ERDA	Ford-MITRE	MRG	WAES	CIA
"Alternative approaches to environmental control should be evaluated on a risk/cost/benefit basis to ensure that funds are not expended for control in instances where neither the potential effects nor public concerns warrant such expenditures" (p. IX-4). Appendix B measures quantities of effluents, but—except for occupational safety—does not translate energy scenarios into biomedical consequences.	"Despite these large uncertainties, the general conclusion is that on the average new coal-fueled power plants meeting new source standards will probably exact a considerably higher cost in life and health than new nuclear plants" (p. 196).			
			"Petroleum demand could exceed supply as early as 1983 if the OPEC countries maintain their present production ceilings because oil in the ground is more valuable to them than extra dollars they cannot use" (p. xi).	"The rising pressure of oil demand on capacity in the early 1980s is bound to cause oil prices to rise well in advance of any actual shortage. Saudi Arabia's ability to moderate OPEC price decisions will be weakened as Saudi excess capacity is used up... By 1982 or 1983, sizable price increases are inevitable unless large-scale conservation measures cut demand sharply" (p. 18).
	"Our net conclusion is that reprocessing and recycle are not essential to nuclear power, at least during this century... In addition, there are potentially large social costs, including proliferation and theft risks in proceeding. A U.S. decision to proceed despite disincentives would induce other countries to follow suit and undermine efforts to restrain proliferation" (p. 321).			

ly. The quotes excerpted in Table 9-5 suggest how many of these issues remained unresolved. Environmental, health, safety, social, and political factors lay well beyond the scope of the technoeconomic models. The externalities were inadequately handled, partly because they were poorly understood and partly because they led to value conflicts outside of the analyst's purview. Yet analysts could not, except at their peril, ignore these factors that were often at the very heart of the controversy. It was a central dilemma for the energy studies of the seventies.

10 THE ROLE OF ANALYSIS IN PRINCIPLE AND PRACTICE

To those who identified it with science, analysis offered the road to rational decisionmaking. When neat solutions eluded energy policymakers, analysts often took the rap. Had they accepted a facile claim to the powers of science in an objective search for truth, analysts would be guilty as charged. For the policy environment does not hold still, and scientific objectivity is a delusion in the world of politics. Furthermore, analysis is not ideally suited to the pluralistic and fiercely political nature of life in a democratic society. Its application to policy issues in practice has been sharply different from its promise in principle. In this chapter we outline the fundamental character of the analytical process, sketch its policy potential from the analyst's perspective, discuss its shortcomings, and examine its use in the pursuit of political ends. This sets the stage for consideration of how analysis might better serve our plural democracy in future years.

Analyses will pour out, the computers will burn up, but the issues are mainly political.[1]

Washington is an "analysis town." The analysis wars were fought and won by McNamara and the systems analysis people years ago. Numbers just have to be produced in this town.[2]

These contrasting comments depict the curious blend of cynical resistance and urgent imperative that pervades efforts to apply analysis to the political enterprise. The actual application of analysis to policy matters is very different from the utopian image—a point made repeatedly in observations from the energy studies and interviews with participants. Analysis is inevitably subordinate to the very political process that provides it with raw material. It is itself a component of that process and can only work its influence through it. Such interlocking dependence makes the role of analysis in politics complex and varied.

ANALYSIS AND SCIENCE: AN EASY CONFUSION

Analysis by the dictionary definition is the examination of a subject of study to distinguish its component parts or elements. In a policy context, analysis often includes a consideration of alternative courses of action (including inaction) and attention to the indirect consequences of these actions. Policy analysis is not necessarily quantitative in nature but does typically involve numerical calculation, physical measures, and empirical data. The form policy analysis takes varies from simple projections and comparisons of magnitudes by individual experts to extensive organized group endeavors employing elaborate computer models, as in the case of PIES. The results of policy analysis may be presented in offhand remarks to make a point, in project reports, journal articles, or published volumes complete with press releases and author interviews. We have seen all such modes of dissemination, and others, in our review of the energy studies of the 1970s.

To qualify as policy analysis in the sense we have in mind, there need not be exhaustion of all ramifications of a policy question, but there should be clarity on what has been included and what has been omitted. Reviewers should be provided with the basis for determining for themselves the adequacy, legitimacy, and importance of the analysis. They should not be left guessing about what the analysts did or did not assume or the meaning of what they have or have not discovered.

Analysis is an appeal to reason rather than to emotion or credo. In this respect, it differs from oratory and is similar to science. Indeed, analysis shares with science a good deal of methodology and approach, including certain tools of mathematics, use of the computer, formal modeling methods, technical terminology, and both the inferential and deductive aspects of empirical investigation. Analysis is, in fact, an important part of the scientific process. *But analysis, especially policy analysis, is not in itself a science.* We make this obvious point with emphasis, though it has been made many times before,[3] since misunderstanding and mistrust often come about from an easy if not deliberate confusion on this matter.

Policy analysis has no experimental controls and no opportunity for independent replication or precise validation of results. Neither does astronomy, geology, or meteorology; however, there is a crucial difference between policy analysis and these observational sciences. The economic and political environment studied by policy analysis is constantly shifting as it adapts to changing conditions and surfacing problems. The results of policy analysis are heavily dependent upon the assumptions and simplifications it makes and the conventions it adopts, none of which are guaranteed to be valid for very long.

Furthermore, policy analysis, like engineering, includes a design function. Unlike engineering, this entails estimation of the consequences of deliberate intervention into the behavior of sociotechnical systems.

Science has to do with the discovery of general truths or the operation of universal laws discerned over time through empirical observation. Science strives for consistent explanation. Analysis, by contrast, seeks to apply the mind's resources to empirical data to reveal the contours and dimensions of a problem in a systematic and logical fashion. A phenomenon is resolved into constituent parts, and each part is examined individually and in relation to the whole. Analysis brings forth ideas that help in the dissection of a problem. It is primarily a contemplative activity. It may be illuminating; it may be useful for understanding the reasons for disagreement; it may organize information and suggest courses of action; it may employ the products of science; but it is not itself a science.

Many of the people who do policy analysis have scientific or technical training. For this reason, and especially because science and technology have tended to have a glamorous image and strong positive connotation for much of society, political figures have deliberately imputed to analysis a scientific or technological cast. Richard Nixon's reference to Apollo and the Manhattan Project in launching Project Independence was an example.

Since analysis plays an important role in science, scientists often make good analysts. Nearly everyone involved in the CONAES study had a strong technical background and familiarity with the methods of analysis. But scientists floundered when confronted with the inherent conflicts in values and objectives that dominated the energy policy scene. Analysis permeated the work of CONAES, and its quality improved under the judicious scrutiny of the scientific eye. But good scientists are not necessarily well suited to the subtleties and cross-cutting considerations characteristic of problems in public policy fields.

The policy process involves many players. As a social activity, it adapts to the changing interests and perceptions of its participants. Every step is confounded by new demands and alliances. Unlike the analyst, the policymaker values compromise over consistency. Analysts and policymakers often have very different perceptions of the same policy problem. To see why, we first look at policy analysis from the fundamental perspective of an orthodox analyst—the modeler. We then turn to the domain of politics, where the gears of analysis frequently grind in their efforts to engage with the wheels of political reality.

THE ANALYTICAL PROCESS AND THE FORMALITY OF MODELS

Introspection is never easy, especially for analysts who would apply to self-analysis the same criteria they use in their work. As one reviewer has astutely observed:

> It is paradoxical that experienced policy analysts have their greatest difficulty when it comes to trying to tell us what it is they do. They clearly feel much better when they are just doing it.[4]

From the viewpoint of a modeler—a meticulous practitioner of analysis—examination of a problem is analytic to the extent that it meets the following criteria:

1. Reduction of the problem to primary components of cause and effect to isolate its elements, simplify its investigation, and make it more comprehensible and communicable.
2. Explicit statement of assumptions and problem boundaries to clarify the basic premises of the analysis and highlight its limited scope, specifying both what is and is not included.
3. Explicit and logically consistent reasoning from premise to conclusion to make the analysis accessible to third parties.
4. Ability to be understood, checked, and challenged so as to permit a gradual improvement in the analysis and guard against mistaken or spurious conclusions.

Analysis, by these norms, is a form of disciplined thinking. But it is more. By responding to challenges and questions, a good analysis improves and builds understanding. By reducing issues to their primary components and by laying out the reasoning process, analysis can stand on its own and provide transferable insights. This ability is similar to, but not the same as, the requirement for replication in science. For example, few can share the flash of insight that led Einstein to his theories of time and space, or follow the line of argument that resulted in a twentieth-century revolution in physics. That insight was not transferable (and it was not produced by analysis).[5] Yet many scientists have been able to replicate experiments verifying Einstein's theories. Einstein's results meet the standards of science eminently.

Problem-solving methods applied to policy issues do not permit tests against observation in this same way. A simple mathematical derivation used to eliminate an impractical energy technology from consideration on the basis of prospective costs and benefits can be analysis of the most useful form, even though it lacks a test. For many questions, science is unavailable; only analysis can support decisions. A case in point is the question of how best to dispose of high-level nuclear wastes. We cannot wait a millennium to collect data. The choice of disposal site must be made using complex models of nuclear decay and the leaching of wastes—analysis, not science.

A model is a simplified, idealized representation of reality.

"Model" is a general term that may be applied to many different things, from a toy car to a full-scale prototype of a supersonic aircraft, from the game of Monopoly to a set of mathematical equations that represent the behavior of the national economy, and from an engineering curve to a complex procedure for making long-range projections of future energy resources. A model need not be quantitative or even explicit. One's subjective perceptions of his surroundings form a mental model composed of images of meaningful features and important relationships.[6]

All analysis includes some type of model, even if only a mental model constructed unconsciously to organize and interpret observations about the world. In fact, analysis is almost synonymous with modeling, broadly defined. The more explicit and detailed the analysis becomes, the more the underlying model is revealed and the easier it is to identify and correct errors.

Modelers are frequently criticized for ignoring subtle interactions and important qualitative aspects of a problem during the process of reducing it to a system of equations. Such criticism is directed at formal quantitative modeling. It should not be taken as an objection to analysis per se. To avoid this misunderstanding, it is helpful to narrow application of the term "modeling" to the formal activity of constructing explicit system representations.

A model can aid in organizing hypotheses to be investigated. For instance, we may be interested in assessing the health effects of prolonged exposure to sulphur oxides produced in the burning of high sulphur fuels. A scientific study would require controlled experiments on large populations over many years—an undertaking that would be ethically repugnant, politically infeasible, and very expensive. A more practical approach is to separate the problem into manageable components: the sulphur content of the fuel, dispersion through the atmosphere, natural mechanisms for purifying the atmosphere, intensity and duration of population exposure, and the effects of this exposure on biological processses in human beings. Hypotheses on how these steps occur and interact can then be examined by means of individual experiments and measurements.

Analysis and modeling are specialized activities. They carry their own particular baggage of theory and methodology, filtering issues to be examined, shaping the approach, subtly influencing the choice of questions, coloring the interpretation of results. The control engineer adept at solving stationary linear systems finds that problems tend to look stationary and linear. The decision theorist wears problem-framing glasses etched with decision trees. The economist feels most comfortable referring to elasticities and curves of supply and demand. Yet modelers and analysts who can shed the disciplinary cloak and communicate their ideas and insights effectively to an audience of nonspecialists—those who can rise above the formulas and rites of their calling—are among the most valued.

Three Phases of Analysis

Analysis in the orthodox sense has several phases; we name them *ideation, classification,* and *codification.* Ideation is the preliminary phase. Dealing with ideas, it begins to define and structure the problem. Classification organizes the elements of the problem. Codification, using formal models, describes the relationships among these elements. We shall see that these three phases, though interacting, yield products of very different character and function.

Analysis is informal at the stage of ideation. Many critical assumptions are unstated. The inquiry tends to be exploratory and broad in scope in order to establish a foundation for further investigation. The problem is not yet well understood. Its very nature is called into question. Problem description is an art; skill and experience play prominent roles. Judgment and common sense establish the extent of the investigation, the shape of the issues, the hierarchy of questions, and the terms of reference for choosing among alternative policies. Training and precedent guide design of the study and suggest tools for attacking policy questions. Ideation is the most important phase of any analysis. It is where the payoff for wisdom and good judgment is highest and where precision can be inappropriate. Here it is better to be approximately right than precisely wrong.

The output of the ideation phase is a written or verbal summary—an essay—that outlines the concepts and arguments, the theories and judgments suggesting tentative problem solutions. Such essays deal with the subtle as well as the obvious, the implicit as much as the explicit. Refined and revised as the analysis proceeds, the problem essay can be among the most valuable contributions of analysis.

The overall study may not go much beyond this literary stage. Some would call the Ford Energy Policy Project chiefly a literary effort. The study did draw upon analysis for support (an evaluation of the energy conservation potential of the iron-and-steel industry, for example, and an estimation of the impact of tax subsidies and new financing arrangements on electric utilities), and several of the supporting studies involved formal models. But the final report used the analytical findings selectively. The integration was artful and informal—a persuasive essay unconstrained by the reaches of its analytical underpinnings.

The Ford-MITRE study drew more on the integrated judgment of participants than on the specific results of any model or underlying technical analysis. Similarly, the CIA study consisted of reasoned argument mixed with impressions. Its forecast of crisis in the world oil market from what critics referred to as "wisps of evidence" did not show supply and demand broken down in any detail.

It is clear that judgment, convincing argumentation, and a flair for communication are enough to produce eye-catching essays. This led James Schlesinger many years ago to observe, "only half facetiously, that experience in debate is the most valuable training for analytical work."[7] But essays alone give little basis for replication or critique. If we stop with the essay, we lose important benefits to be gained from the challenging, testing, and reformulating of ideas.

A second phase in the analytical process, and a major step toward more formality, is systematic classification of problem elements, as illustrated in Figure 10–1. By introducing an accounting scheme, analysts create the ability to be explicit and quantitative. For instance, the energy balance tables of the WAES study helped maintain consistency in matching the energy demand and supply estimates of different countries. Similarly, the Brookhaven Reference Energy System became the centerpiece of the ERDA studies and the leading illustration of a network approach to describing and manipulating energy flows.

An explicit accounting scheme can provide an effective quantitative structure for exercising judgment and setting priorities. WAES inferred an impending liquid fuels crisis from conventions

10–1.   Energy Accounting Schemes

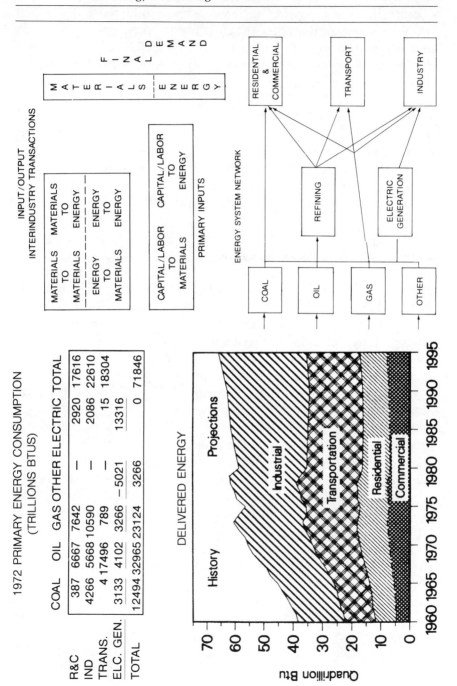

established when it classified fuel balances by region in order to decentralize the analysis. National teams conducted independent supply and demand studies for each region, then put their results in a common format for integration. The ERDA efforts followed similar procedures to obtain fuel-specific imbalances between energy supply and demand.

These accounting frameworks summarized disaggregated results but supported only minimal interaction among component investigations. Analysts discussed anomalies without resolving them. They presented alternative scenarios, yet did not attempt to achieve consistent integrations. ERDA's scenarios for a variety of energy policies left it to the reader to decide which ones were likely or feasible. WAES analysts invented the "notional gap" (a measure of the imbalance between supply and demand) to spot trouble but not to work out inconsistencies among countries. The accounting frameworks permitted a level of detail that would have otherwise been impossible, but there was no attempt to push the formal analysis to the point of total system integration.

The third and most formal phase of analysis is codification of the problem—the description of explicit rules for interaction among system elements. These rules record the analyst's view of the nature of the system and help maintain consistency in assumptions, arguments, data, and results.

Codification is essentially modeling. It facilitates communication beyond the inner circle of analysts in the study group. Equations and computer programs record hypothesized relationships among variables and parameters, ranging from technical descriptions of physical processes (heat rates in power plants, losses in refining, and so forth) to representations of behavioral choices (consumer flexibility in fuel mix or the pace of adoption of new technologies). The equations provide the starting point for empirical estimation and quantitative inquiry.

The theoretical structure of a formal model can reveal the basic approach to the policy issue. The FEA PIES model, for instance, reflecting an economist's view of consumer response to changing levels of energy price and economic activity, portrayed market equilibrium within a disaggregated network of supply and demand centers. Although it provided a sophisticated representation of many facets of the energy system, PIES did not deal with income distribution among competing interest groups. When this issue became a central focus of energy policy debates, PIES was left behind.

The Synfuels Commercialization study emphasized a similar network equilibrium approach, embedding it in a decision-analytic framework. It did not describe income distribution effects either, but did add a capability for systematically investigating the effects of uncertainty on investment decisions, thereby improving the quality of the analysis.

The MOPPS study took an unusual approach. Its model, formal but not automated, had individual actors using judgment and manual adjustment to balance supply and demand across many energy technologies. Participants followed explicit rules iteratively to achieve integration. This human feedback loop served to resolve inconsistencies among studies of individual fuels and technologies. The model was an organizing device.

The CONAES Modeling Resource Group went the comparative route in its utilization of models.[8] It ran six different models of the energy system to test the importance of alternative assumptions and different model formulations. This educational process strongly reaffirmed the need to scrutinize assumptions and demonstrated how models can function both as subjects and instruments in this task.[9] The process has since been institutionalized in the Energy Modeling Forum at Stanford University.[10]

In the idealized case, the three phases of analysis—ideation, classification, and codification—become a whole that is greater than the sum of its parts. The intuitive and artistic phase identifies the key issues; a systematic accounting structure orders the data and spells out the main parameters; and a formal model reveals the central assumptions and offers a tool for detailed examination of the policy issues. Interactively and iteratively, each phase supports the others. Building the model may expose the inadequacy of the accounting structure and help define new data needs. The model provides a mechanism for using available data and deriving analytical results, drawing the analyst's attention to subtle interactions. Occasional discovery of counterintuitive results is an opportunity to review preconceptions and problem formulation. So employed, the model becomes a learning tool and a repository for accumulated understanding.[11]

This perspective of an orthodox analyst does not describe all of policy analysis, and is undoubtedly of more meaning to analysts and especially modelers—than to nonanalysts.[12] Most significant-ly, it omits a description of policymaking and the policy process—the process within which analysis must function—the process that determines how policy studies get designed, conducted, and uti-lized.

THE POLICY PROCESS AND THE POLICY STUDY

In Allison's idealized model of the rational actor,

> A government, firm, or agency behaves as a single unit which acts completely rationally in terms of a definite set of goals. The decision-maker has a payoff function which ranks all possible consequences of alternative policies or possible actions in terms of these goals, taking full account of all possible side-effects. The choice consists of selecting the alternative whose consequences rank highest in the decision-mak-er's payoff function.[13]

Clearly, this is a narrow caricature of actual decisionmaking in a policy setting. It misses the very substance of policy problems—a plurality of partisans with many conflicting, value-ridden objec-tives. Politics is the realm of clashing interests. Its interpersonal dynamics has a chemistry, not a deductive logic. Politics deals in pressure, power, and negotiation, not optimization or rational de-duction. Its instrumentalities are coalitions, compromise, and rec-iprocity, not decision rules, computer models, or payoff functions.

One way to apply the rational actor model to the energy prob-lem is to view the political process as consisting of many interact-ing interests, each with its own self-consistent set of aims. Another way is to invoke the model in the context of what are assumed to be broad social goals. Society becomes the rational actor. Several energy studies took this approach in their efforts to establish criteria for ranking options. The EPP focused on saving the environment and reducing energy use. PIES, in the wake of the oil embargo, centered on the reduction of oil imports. The Synfuels Commercialization study adopted classical cost/benefit criteria within a framework that took account of uncertainties. The strategy of positing overarching goals gave direction to these analyses and clarity if not legitimacy to their results.

What is the nature of a policy study? Its first stage resembles the ideation phase of analysis. It begins with recognition of a problem, the formation of new ideas, and the generation of alternative proposals for action. Success depends upon a complex interplay of intuition, judgment, ideology, values, and analysis.

In this critical phase, predisposition and focus are all important. The EPP, led by a lawyer with a background in regulation, emphasized the role of government intervention. ERDA analysts, working in an agency populated with engineers, saw energy policy in technological terms and searched for the most efficient means of increasing supply. PIES, whose economists were surprised by the sudden rise in energy prices, fixed on the economic issues of pricing and consumer response.

Meanwhile, across the Potomac, CIA analysts, schooled in the intelligence tradition, saw energy problems in terms of institutions and international power. WAES measured world supply/demand balances. The Synfuels Commercialization study formulated event trees laced with compound probabilities. Rasmussen's risk analysis of reactor safety took a similar approach. Ford-MITRE had a disarmament orientation. Stobaugh-Yergin stressed opportunities for conserving energy. CONAES and MOPPS brought interest groups with obviously divergent views into their analyses. Ford-RFF and RFF-Mellon had strong economic bents and placed much weight on issues of technoeconomic efficiency and environmental impact. Amory Lovins, separating himself from the majority of these studies, questioned the basic social structure he charged the others implicitly assumed.

The most pivotal points of the energy studies were not in the forms of their analyses as such but in the questions they asked, the problems they framed, and the unspoken restrictions they imposed on possible solutions. These aspects, constantly under revision during the course of the studies, were decisive in their outcomes.

Another stage in a policy study is the evaluation of policy alternatives from many perspectives, measuring anticipated results against policy objectives. Policy evaluation can be a study's main activity and consume much of its resources. The FEA organized the entire PIES effort around models and data-collection procedures for evaluating alternative equilibrium market solutions. Analysts compared large numbers of alternate paths for expanding supply or reducing demand—expressed within model equations solved by speedy computer algorithms—and examined thousands of possible energy flows among hundreds of regions and fuel types. In these simulations, modifications in the construction schedule for the Alaskan oil pipeline would cause adjustments to estimated electricity prices in New England, and changes in environmental standards in New York would alter the projected demand for Colorado coal.

Some early PIES runs suggested counterintuitively that more stringent "new source performance standards" for burning coal would *increase* the total level of sulphur pollution. Reexamining the model solutions, analysts found that the tighter standards on new sources made older plants more attractive economically, thereby delaying their retirement. The older plants were greater polluters. The increase inferred in total insult to the environment had been logically correct.

Such insights, along with new questions and new possible solutions, arise in recycling through the ideation phase. Occurring repeatedly during the course of a study, this natural iterative process can be a valuable learning experience for participants.

Evaluation does not always lead to policy recommendations. The Rasmussen study estimated the likelihood of a variety of failure events occurring in a nuclear reactor and characterized the nature of these failures but did not propose means for their avoidance. PIES analysts held back from suggesting action after describing the implications of a range of government regulations and initiatives. The CIA study also made no explicit recommendations, although its gloomy picture of the future if the world failed to reduce the demand for oil had obvious policy implications, and was so used. A similar message came from WAES. The RFF-Mellon team pointedly abstained from proposing policy solutions.

Many energy policy studies do make policy recommendations, and seek to publicize these recommendations through press conferences, talks, and testimony to Congress. The EPP, Ford-MITRE, Lovins, and Stobaugh-Yergin gave strong public advocacy to their conclusions. Members of CONAES, Ford-MITRE, and the EPP served after their studies in senior government posts where they were able to execute policy initiative and help implement the results of their analysis.

The Central Question

Observing study group members in the roles of policy advocates and implementors brings the discussion down from the rarefied world of the orthodox analyst to the pragmatic world of the political practitioner. It is at the transition from policy evaluation and recommendation to advocacy and debate—moving from the inward focus of searching for attractive policies to the outward focus of dealing with those who think differently—that the analytical art begins to have trouble meshing with political reality.

Even good analysis can get in the way of political accommodation—a fact that can be difficult for the conscientious analyst to accept. From the perspective of the rational actor, more and better analysis should only help, but it is another matter when the analysis goes public. Analysis that is openly explicit in identifying alternatives, arguments, and tradeoffs can create problems for the policymaker. To one involved in delicate negotiations, displaying your hand before the bidding begins can undermine the negotiating position and hinder formation of a political coalition.

If analysis benefits negotiators most when done only for their own interests, is it then to be confined largely to a proprietary rather than public role in the political process? Would not that limit its usefulness as an instrument for social progress? Not necessarily, for two reasons.

First, it is a cardinal maxim of policy research that analysis is at best only one of many influences on policy and is dominated typically by values and ideology. Other factors, such as intuition, judgment, and appeals to authority and precedent, interlace with analysis and can be affected by it. Analysis provides a language for politics and a paradigm for political discourse. In a policy setting, analysis is not a sterile, encapsulated activity that stands alone. It is a process that propagates, interacts, and diffuses. Policy analysis can be useful in many subtle ways.

Second analysis does have an important nonproprietary function to perform in a democractic society that is different from what it might be expected to perform under another form of government. Without meaning to detract from the contributions analysis has made, we feel this function can be done better than it is being done at the present time. Before being more explicit, we review the roles policy analysis has served in our society and the difficulties it has encountered in the process.

DEMOCRACY AND ANALYSIS: AN ESSENTIAL TENSION

The emergence of expert analysts in government over the past two decades has thrust yet another interest group into the political cockpit.[14] Since a democratic society entails conflicting objectives that seek resolution through the political process, analysts are frequently called upon to help clarify tradeoffs among the separate interests and thereby contribute to the evolution of policy. But they end up resembling the interests themselves when they turn to promoting their own view of the good or just society. They are then entitled to no greater consideration than any other advocacy group.[15]

> If, in the guise of analysis and exposition, an expert becomes an advocate for a particular position, he sometimes may have his own way, but only by substituting his own judgment for that of people who have the responsibility for decisions and who might weigh values differently if given all the facts, and whose judgment may be better.[16]

Decisionmakers know the difference between analysis and advocacy, and soon learn how analysis is invoked to further political ends. This can lead to a form of analytical inflation—to cynicism and a devaluation of analysis. The doubt is not so much over *whether* analysis should be used as it is over *how* it is used and what its products mean.

Analytical results mean different things to different people, especially given a tendency to "leave goals phrased in ambiguous language, hence open to misunderstanding."[17] Goals enunciated by the makers of policy can be inconsistent or even contradictory, as in the design of government programs where compromises and coalitions among diverse interests are required to secure passage of legislation.

Analysts yearn for a clear statement of purpose, yet the political process rarely provides such clarity. Decisionmakers and their aides have different yearnings from analysts and different agendas. Said one congressional staff analyst:

We need discrete and simple answers, and no model is going to give you that ... We just ignore [the analysts] unless they seem to be saying something we need. Staff people are not influenced by studies, except if a fact or conclusion conforms to a point of view they need to promote.[18]

We heard such cynical comments frequently in our interviews. They confirm the worst fears of those who value understanding for its own sake.

Analysis has limits. In selecting a site for construction of a nuclear power plant, rationality can help in devising an efficient plant design or determining the amount of energy to be produced. But when environmental or safety considerations enter the picture, all bets are off.[19] Facts alone cannot resolve controversy rooted in subjective values and ideological positions. Indeed, facts can exacerbate the controversy.

The tension between politics and analysis is paradoxical.[20] We require consensus, but the very act of exploring options and trying to attain consensus heightens our awareness of the many interests and values at stake. As a result, it becomes nearly impossible to produce a logically consistent set of policies and programs. Attention to the "big picture" gives way to "fragmented and personalized interests and values ... difficult to reconcile with integrated and rational planning and foresight."[21] The "correct" answer is seldom known, and often "no solution ... can be proved to be right or even to be the best."[22]

Analysis that recommends a hike in the price of oil to force conservation may show that oil consumers in New England—a politically powerful group—will lose even though the country as a whole will gain. Different interest groups will interpret and act on the same analytical results in different ways, according to their values, interests, and perceptions. The recommendation may not make for politically palatable policy.

Many fundamental energy policy questions have so far failed to yield to analysis: questions about the legitimacy of authority, political accountability, guarantees of energy security for the general populace as well as local citizens, and private ownership versus public needs. Ethics, politics, economics, and other considerations are involved. Policy is determined by resolution of all the relevant factors. Analysis deals with but a few.

In examining the uses of analysis, we first need to understand the role that ideology and values play in politics.[23] "Prior to politics, beneath it, enveloping it, restricting it, conditioning it, is the underlying consensus on policy ... among the ... politically active members ... Without such a consensus no democratic system would long survive."[24]

One might expect that analysts could foster the essential consensus, and hence sustain the democratic process, by facilitating full and fair consideration of issues from all politically active points of view. One might further hope that through analysis society would be able to reconcile conflicting ideological and technical information. Ideological information deals with the thoughts, feelings, attitudes, and conduct of human beings; technical information deals with the material and measurable aspects of an issue. Policy goals almost always have intertwined ideological and technical elements. Analysis so far has been applied successfully only to the technical side, although insights from analysis have helped explain behavior and reveal inconsistencies in attitudes and premises. Still, ideological matters are generally considered within the domain of politics.

Energy independence, for instance, is an ideological goal encompassing thousands of technical questions on current and projected patterns of energy consumption, worldwide distribution of energy resources, the use and availability of technologies, and future scientific and engineering advances. Consider the subgoal of producing enough synthetic fuel from nonpetroleum sources to maintain current levels of consumption without increasing dependence on foreign oil. Even this technical-sounding subgoal implies feelings about being dependent on Persian Gulf oil, about sinking huge amounts of capital into uncertain synthetic fuels technology, and about having to pay a larger share of family income for something hitherto taken for granted.

Can the analyst work back and forth between ideological and technical facts to illuminate all sides of such a policy matter? That would require the analyst to:

1. Define the problem according to various points of view;
2. Collect technical facts bearing on the problem;
3. Explore the attitudes of those most affected by it;
4. Integrate the definitions, technical facts, and attitudes into alternatives for decisionmakers to consider.

These analytic tasks have generally been given short shrift in the emphasis on technical factors over ideological ones. Policy analysis gives top billing to economics and systems analysis—by virtue of the relative power of their normative structures—and second-class status to the softer, less quantitative disciplines that are more attuned to ideological aspects. This shows up in a preference for energy-economic models despite widespread reservations about their use.

Said one critic,

> Models are a symptom of a belief system especially evident in the economics profession. Economists believe them, even though they don't work.[25]

By "don't work," the critic meant that they are not really relevant to policy. But what is policy?

> Policy brings to statement what is judged to be possible, desirable, and meaningful for the human enterprise. In this sense, policy is the nexus of fact, value, and ultimate meaning in which scientific, ethical, and theological-philosophical reflections meet.[26]

Public policymakers are engaged in orienting society to "an image of the future filtered through an elaborate value system." This "image contains not only what is, but what might be. It is full of potentialities as yet unrealized. In rational behavior, man contemplates the world of potentialities, evaluates them according to his value system, and chooses the 'best' "[27] according to an abstract, highly simplified conception of the world.

Time complicates the picture. As a problem setting moves from the immediate into the future, the number of ways it might change increases; so does uncertainty over the ways it is most likely to change. As one peers further into the future, rationality yields to intuition.[28] The result is that policymakers have "muddled through," doing their utmost in incremental steps at the margin of time.[29] Expert forecasts have fallen short of expectations.[30]

The simplifying tactics of analysis cannot do justice to the complexity of the future. Nor can statistics and indices such as money and efficiency handle the crucial qualitative imponderables[31] that stubbornly resist quantification.[32] It has been charged that political factors are for this reason frequently "left out of assessments. Thus some analysts have ignored the fact that the OPEC price is a political price [and] that an international oil system that depends so heavily upon a few fragile Middle Eastern countries is a crisis-prone system."[33]

Lacking adequate ideological facts about the images of the future held by political authorities and affected parties, analysts inject their own simplified, explicit, orderly preferences. They do so either in the interests of technical efficiency or in hopes that complications will work themselves out. But "when the theorist attempts to make policy judgments as he would theoretical judgments, he simply becomes a naive policymaker ... The theorist who fails at policymaking operates with a bad theory, and it is bad theory partly because he has tried to keep it explicit, articulate, conscious, and orderly."[34]

The reductions of analysis are further simplified by interpreters trying to convey technical results to a general audience. Interpretations may be made by specialized briefers and consultants who did not participate in the analysis themselves or, as described in Chapter 7, by newspaper reporters working under tight deadlines. Interpreters dramatize and condense for the benefit of busy decisionmakers and an easily distracted public. They translate analysis into terms their clients can understand, resulting in what one person we interviewed called "the interpretation game."[35]

Further compounding the difficulties, analysis takes time, often years for problems of great moment. Meanwhile, the problem or the decisionmaker's perception of it may have changed. There may be new world conditions. The issue that prompted the analysis may no longer be germane. Other problems may have taken precedence, or a solution may have evolved naturally. The political clock ticks much faster than the timepiece of analysis. Decisionmakers are impatient for data. The upshot is "selective picking and selling of one number or result."[36]

Politics: Comprehensible If Not Rational

Many students of politics have stressed the irrational, nonobjective, illogical nature of the political process.[37] Political scientist Harold Lasswell called politics a "sphere of conflict," an open arena into which all the vanity, venom, narcissism, and aggression of the contending parties and all "the irrational bases of society are brought out." When in the process the moral order is devalued and challenged, pressure builds up for the political authorities to find "a reflectively defensible solution."[38]

Lasswell's characterization of politics, written in the thirties, rings true when applied to the events of the seventies. Consider the following comments made in mid-1979 by William F. Kieschnick, Atlantic Richfield executive. Kieschnick was reflecting on the effects of the energy crisis on the American people:

> As energy upsets increase, more consumers are experiencing these upsets personally and seeking to understand them. They are at the beginning of their learning curves, and their primary sources are television, politicians, and advocates, all of whom produce unsettling and contradictory answers—rip-off, conspiracy—that anger them. The net effect is a growing incredulity about the performance and the ability of energy industries, government, and large institutions generally.[39]

Consumer anger (as Lasswell would have expected) put the government on the defensive. Said Kieschnick:

> The government has become accusatory in order to justify itself. This was evidenced in Carter's [May 1979] speech on decontrol and windfall profits, where he found it politically necessary to make unsavory remarks about industry. It is a psychodrama in which everyone—consumers, industry, government—is estranged from everyone.[40]

People no longer accepted the old order, with all its irrational and nonrational sanctions. Political initiative was necessary to reestablish a new moral consensus.

> Politics is the transition between one unchallenged consensus and the next. It begins in conflict and eventuates in a solution. The solution is not the "rationally best" solution, but the emotionally satisfactory one. The rational and dialectical phases of politics are subsidiary to the process of redefining an emotional consensus.[41]

Intellectually, a politician may appreciate an analyst's rational justification for change but oppose it nevertheless if it appears to create inequities or otherwise threaten the underlying or emergent consensus.[42] In order to remain effective and in office, the politician must work to support and sustain the consensus, especially among his constituents. He will use decisions on allocation and distribution, the main tools at his disposal, to serve this purpose more than to improve government programs. He will be naturally suspicious of any redistribution affecting wealth or power. It is a rare political change that benefits everyone at a cost to no one.

Those with heavy investments in the status quo were reluctant to admit during the 1970s that a new public mood was developing, a mood that was introspective, skeptical, distrustful of established institutions, antagonistic toward authority, and very concerned about the problems of environmental impact, waste management, and nuclear safety. "The man in the street has not shifted his views very much," said one industry official. "The public has new cliches, not new views."[43] Contrast that statement with the assessment of a CONAES participant: "There is a paradigm shift; people are fighting for their lives."[44]

The EPP was a harbinger of the political changes taking place. One caustic critic of the project, Lewis Lapham, acknowledged that if it "accomplished nothing else, it testified to the lack of leadership in the country and to the collapse of an idea of legitimate authority." Observing "the way in which the old and the new class preferred to conceive of the other as malevolent conspiracies," Lapham charged members of the "old class" with having "affected an air of exaggerated certainty" and with having "pursued their objectives by means of indirection."[44] The effect was to increase distrust, convince the populace that it was being lied to, and further undermine respect for government.

The views of counterelites and utopians flourish in times of mass unrest, cynicism, and introspection.[46] That phenomenon accounts in part for interest in the soft path of Amory Lovins. On the other hand, when the populace is reasonably well satisfied and trusts and respects the governing authority, there is apt to be greater acceptance of government policies. The enabling consensus is left unchallenged.

Taken individually, the events that capture public attention may be insufficient to disturb the consensus. Cumulatively, however, they can cause widespread malaise and anxiety. Gas lines, soaring fuel prices, fears about reactors threatening the health and lives of millions, power blackouts, polluting firms, foreign insurrections, military adventurism, government scandals, and a seemingly ineffective president all add to growing political turbulence and unrest. Further eroding public confidence in the reigning expertise are: planes whose engines fall off, drugs that harm rather than heal, popular food items that contribute to cancer and heart disease, cars that fail and maim, and an economy that defies the ministrations of financial and economic authorities.

Not knowing whom to trust, the public retreats into cynicism. "What energy crisis? The oil companies are just trying to rip me off. Look at those obscene profits." Energy looms as a symbol for all that ails people, spurring the drive to reach a new consensus and political equilibrium. Pressure mounts to apply more analysis to clarify issues and set matters straight. But analysis may only make matters worse, emphasizing rational argument and reason at a time when irrationality and nonreason dominate.

Needing to maintain a consistent orientation, analysts are likely to cleave to the familiar, even in the face of signals from the environment that the old reality no longer exists.[47] CONAES tried to take account of the social dimension, but its review committee, composed mostly of mathematicians and scientists, believed the report should concentrate on technical issues. It objected to an overview chapter being written for a lay audience, thus giving support to those who said the staid National Academy of Sciences was unsuited for a problem so deeply woven into the social fabric.

When research and analysis threaten to undermine comfortable habits and patterns of thought, individuals will go to great lengths, including unconscious distortion of facts, to avoid confronting the new reality.[48] Organizations can also react irrationally.[49] To illustrate, let us look at how analysts and politicians deal with the central economic concept of price in times of instability and turmoil.

To the economic analyst, prices are market equilibrators, mechanisms that promote allocational efficiency. A rising price is not a conspiracy but a natural signal of demand outstripping supply. Yet the public regarded the pre-1973 period—inexpensive gasoline, no lines—as the familiar and fair situation. It viewed the rise in prices as unjust. The spoils appeared to be going to some unseen force through a manipulative scheme. To the economist trained to accept economic realities, the public's attitude was naive and misguided. It could largely be ignored. To the politician, whose business was to decide who among contending parties gets what, the public's attitude was crucial. It was the economic reality that should be ignored if doing so would preserve the consensual foundation that makes politics possible.

Democratic politics is the struggle for advantage among society's competing factions. When a technical issue enters the political arena, the politician treats it more or less like any other issue. While the technician's arguments will be heard, they will be received in the style of politics: believed if convenient, ignored if not. We have seen in the debate over natural gas deregulation how analysis was accepted or rejected depending upon whether it confirmed or contradicted the preconceptions of the adversaries. Confided one congressional user: "Let's face it. We do quick hits around here. Pick your analyst to suit your political needs. Any analysis can be useful if it produces something that supports a point of view. Otherwise, forget it."[50]

That had to be one of the most cynical of the many cynical comments we heard during the course of our interviewing. It reminded us of George Orwell when he wrote that the language of politics was "designed to make lies sound truthful ... murder respectable ... and give an appearance of solidity to pure wind."[51]

PURPOSES AND USES OF ANALYSIS

It was striking in our review of the energy studies how often there was disparity between intention and result. The purposes of the studies—as gleaned from project contracts, formal work specifications, planning documents, and personal recollections—were one thing, the products of the analyses and ultimate outcomes, quite another. Here are a few of the actual uses.

Talent Shows. A politician confronted with a thorny problem might commission an analysis to locate knowledgeable people, but more often a study commissioned for other reasons will groom or identify talent as a by-product. An analytical exercise can expose participants' strengths and weaknesses, including their ability to withstand the rigors of political life. While providing a valuable learning experience, it can also test one's mettle and provide a qualifying round for high office in public service or industry.

David Freeman became a member of President Carter's energy planning staff a few years after leading the Ford EPP, and was subsequently made chairman of the Tennessee Valley Authority. Joseph Nye assumed policy responsibility within the State Department for the nuclear proliferation issue he had been addressing as a member of the Ford-MITRE study group. John H. Gibbons, chairman of the CONAES Demand and Conservation Panel, became director of the congressional Office of Technology Assessment. Floyd Culler, acting chairman of the CONAES Supply Panel, was appointed to the presidency of the Electric Power Research Institute. Kenneth Davis, initial Supply panel chairman for CONAES, became deputy secretary of energy in the Reagan Administration. These appointments may have been made anyway, but participation in the studies did provide useful background.

Personal Education.   A common reason why people agree to engage in or finance policy analyses is to acquire knowledge. This educational benefit will often accrue not only to those centrally active in the analyses but to others more peripherally involved as well. Decisionmakers exposed to the energy studies incorporated insights from the studies into their thinking. The shift in attitudes about energy conservation reflected the cumulative educational effect of many analyses that progressively established the potentialities for greater energy efficiency and questioned the notion that moderating energy use would bring harm to the economy.[52] "Some of the most deeply held beliefs of the participants on what is possible with energy have been shaken," said one involved in CONAES. "I learned a lot, although if I had thought of it as continuing education, I would not have suffered as much agony.[53] In MOPPS, ERDA managers were said to have been forced to think for the first time about how each other's technologies would gain acceptance and penetrate the market, helping ERDA solve a touchy management problem.

Communication.   When a problem is being analyzed, word often spreads on the issues being discussed and the preliminary findings. Such communication can lead to suggestions, ideas, and information from the outside that feed back into the analysis, and it can stimulate helpful complementary work. (An example is the productive working relationship that the CONAES Modeling Resource Group had with the Energy Modeling Forum.) It can also gain recognition for the analysis and pave the way for acceptance of its conclusions and policy proposals. Candidate Carter's views on nuclear policy were expressed before the 1976 election in his May 1976 speech to the United Nations, drafted by members of the Ford-MITRE study group. The president's later espousal of the Ford-MITRE recommendations was clearly not fortuitous.

Mass Education.   An analysis may be undertaken primarily to inform and influence the thinking of the general public. This is considered especially appropriate for highly complex, universally applicable issues like energy policy and nuclear concerns where a consensus is lacking and must be painstakingly built. Wide dissemination of the analysis comes about through active distribution of the final report, press conferences, general publicity, and special briefings. The EPP promoted its environmental and conservation values even as the study was in progress. A preliminary report on the study became a Book-of-the-Month Club selection, circulating over 300,000 copies. Also achieving public prominence were the Rasmussen study of reactor safety and Stobaugh and Yergin's *Energy Future.*

Forum for Debate.   If politics is the art of compromise, there is always the hope that analysis, in highlighting contending points of view, can lead to better balanced, more solidly grounded agreement. But it may not. A model, for example, can provide a very useful simplifying device and a tool for sharpening issues and exposing their roots. But rigid adherence to the model can distract attention from the broader perspective needed for compromise. Careful analysis may merely harden the positions of opposing factions. Indeed, for purposes of negotiation, ambiguity can be preferable to explicitness.[54] An analysis designed to illuminate disputed aspects in a policy debate can turn into an arena for the debate, as in CONAES. It is easy to overlook the significant learning and reconsideration of positions that can also take place.

Scapegoating.   When a policy decision does not work out as well as expected, a politician may try to avoid personal criticism by implicating the analysts who advised him. Officials have been known to take credit for successes and disavow responsibility for failures. Analysts, models, and studies make convenient targets, as we have seen. One Department of Energy administrator put it bluntly: "Analysts must learn there is no fame for them in this business."[55]

Delaying tactic.   Some complex issues pose so severe a dilemma that any response seems bound to be a no-win proposition. A decisionmaker facing such an issue may commission an analysis to gain time and maneuverability.[56] As additional facts come to light, the problem could resolve itself, other options could arise, or a compromise could be arranged without injuring the politician's reputation or chance for reelection. In many policy areas, there is a time-honored tradition of referring recalcitrant problems to blue-ribbon commissions for extended analytical treatment. One social critic, alarmed at society's appetite for information, observed that "there are vast realms of the bureaucracy dedicated to seeking more information, in perpetuity if need be, in order to avoid taking action." The fear is that the quest for ever more information has "destroyed our ability to make use of it."[57]

Posturing.   An analysis may be used by a politician or the head of an agency to provide an appearance of concern and attention—or otherwise carry a positive symbol—for the benefit of constituents and the general public. Such posturing is normal practice in the world of politics. It may or may not be accompanied by substantive action.

Eyewash.   Results of an analysis that support a program or policy are fair game for politicians to refer to selectively for promotional purposes. This can catch the analysts by surprise, as was the case with the CIA study of world petroleum. Commented one CIA analyst, "We try to do research that is useful, but this time we were unusually successful. We just happened, by accident, to produce something that buttressed the Administration's position. It wasn't anticipated."[58] After Carter cited the findings of the classified study to underscore the need for his energy program, there was instantaneous demand for the report. The unprepared analysts had to work under pressure over a very long weekend to get it out.

Emperor's New Clothes. In examining the use of analysis by Congress in the natural gas deregulation debates of 1977, political scientist Michael Malbin asked himself why analysis was used so heavily.

> Part of it was public relations, but there were two deeper reasons as well. First, some members hid behind the studies: even though they may have made their decisions on other grounds, they felt more comfortable knowing that if something should go wrong, they would be able to say they simply went along with the experts ... The second reason is evident from something Representative Stockman said ... "Nobody wants to say it's all based on a basic value. My position is that the free market does work, but that is a pretty thin fig leaf for something as important as this."[59]

"A fig leaf!" exclaimed Malbin. "Members of Congress apparently are ashamed to discuss public issues without hiding behind the numbers provided by economists." Malbin did not think much of these numbers. So far as he was concerned, all the analysis did for Congress "was to cover the fig leaf with the emperor's new clothes."[60]

Moderator of Goals. Government may commission a study to define or reconsider a program declared by an act of Congress or presidential proclamation. Such were the motives of the Project Independence and ERDA studies, and also of the Synfuels Commercialization study, all of which served to moderate overly ambitious goals into more reasonable expectations. A similar purpose was served by the three studies discussed in Chapter 8. They implied disapproval of President Carter's post-Iranian call for aggressive government promotion of synfuels. The country and Congress were very impatient at the time. Carter was responding to public pressure. Analysis, time, and the flow of events helped moderate unrealistic aspirations.

When Analysis Does and Does Not Get Used

Professors of policy analysis would like to tell their students that when an analysis is done professionally, it attracts attention and gets used and when it is done carelessly or irresponsibly, it does not. Alas, according to our interviews and study questionnaire, the highest quality analyses received least attention and exerted the lowest immediate influence on policy. From a longer term perspective, quality may contribute more positively to effect, but that is hard to determine. It is difficult enough to trace the influence of an analysis over a year or two.

If quality does not explain acceptance, what does? Timing, for one thing. An analyst with long Washington experience said that policy studies can be very useful to a young administration with few public positions, vested interests, or political investments. "The new guys are going to be looking desperately for a package they can pick up, massage, and call their own. If someone is ready with that package, his chances of having an impact are better.[61] In 1977, the incoming Carter Administration embraced the Ford-MITRE report, the CIA analysis, and the ideas of Amory Lovins. Findings coming out of CONAES, particularly on conservation, also got a hearing.

Analysts and politicians march to different drummers. By the time an analysis is complete, political attention may have turned to matters other than those pressing when the analysis began. An analysis not ready in time to serve the purpose for which it was intended is taking a chance of losing its audience. In 1975, Norman Rasmussen was buried under an avalanche of responses to the draft version of his Reactor Safety Study which he had circulated the previous year. Nuclear safety was a central issue in congressional review of the Price-Anderson amendment about to get underway. Rasmussen succeeded in getting the final version of his report released just in time for the review. He was later criticized for responding inadequately to peer criticism; but his study got used.

Also important in determining acceptance is whether the results of an analysis accord or clash with the beliefs of those in power. Ford-MITRE was in harmony with the Carter Administration's position. MOPPS collided with it. The president warmly received the Ford-MITRE report and presented it proudly to the Japanese prime minister. The MOPPS report was recalled and studiously ignored.

Research is more likely to gain approval and be put to use when presented in a concise, comprehensible format than in a voluminous, highly technical tome. The complex and massive Rasmussen study, which its critics called "inscrutable," was never really applied operationally, even though its popularized executive summary was widely cited.

It has been said that policy analysts do best to stay away from wholesale changes in social values if they are to have an impact.[62] Though he would disagree that his ideas required a revamping of social mores, Amory Lovins could be considered a counterexample. He was the exception that proved the rule. Analysts generally do assume the status quo and tend to concentrate on matters where consensus seems feasible.

There never was a period of true national agreement on energy during the seventies. None of the energy studies, singly or in combination, whatever their virtues, inspired a real consensus, despite a felt urgency within the country to reach one. Energy problems, "marked by a high degree of uncertainty and large political or economic stakes," were an example of the "kinds of political circumstances in which policy modeling has risen to prominence: the big, unprecedented issues, the crises in public policy that tend to create a market for modeling."[63] A paradox of the democratic society is that these are just the conditions under which analysis is most likely to be disputed or ignored. Our investigation into the energy studies of the seventies reconfirmed this lesson from earlier research.

Pro Bono Publico

One of the more troubling dilemmas of a democratic society has to do with how much individuals should relinquish decisionmaking powers to those with presumed special competence.[64] Such relinquishment is commonplace in certain areas of life. We routinely submit ourselves to the care of airplane pilots and surgeons, for example. But in a democracy, delegating policy decisions to those with technical, moral, functional, or other supposed expertise runs the risk of allowing political equality to be undermined.

Is the ordinary person competent to judge difficult technological questions?

If you believe ... that on the whole the ordinary man is more competent than anyone else to decide when and how much he shall intervene on decisions he feels are important to him, then you will surely opt for political equality and democracy. But if you believe that he is less competent in this fundamental way than some particular person or minority, then I imagine that like Plato your vision of the best government is an aristocracy of this qualified person or elite.[65]

Institutionalizing a technological elite would be one means of responding to those who "find the requirements for democratic participation more and more in collision with requirements for internal consistency in the social management of technology."[66] But then how does one control such a presumably wise and righteous aristocracy and keep it from becoming "a cunning and voracious oligarchy."[67] Quis Custodiet Ipsos Custodes? "Who will watch the guardians themselves?"

Part of the energy problem has been represented as requiring superior technical knowledge and special responsibility for the crucial decisions that need to be made.[68] Plato insisted his guardians possess an excellence of character and deep philosophical understanding "in order to justify their rule."[69] Are our technical experts up to those lofty standards? The great minds among us are no priesthood. For one thing, they disagree among themselves on technical issues as well as values. For another, they are human, thus fallible.

We do not want a technological aristocracy making decisions for us. But we do have need for trained professionals and for nonpartisan analysis of the complex problems facing society. That is not a conclusion, it is the fundamental assumption with which we began this book. These nonpartisan analysts have the responsibility to present arguments on all sides of the debate and communicate this information widely.

The task ahead for analysts and the designers of analysis is not to preempt normal decisionmaking processes or strive for comprehensive solutions. It is rather to illuminate policy issues, evaluate feasible alternatives, help society understand how values relate to policy, and identify tradeoffs between social and individual priorities. The special challenge is to find effective means for conveying these insights clearly to decisionmakers and the citizenry at large. Over the long run, analysis that achieves these ends will serve both society and the professions—economics, political science, engineering, science, operations research—from which the legitimacy and credibility of analysis must ultimately derive.

Here, then, is a mission for analysis. It leaves unanswered the question of how analysis can be used to build the political consensus needed for policy decisions. That is the business of the political process. We have seen how this process can obstruct and frustrate orthodox analysis. We also know that no one can hope to comprehend fully the intricate systems that routinely condition our lives. No one can engineer them to work most efficiently, as well as most equitably and safely. Analysts must be humble.

Recognizing the limits of the human intellect in the face of modern complexities helps us view policy analysis more realistically. Policy analysis is not an oracle or finder of best solutions. It is fallible and has definite inadequacies. Yet, together with wisdom and the freedom to question and disagree openly, analysis conducted in the public interest and communicated effectively is among the best hopes society has for dealing with its problems.

Analysis brought reason to the rampant blaming and scapegoating of the seventies. It contributed impressively to policy discussions. Without minimizing the meaning of this contribution, we would ask whether analysts might aspire to do better still. We suggest an answer in the concluding chapter.

11 ASKING MORE FROM ANALYSIS WHILE CLAIMING LESS

In this chapter, we present several perspectives on issues central to the use of analysis in policy, along with our own considered position. Sometimes we lean more to one side of an issue than the other. Most often, we seek a middle ground or reasonable resolution. In examining attitudes harbored about analysis and the stereotyped roles it is assigned to play, we draw some conclusions and make some recommendations. In particular, we suggest specific actions we believe can improve the payoff of analysis in policymaking. We also seek ways of fostering a better, more realistically based relationship between the analytical and decisionmaking communities. These, in our opinion, are goals worthy of the country's best efforts.

On the first day, the Lord reviewed two alternate designs for the human race. Results of this comparison of a low-growth, divine-control versus high-growth, free-will scenario have never been revealed. There is no way to determine its influence on the Almighty's decisions over the next five days. Rumor has it that the archangel who prepared the analysis was cast from heaven afterwards for obfuscation, presumption, and irreverent pride.[1]

This biblical invention makes three points: the possibility of analysis having an influence on decisionmaking is as old as decisionmaking itself; we cannot presume to know the extent of its influence; advisers are not always appreciated.

Isaiah suggests that God takes no counsel and needs no analysis save his own.

Who hath directed the Spirit of the Lord, or being his counsellor hath taught him? With whom took he counsel, and who instructed him?[2]

Yet, according to Bacon,

The wisest princes need not think it any diminution to their greatness, or derogation to their sufficiency, to rely upon counsel. God himself is not without, but hath made it one of the great names of his blessed Son, *The Counsellor*.[3]

Bacon wrote of "the inseparable conjunction of counsel with Kings, and the wise and politic use of counsel by Kings."[4] Historical documents show the counsellor "as much a part of the political structure as the king himself."[5] Over the years, the role of advisers and the analysis they employ has been central in the rule of nations and the conduct of world affairs. It is remarkable how little systematic attention has been paid to it.

The federal government goes to quite a bit of trouble to check if economic markets are operating freely and efficiently. Members of Congress return to the voters every two years to be scrutinized and revalidated on their performances. Professors and university departments are constantly being appraised in an unending competition for intellectual excellence. What about analysis? It has been observed that "knowledge about the analytical process is just as important as knowledge about policies if the effectiveness of public policy is to be improved."[6] What does the nation do, in its search for wise and discerning policy guidance, to check the soundness and responsibility of its analysis? For the most part, it relies on an unstructured and imperfect system of professional review, media reports, and informal word of mouth.

MARKETPLACE OF IDEAS

The results of analysis are transmitted to policymakers through an undisciplined marketplace for ideas. This marketplace, replete with false advertising, wide gradations of quality, conflicting claims, and growing consumer cynicism, suffers from a Gresham's law where quick explanations and appeals to emotion tend to drive out research results carefully derived and documented. Timing is of the essence in policy encounters, and policymakers typically "have to reach decisions before all the evidence is in."[7] Besides often taking too long, studious research may seem too politically innocent to be taken seriously. There is little organized effort to review and compare analyses critically or explain their findings in politically meaningful terms. The upshot is that policymakers and their staffs, given their limited time, have inadequate grounds or motivation for sifting through the many studies available. Confidence in analysis and its effectiveness in application inevitably suffer.

As we began to address problems such as this in the early stages of our study we became acutely aware, as collaborators in a team effort, of the significant differences of opinion among us—a product of our varied backgrounds and staunch individualism. There were five of us: Garry D. Brewer, Martin Greenberger, William W. Hogan, Alan S. Manne, and Milton Russell. Two of us (Hogan and Manne) were modelers in economics and operations research. A third (Russell) was an economist with an institutional orientation. Another (Brewer) was a political scientist with healthy disdain for the pretentiousness occasionally found among modelers, economists, and operations researchers. And one (Greenberger) was a bit of each.

Early in the project, we subjected ourselves to our own attitude questionnaire to see where we lined up among those whom we were interviewing. One of the modeling economists (Manne) turned out to be almost all traditionalist. Brewer, our political scientist, was primarily reformist, but not nearly as much a reformist as Manne was a traditionalist. The other two economists, Hogan and Russell, were largely traditionalists, although Russell showed significant weighting on the reformist scale as well. Greenberger, the eclectic among us and leader of the team, was traditionalist and reformist in equal measure.

We did not strive to suppress our differences, nor did we attempt to unify our individual stands. It would not have worked had we tried. Instead, we identified for each problem area one or more outlooks on the role of analysis in policy that spanned our intellectual divide without necessarily conforming to a traditionalist/reformist breakdown. We then proceeded to look for a reasonable resolution of the issues with which most of us could agree. That approach did seem to work, although we took no votes. As in the energy debate, real consensus was elusive. We settled for friendly reconciliation.

With respect to the marketplace of ideas, we see society beset with a confusion of analyses and contradictory pronouncements that leads to failing trust in the statements of experts and findings of studies. Some believe that a research institute or review board of high repute and integrity is needed to critique objectively the plurality of available studies and distill out responsible results, making them credible, understandable, and meaningful to policymakers and other interested parties.

> There needs to be an institution something like a technical-analytical court, an organization with the technical competence to deal with the various adversary analyses on their own terms, comparing their assumptions, their formulations of the problem, their methodology, the boundaries of their analyses, and their data. In the language of technology assessment, the institution needed is an assessor of assessors.[8]

Others insist that the marketplace of ideas, with all its problems and haphazardness, is vital to a participatory democracy and open society. It ensures the ability of society to entertain unpopular and minority opinions. An aristocracy of analysts, by this view, is a dangerous notion that should be discouraged. There is no such thing as an unbiased expert. Nonexperts need to be heard.

Considered Position. We have strong sympathies with this second position and recognize merits in the first one as well. We believe the open marketplace should be supported, but we feel it must also be strengthened. We would like to see the functioning of this intellectual exchange improved so as to reduce the possibility of important ideas and careful research being neglected. More mechanisms for the critical comparison of alternate analytical approaches and results would help. This is especially important given the expediency that characterizes much policy analysis. The reasons for rejecting ideas that do not stand up to scrutiny should be noted and widely publicized to expose misconceptions that obstruct constructive action. We recommend a concerted attempt to accumulate understanding and insights (as in the social sciences) toward building a broadly based discipline of policy analysis. This is one way analysis can legitimately and usefully seek to emulate the physical sciences, despite the fundamental differences between them.

What is needed, we believe, is not a single review board but multiple centers of excellence firmly rooted in the political system. These centers should work to enliven rather than dampen debate. Their activities, extending beyond the boundaries of traditional policy analysis, might include comparing studies, revealing hidden assumptions, clarifying the substance of policy arguments, and framing issues for discussion without taking political stands. Clarity and simplicity of presentation should be emphasized in the execution of these functions so as to promote communication with both technical and nontechnical communities.

The unwillingness of society to rely on experts comes from an instinctive reluctance to lay one's fate in uncertain hands at a time of heightened peril. It is a sensible attitude. We need to examine whether research results affecting social concerns are being developed and presented in a way that facilitates the nonexpert's ability to judge. We feel the job could be done better.

PUBLIC ANALYSIS

At the same time that cynicism about its use was growing during the seventies, policy analysis was having a heyday with the blossoming of commercial firms doing economic consulting, forecasting, and systems modeling. Analytical results entered into numerous policy debates, influenced key decisions, and were appealed to repeatedly in partisan and advocacy frameworks to make a political point, defend a position, or take a side.

But does the commercial acceptance and popular political currency of analysis necessarily constitute benefits for national decisionmaking? The mounting cynicism does not seem to auger well for the contributions of analysis to policy in the future. We asked ourselves whether the nation in its long-run interest should be setting a higher goal for policy analysis. One of us in congressional testimony has advocated a role for analysis insulated (but not divorced) from partisan politics and commercial enterprise, one that underpins the advocacy roles it is now playing while reaching out beyond them.[9] Such a role requires that analysts gain the confidence and respect of policymakers without becoming beholden to them. Given the desirability of this goal, the question is, how can the nation help bring it about?

The Ford Foundation tried to tread the slippery path between apolitical analysis and policy relevance with its idea for an Environmental Institute. It became disillusioned and backed off when the White House staff got defensive about policy overtures from outside and insisted on playing politics in the appointment of a director. The RAND Corporation, after doing independent research for the Air Force on problems of strategic defense in the years following World War II, fashioned an applied research office in its own image (in the late 1960s) to work on the problems of New York City. Sponsored by Mayor John Lindsay, the New York City-RAND Institute had its share of successes but ran into repeated head winds of political resistance and was never fully accepted by many of the municipal agencies intended as clients. Seven years after its creation, the institute was terminated under a less friendly mayor and conditions of budgetary stringency.[10] A reasonable conclusion to draw from these two examples is that it is impossible to insulate a politically meaningful policy center from partisan politics.

Yet the role performed by the RAND Corporation in defense is very much needed in government policy generally. RAND was not afraid to question conventional military wisdom or make unpopular suggestions, even some running counter to the immediate interests of the service from which it derived its support. By assuring RAND's independence, the Air Force gained for itself and the country an enviable reserve of analytical talent and a stream of ideas and critical assessments that worked to strengthen military preparedness and cut costs in doing so. RAND furnished many experienced experts for high government posts: among them, Alain Enthoven, Charles Hitch, Henry Rowen, and James Schlesinger. Secure funding and autonomy in research were key to its success.

Considered Position.   We have seen repeated instances of the partisan uses to which analysis has been put in a "search for the winning argument as opposed to the correct conclusion."[11] Yet most of the studies we reviewed were nonpartisan in origin. Unfortunately, the political cast they acquired in application often detracted from their value and beclouded the insights they had to offer. The highly publicized gas curves of MOPPS gave the effort an unintended maverick tone that caused the bulk of its findings to go unnoticed. A similar fate befell the Ford Energy Policy Project and the AEC Reactor Safety Study. Because of the populism of *A Time To Choose* and the popularization of the WASH-1400 Executive Summary, many overlooked the serious research that went into these two studies

Words can be used for salesmanship and solicitation or they can be used for conveying information and the lessons of the ages. Most of us can tell the difference between promotion and knowledge. Certainly no one would propose to equate language with persuasion even though persuasion is indeed an important use to which language is put. Similarly, advocacy is an important use to which policy studies are put, but analysis is not thereby advocacy.

How can the separateness of analysis and advocacy be protected? In radio and television, the boundaries between commercial and noncommercial have been marked out with the establishment of public broadcasting. This analogy, though imperfect, prompts us to coin the phrase "public analysis" to designate the kind of effort in policy analysis that we would like to see encouraged and supported. By public analysis we mean analysis in the service of the country at large. We also mean analysis that explains the substance of the debate—including private positions and special interests—in straightforward terms that can be generally understood and appreciated. This could raise the level of debate and might just also aid in keeping the overall volume of analysis within manageable bounds.[12]

One approach for fostering such a concept would be for government to fund new organizations or expand existing ones already doing something close to what we have in mind. A type of noninterference agreement would have to be worked out to give the organizations continued financial security without compromising their operational independence. As first approximations, the Office of Technology Assessment and the Congressional Budget Office, helped by politically heterogeneous congressional bases, have been able to maintain a high degree of neutrality and critical objectivity in their studies and in testimony before Congress.

A concerted attempt was made in designing these two offices to insulate them from political pressures. They have been careful to preserve that immunity. Yet they receive heavy criticism on occasion, and they can never be certain that their leadership, budgets, or autonomy are secure. Safer by nature of its charter is the General Accounting Office, also a creation of Congress.

The Brookings Institution and the American Enterprise Institute are different kinds of organizations that draw their funds largely from private sources. They are often seen as having political leanings, yet many of their activities resemble what we mean by public analysis. So do the activities of the National Academy of Sciences, Resources for the Future, and numerous other research centers and university groups. Another approach to public analysis would be through the expansion or revision of programs in such already functioning private and semiprivate organizations.

The route taken, then, could be either programmatic reorientation or organizational innovation. Either way, we believe that major financial support must come from private foundations and private firms. They could be instrumental in making possible the initiatives and changes we intend. It is largely to their executives and program officers, along with government officials, that the appeal should be directed.

We do not underestimate the magnitude of the problems confronting the development of public analysis. The complicated history of public broadcasting suggests some of the difficulties that would have to be faced, not the least of which is the securing of long-term funding. A proper mix of impartiality and relevance and a highly motivated staff of top quality would have to be sustained, even though staff members were removed from the front lines of policy action. This would necessitate a well-planned system of "institutional and career incentives for remaining above the battle."[13] Most essential, a reputation for excellence and an institutional character that served as unmistakable demarcation from the work of the advocacy groups would have to be cultivated and jealously protected. Once that were achieved, meeting the other requirements would be easier.

Although government might have only a limited role in the development of public analysis, it would have a major stake in the outcome. In providing incentives, regulating activity, collecting taxes, and authorizing expenditures, government is the conditioner of the social and economic environment. Its decisions are prime movers. It has a responsibility to see to it that there is available a sound analytical basis for policymaking. A traditional role of government is to guide and bolster the economy while participating in a minimum number of economic transactions. In the same way, government should strive to enrich the process of public analysis by encouraging high standards and a lively competition of ideas while taking care not to dominate the process itself.

ASSUMPTIONS, INCLINATIONS, AND IMPLICATIONS

In distinguishing in the last chapter between policy analysis and science, we observed that the results of analysis are heavily dependent upon the assumptions and simplifications made. We also noted the importance of predisposition and focus—the entrenched premises, underlying all others, that express from where the study comes and determine to where it will lead.

When inadequately isolated from political influences, analysis may "only play back to the decisionmaker a more sharply defined version of what was already implicit in his assumptions. The role of analysis then becomes not so much to *sharpen* the intuitions of the decisionmaker as to *confirm* them. Under these circumstances analysis is not being used in in its most fruitful form, that of raising questions."[14]

Such political permeation can occur unwittingly among policymakers of good will but little time, served conscientiously by well-meaning analysts lacking more in detachment than integrity. The pity is that assumptions stated explicitly and set consciously—the handles and knobs of analysis—are the best means by which to begin to understand the process wherein results are obtained. Tracing the effects of assumptions and predilections can be a valuable, mind-expanding exercise, though not an easy one.

> Even when we have correct premises, it may be very difficult to discover what they imply. All correct reasoning is a grand system of tautologies, but only God can make direct use of that fact. The rest of us

must painstakingly and fallibly tease out the consequences of our assumptions.[15]

Assumptions, sometimes buried deep within an analysis, may themselves have to be teased out in an inductive process more exacting than was the analysis. Harder still is the effort, first to identify, then to understand the significance of one's preconceptions and intent. Not generally expected, such demanding exertion is rarely undertaken. The opportunity for insightful introspection, missed by analysts, is lost to their clients as well.

Assumptions are often made—consciously or unconsciously—with a conclusion in mind. The world of politics presents an enticing stage for analytical improvisation. A part of politics, it could be said, is the art of translating a complex (many-variable) problem into a simple (single-variable) version that—with some congenial assumptions—best supports a point of view. This is not as cynical as it sounds.

Take the issue of natural gas deregulation. The petroleum lobbyists favoring higher prices, joined by free-market economists, focused on supply elasticities. Their single-variable rendition of the gas markets had rising gas prices increasing supplies and alleviating energy shortages. In the 1977/78 congressional debates, David Stockman, then a Congressman from Michigan, argued the cause of deregulation with a small model run on a hand calculator.

> The crucial first assumption for Stockman ... concerned the amount of gas that would be produced under deregulation. Instead of attempting to assess the supplies on his own, Stockman simply assumed that the quantitative elasticity index the administration used for prices of $1.45 to $1.75 would apply for prices above that as well. The assumption was made with literally no analysis of the underlying geology. Stockman might as well have been analyzing soy beans or any other commodity whose supplies are not strongly limited by physical constraints ... Everything else in Stockman's analysis followed from these basic assumptions about the supply elasticity of the resource base.[16]

Meanwhile, consumer interests wishing to keep prices down emphasized the plight of the poor, the adverse effects of higher gas prices on the farmers, the impact on New England, and the costs to the nation. Senator Edward Kennedy was a vigorous advocate of continued regulation. By one account, members of his Energy Subcommittee of the Joint Economic Committee asked the Congressional Research Service for an analysis of projected costs "to be done by someone who they knew would oppose deregulation. They were looking, in other words, for material to use in their speeches, not for analysis to help them make up their minds." In the rush to present his case to the Senate, Kennedy is said to have incorrectly combined the results of the Congressional Research Service's analysis with a contrary study by the Congressional Budget Office.[17] Such acts are troubling features of the political process.

Both sides of the deregulation controversy searched for elementary, forceful, single-variable interpretations, and waved aside dimensions of the problem that would complicate or vitiate their arguments. They were playing to a public that felt injured, angry, and frustrated; a public that had little appetite for qualifications or convolutions; a public that wanted black or white, not some shade of gray.

Another example of a single-variable argument occurred in the debate over whether economic growth depends on energy consumption. For years, industry spokesmen contended that energy and the economy were in lockstep. They claimed that conserving on energy could only harm the economy. GNP was the one variable they were considering. Neglected were energy price, the trade deficit, and other critical variables.

Too often the response to an oversimplified argument is an equally misleading counterargument. Exclaimed one analyst, deploring what he considered to be another's constantly irresponsible pronouncements, "Sometimes I feel I could get up and say anything, and get away with it for years!"[18]

In an attempt to avoid misunderstanding and misrepresentation, Robert McNamara tried to promote a procedure of "open and explicit analysis" in the Department of Defense in 1961, designed so all interested parties could "see exactly what assumptions were used and how the conclusions were reached."[19] Traces of McNamara's innovation are still visible in parts of the government today, but a tradition for analytical accountability and follow-up has never fully taken hold.

Many of the energy studies were charged with omissions or biases in their assumptions and priorities. According to critics:

Ford-MITRE neglected international considerations;[20] the Rasmussen study concentrated on risks to public safety, ignoring the (not-unrelated) risks to private enterprise that eventually stalled orders for nuclear reactors;[21] the EPP overemphasized regulatory remedies and slighted economic factors; the CIA underrated the Soviet Union's natural gas potential and its ability to shore up petroleum production; WAES missed the major cutbacks in world energy use and the sharply rising oil prices; and most of the studies, in their traditionalist orientations, took for granted the continuity of existing social structure and neglected institutional and political forces for change. Questions of equity, power, money, lifestyles, and regional impact were at the heart of the energy debates of the seventies, yet were not represented, the critics say, in the formulations of most energy studies.

Such omissions and biases, a natural part of learning, can be much more instrumental in determining the outcome of an analysis than its methodological approach. Methodologies are vehicles for tracing the implications of assumptions. When the assumptions are invalid, methodology cannot save the analysis. The imposing machinery of a major analysis, in fact, may serve only to obscure the fact that certain assumptions are weak and need to be exposed to critical review.

Considered Position. We found that few energy studies made a concerted effort to detail their assumptions and explain how they affected the conclusions reached. Such detailing of assumptions and tracing of their implications is the exception rather than the rule in policy analysis. We think it ought to be expected of all major policy studies—initially in the public analyses we have suggested, eventually in partisan-based studies as well.

We know the difficulties this requirement would pose for analysts. Identifying assumptions can be hard work, especially the deep, unconscious assumptions that are better termed inclinations. Pinpointing their meaning would be harder still. Be that as it may, we see no shortcuts to making analysis more responsible and responsive to national interests. Zeroing in on these most basic determinants of policy conclusions would assist policymakers and third parties to be more critically discerning in their use and assessment of the studies.[22] It would be salutary for analysts as well, encouraging a level of questioning and discipline of thought that could expose fallacies of reasoning and clarify relationships. These, after all, ought to be the primary benefits of analysis.

REMOVING THE BLINDERS

Public analysis was to a large extent an objective of the Ford Foundation in each of the three energy studies it sponsored during the 1970s. The EPP did not achieve this objective because of the advocacy position it espoused. Ford–MITRE came closer, but it too was charged with taking sides. Ford–RFF came the closest of the group. RFF–Mellon came close too, as did CONAES, despite its turning into a heated debating forum among experts of different persuasions. If these three low attention/influence studies are the best examples we can cite of public analysis, we face a dilemma. The same fate could befall the institutionalization of policy analysis.

We referred earlier to the "blinders syndrome." As though by an agreement among scholars, many analysts tend to follow the guidelines of their professional disciplines and concentrate on the technoeconomic aspects of an issue, leaving political considerations and questions of values to politicians, decisionmakers, and society at large. These analysts put themselves at a disadvantage in the contest for attention and influence, and their work loses the impact it might otherwise have.

The National Academy of Sciences has for some time been in a quandary over how to get the country's scientists and other experts to contribute input to national decisionmaking without going beyond the bounds of their special skills and knowledge. CONAES was a great trial of conscience for the Academy. Its final report was delayed for well over a year because the Academy's review committee objected to the study's repeatedly moving outside the domain of what the scientist-reviewers considered its technical competence. Yet the study group had been manned (literally, the sole female member would point out) not only with engineers and physical scientists but with lawyers and social scientists as well. The study director was convinced he could not adequately treat the substance of the issues without taking account of a broad range of nonscientific, nontechnical considerations. The strain was a severe test for the stately, marble-walled Academy.

Philip Handler, former president of the Academy, personally experienced much of this strain. A year and a half before his death at the end of 1981, Handler reflected on the public's seeming loss of "confidence in the ultimate value of the scientific endeavor," venturing that the image of science had "been distorted by the participation of scientists in public policy formation." By his perspective, "scientists best serve public policy by living within the ethics of science, not those of politics."[23] Can they do this and still exert a constructive force in the hurly-burly world of power, influence, and tactical manuevering? Handler would answer in the affirmative.

By another perspective, restriction of analysis to matters of technoeconomic efficiency misses the substance of the policy debate and largely explains the mismatch between analysis and politics. The public is concerned with issues of self-interest, equity, regional impact, lifestyles, freedom, and democracy. When analysts behave as though such factors do not exist, they do not thereby gain an image of greater objectivity for their work, but one of naivete—and possibly of callousness as well. Better to remove the blinders and make a candid appraisal of the commercial and political interests at stake, taking a perceptive look at the contention points and contending parties. "Analyzing the *interests* and the *participants* may be as important as analyzing the *issues*."[24] Building this examination into the analysis would improve it, make it politically more relevant, and thereby increase its usefulness and the respect with which it is received.

Considered Position.   The current marketplace for analysis and ideas does not offer equal opportunities to be heard, and analysts cautiously adhering to the standards of their professional disciplines can be at a distinct competitive disadvantage. We take this as reason to improve the marketplace rather than as cause to relax intellectual principles. Analysts need to realize that the policy questions they are examining are going to be decided within a political framework in which clashing interests can upset the most carefully laid designs and most reasonable of logical arguments. We feel they should take this political framework into account in presenting their conclusions.

We would have public analysis delineate the points of view of each of the parties in dispute, define the problem according to these points of view, collect technical facts bearing on this problem, explore the attitudes of those most affected by it, and then integrate the definitions, technical facts, and attitudes into alternatives for decisionmakers to consider. Analysts would thereby be supplying the political dimension now missing in much policy analysis. In effect, we would join political analysis with techno-economic analysis much as economic analysis was joined with systems analysis at the RAND Corporation and in the Department of Defense many years ago. That would provide valuable links for public analysis with partisan-oriented studies.

We would have public analyses compare studies critically and explain why they reached different conclusions.[25] Public analysis must not be done primarily for elites. That was a criticism lodged against many of the energy studies. Public analysis should search for effective ways to convey key insights to the media and general public as well as to decisionmakers and other analysts. Its purpose should be understanding and communication, not persuasion or prescription.

PISTONS OF PUBLIC POLICY

Policymakers are the pistons or pressure chambers through which propulsive social and political forces get transmitted to the wheels of policy action. We choose the power metaphor for a reason. Policymakers operate under pressure, respond to pressure, and exert pressure—a great deal of it. Their hectic schedules and political preoccupations leave them little time for reading and reflection. We find it remarkable how well many of them stay abreast of technical issues under the circumstances. Analysts function differently. Yet if analysis is to serve a useful national purpose, it must somehow make productive contact with those hard-to-reach, fast-moving pressure pistons.

Analysis is most needed and potentially most useful when it helps in making or changing complex and important decisions. Since these decisions are often those for which the political stakes are highest, as in the case of energy, they are likely to be occasions of greatest disagreement among policy studies and greatest confusion among those trying to make sense of these studies. Policymakers must be more like athletes than professors, says one former professor turned policymaker.[26] They must train themselves to act and respond instinctively. According to this perspective, people in high office have an obligation to the public to show confidence and disguise uncertainty. Analysis cannot really be useful to them unless there is a way they can understand and internalize it. Analysts must make their results transparent and intuitive so that decisionmakers can check them against their own mental models with simple back-of-the-envelope calculations. That, by this view, is the best way for analysts to make their work meaningful to policymakers and gain for it greater appreciation and acceptance.

Analysts and politicians look at the world differently. They are often seen as dissimilar cultures shaped by very different sets of experiences and incentives. They may use similar terms without speaking the same language, even in areas where they have interests and concerns in common. These differences, to the extent they are real, can result in misunderstanding between the cultures, mistrust, and misapplication of the benefits analysis is able to provide.

Analysts, according to a common caricature, thrive on harmony and regularity. They prefer to work in tranquil settings and contemplate the longer term picture. Because they are able dispassionately to observe, order, and measure—using procedures rooted in scholarly traditions—they can bring a much-needed degree of rationality and precision to political deliberation. Analysis can help define and clarify a problem, locate boundaries of feasible and appropriate action, and inject factual substance into what might otherwise be hollow rhetoric and emotional political debate.

Politicians, in contrast, have to be aware of many features of the problem not amenable to analysis. They must take stock of the likely effects to their constituents of a range of policy options, stressing the immediate here-and-now and heeding symbolic and subjective meanings along with those of a more quantitative type. Miscalculation threatens their political survival. This fact, above and beyond all others, enters into every move, every decision. Under such circumstances, the politician is apt to use analysis and analysts in ways analysts think improper. But analysis is one kind of activity, politics another. East is East and West is West, and many are convinced the twain shall not soon meet.

Considered Position.   To acquire analytical skills for their office, members of Congress have hired analysts and investigators in recent years, instructing them in constituent politics and appropriate priorities. The analysts hired were not so much of the "technician" kind as the "political" and "entrepreneurial" varieties.[27] By employing these analysts, the members may have hoped to gain time for more diligent delving into policy questions themselves. This rarely occurred.

> If anything, the use of entrepreneurial staffs has meant an increase in the numbers of hearings and amendments considered ... While representatives as recently as 1965 spent almost one full day every week on 'legislative research and reading,' by 1977 the time spent on reading was down to an average of eleven minutes per day. In other words, instead of freeing the members to concentrate, the staffs contribute to the frenetic pace of congressional life that pulls members in different directions, reduces the time available for joint deliberation, and makes concentration all but impossible ... The situation feeds on itself. The members need staff because they have so little time to concentrate, but the new work created by the staff takes even more of the members' time, indirectly elevating the power of the Washington issue networks in which the staffs play so prominent a role.[28]

It would be unrealistic to ask politicians to use their budgets to hire reflective, dispassionate conveyors of the results of analysis, and it would be unrealistic to expect staff assistants to abstain from political zeal in trying to be helpful. Better to settle for strengthening the marketplace of ideas, improving communications with the general public, supporting the development of public analysis, and making analysis more understandable and politically germane. Politicians are born of the people, put in office by the people, and committed to work for the people. As the public becomes better and more responsibly informed in the lessons of policy analysis, politicians will too. That is the sure way for public analysis to make contact.

Public analysis, if it meets its goals, is certain to reach decisionmakers and help shape or reshape the mental models they carry with them to policy deliberations. When the new fellow on the block (public analysis) begins to make friends, policymakers are not apt to overlook or snub him. They may challenge him, they may be inclined at first to minimize him, but there would be little political gain from doing so. Better to listen to what he has to say, or at least have the analyst on the staff listen to what he has to say, and then listen to the analyst on the staff.

ENERGY STUDIES AS THE THIRD WAVE

The active use of analysis in public policy that accelerated with the energy studies of the 1970s includes in its still-recent history the cost/benefit analyses of the 1940s and 1950s. But it was not until the vigorous introduction of analysis into government by the Kennedy Administration in 1961 that analysis came into its own. With the arrival during the Kennedy years of Robert S. McNamara from Ford Motor Company, Alain C. Enthoven and Charles J. Hitch from the RAND Corporation, and a host of intellectual elite from Harvard and M.I.T.—McGeorge Bundy and Carl Kaysen among them—a new era in the application of analysis to policy problems got under way.

This era had two main thrusts: economic analysis in the Council of Economic Advisers and other government offices, and systems analysis in the Department of Defense under Enthoven, Hitch, and McNamara. For economic analysis, it was the beginning of a steady growth in volume and visibility that continues today. For systems analysis, based in economics and operations research, it was the start of a fast climb that reached its peak and topped out by the end of the decade.

Hitch wrote, at the start of the climb:

> The Secretary and I both realized that the financial management system of the Defense Department must serve many purposes. It must produce a budget in a form acceptable to the Congress. It must account for the funds in the same manner in which they were appropriated. It must provide the managers at all levels in the defense establishment the financial information they need to do their particular jobs in an effective and economical manner. It must produce the financial information required by other agencies of the government, [and it] must also provide the data needed by top defense management to make the really crucial decisions, particularly on the major

forces and weapon systems needed to carry out the principal mission of the defense establishment.[29]

Thus was born, based on work Hitch and others had done at the RAND Corporation, a programming function that came to span the gap between planning and budgeting in Defense. This led to the three-phase operation known as the "planning-programming-budgeting system" or PPBS.

Use of PPBS blossomed in Defense during the early and middle 1960s. Impressed, Lyndon Johnson, promoted a massive effort to introduce the methodology throughout the federal civilian agencies. But the wholesale assaults made on the traditional ways of conducting government's business were hasty and not well conceived. Meeting great resistance, the use of PPBS most everywhere except in Defense waned by early in the 1970s.

Systems analysis suffered from its association with an unpopular war and from the resentments it stirred up in the military services whose programs it was being used to monitor.[30] Beginning about 1968, in the wake of disenchantment with Vietnam, there were revelations about egregious mismanagement of defense procurement programs and subversion of the planning process. That was the end of the first vigorous use of policy analysis by government. A second wave was about to begin.

The second wave developed in the sphere of social and urban problems, reflecting President Johnson's emphasis on the Great Society. As if to signal and underscore the shift, principals in the first wave moved away from military preoccupations into civilian endeavors in the private sector, typically in social or educational fields. Over a period of years, McNamara became president of the World Bank, Bundy president of the Ford Foundation, and Kaysen president of the Institute for Advanced Studies at Princeton. Hitch took a financial post at the University of California, later to become president of that institution, and Enthoven, after being nominated as director for the abortive Environmental Institute, first went to an electronics firm, then became a professor of public policy at Stanford, where he conducts broad-ranged studies in the economics of health care.

The second wave of analysis had some carryover from the first, such as PPBS's systematic attention to policy alternatives.[31] A variety of quantitative methodologies were applied to problems of the city, social welfare, health, and the environment.[32] This wave, with its environmental impact statements, social research, and ambitious urban models, extended from the late 1960s to the middle 1970s. Its decline was marked by growing disillusionment with large social programs and by the demise of the New York City-RAND Institute, the intrepid organizational experiment noted earlier.

Once again, as one wave was coming to a close, another was already taking form. The third wave included the energy studies that are the subject of this book. Like the second wave, it rode above the sea of numbers and analyses that was by now commonplace throughout government.

By 1980, the third wave was itself subsiding. As energy use fell back and the budget-minded Reagan Administration took office, there was more oil on the world market than interested buyers at then current prices. The Department of Energy began to be phased down, and government spending was sharply curtailed on solar, conservation, and other energy programs. It looked like the start of a dry season for energy research and energy policy.

CONCLUSIONS

We began our inquiry into the energy studies with two overarching questions. First, what has been the role of analysis in public policy? Second, what should it be, given the needs and nature of a democratic society? The energy studies provided a rich laboratory for examining the role of analysis in policy and also for observing the effect that the political process in a democracy has on the use of analysis. Energy as a policy area was so technically demanding and beset with unknowns and uncertainties on the one hand, and so politically divisive and fragmenting on the other, that analysis alternated between soaring to prominence and being ingloriously ignored. It was highly desired as a way to understand what was happening and what to do about it, and it was just as highly resisted for being too simplistic or for appearing to support an opponent's position. The studies were like clay pigeons, launched only to be shot at and brought down.

It is abundantly clear that there is not just one role for analysis; there are many. As we did with the viewpoints of the energy elite, we may roughly divide these roles—again at the risk of oversimplifying—into two main categories. Let us call these categories the "public" and "partisan" uses of analysis. The first is analysis meant to serve national interests; the second is analysis adopted by advocates to further or defend their positions in policy debate.

These terms help us portray what was happening with energy policy analysis during the 1970s. Many studies designed to serve a role of public analysis were initiated. One after another, they were drawn away to serve partisan purposes by the strong adversarial forces the debate unleashed. Political partisanship is a powerful magnet. It can shape an analysis, lure it off course, or give the analysis unintended meanings. It pulls in many directions. This is why analysis often ends up mirroring or widening a conflict rather than resolving it.

Serving political objectives is a legitimate and natural use for analysis in a democratic society. It can sharpen arguments, provide healthy critique, stimulate new ideas, and (as in the case of the soft path) present additional alternatives for consideration. But the strong pull of partisan analysis can endanger public analysis. This is a menace a democracy must guard against, just as it must guard against the threats to its constitution and basic freedoms that come regularly from the power blocs and special interest groups within society.

Public analysis is an ideal. Its constituency, such as it is, does not have the incentives, drive, and determination possessed by customers for partisan analysis. Public analysis is subjected to an unequal competition within the same democratic process whose interests it is called upon to serve. It has a vital role to play in that process: promoting rationality, correcting misconceptions that obstruct or deceive, offering paradigms for political discourse, revealing inconsistencies in premises and attitudes, providing frameworks for policy debate, suggesting procedures and rules of evidence for weighing opposing positions, producing forums for discussion and for assimilating knowledge, and giving focus to the common objectives that sometimes seem to dissolve in the heat generated by the adversarial system.

If all analytically based arguments are treated the same, if there is no process able to expose or dispose, if conclusions are intended just to score debating points and not stand challenge, and if political ends are made supreme, it is hard to see how analysis can ever give society full value. If the country chooses to view analysis as all partisan, as several of those we interviewed have done, then analysis is easy to dismiss. If the attitude, "Well, everyone does it," in response to a discovered deception is considered perfectly OK, policy analysis becomes no more than a pretense and a sham, and we should not be surprised when we hear from the head of the Office of Management and Budget that "none of us really understands what's going on with all these numbers."[33]

Our culture values numbers more than most—often more than it should. Despite the skepticism they inspire, statistics have become an important currency of political exchange. "We had the facts and figures to combat their dazzling rhetoric," rejoiced a Congressman from an oil-producing state in commenting on his part in the gas deregulation controversy.[34] Facts and figures, as we have seen, were central to the energy debates.

Much of what we saw in the energy studies was excellent work. We believe it had a constructive influence in helping the nation deal with the problems that arose unexpectedly and abruptly in the early seventies. We do not view the fits and starts in energy policy as a black mark for analysis. "Inertia is greatest where the options are broadest. Where there are no choices, it is easy to be decisive."[35] For some options, vigorously promoted at the crest of the debate only later to be rejected, most people now agree the outcome was fortunate. Analysis was useful in that respect. It lit the country's way to realism.

Our investigation of the energy studies encourages us to view the potential role for policy analysis, on balance, optimistically. This does not alter the fact that we were disturbed by the cynicism and alienation we found. We hope we have expressed this concern clearly. The centrifugal force of partisanship, so evident in the case of energy, requires a countervailing centripetal force to keep the analytical enterprise from flying apart. That centripetal force is underdeveloped and undersupported at the present time.

On the time scale of the nation's history, policy analysis is still a youngster. Judgment would be premature. The country has yet to generate the institutional mechanisms and supportive environment necessary to nourish and sustain the kind of activity we believe is needed—the activity we have called public analysis. The pause between the third and fourth waves would be an opportune time to get started.

APPENDIX A
ELITE VIEWPOINTS ON ENERGY

In the energy studies of the seventies, discussions tended to re-volve around interests, issues, theories, the economy, and socie-ty—abstractions at best. Yet energy problems concerned real people. Examining their attitudes and opinions sheds light on the causes for the energy debate and the obstacles that compli-cated the constructive use of analysis in policymaking. In this appendix, we examine the attitudes and opinions of energy ex-perts, as elicited from interviews and an attitude question-naire. We compare these findings with observations about the feelings of the general public from earlier studies of citizen groups. No all-inclusive way of classifying expert opinion can possibly be valid; still, two central viewpoints did emerge in our investigation of what we dubbed the energy elite. We call these viewpoints "traditionalist" and "reformist" for ease of reference. These two polar stereotypes thought differently about most issues, yet in some respects were closer to each other than either was to the public at large.

How odd it is that anyone should not see that all observation must be for or against some view if it is to be of any service.[1]

For energy more than any other problem area in the history of policy analysis, analysts came forward readily and abundantly to work on issues affecting the interests and well-being of every citi-zen. Unavoidably, they brought with them attitudes and opinions on these issues. We wanted to know what these predispositions were. We were curious about how analysts and policymakers differed in view from one another and from the average citizen, and the degree to which their views permitted accommodation and the formation of consensus. To pursue these interests, we compiled an attitude questionnaire using a technique from the social sciences that helps identify underlying viewpoints.

We distinguished in our investigations between the "energy elite" (experts on energy and well-briefed policymakers) and those forced to rely on these elite (the mass or general public). How effective was communication between these groups? To what extent was the public misinformed? What was the balance between emotion and reason in the popular mood? Without sufficient popular support, there is no way to gain acceptance for policy. Could the disagreement and its consequences be mitigated? Was there a basis for agreement between the mass and the elites, or among the elites themselves? Was there any consensus among the experts, or the possibility for it? These were some of the questions we sought to explore.

OPINIONS, ATTITUDES, AND BELIEFS

For much of the seventies, the apparent mood and climate of public opinion consisted of:

- Unrest and a lack of trust in the leadership;
- Introspection;
- Pent-up frustration and a tendency toward aggression;
- Cynicism about authority figures coupled with a contradictory desire for someone to take charge.

Historically, in periods such as that attending the oil supply interruption, when the mass mood was unsettled and authority's ability to govern doubted, antiestablishment views abounded and the status quo was likely to get disrupted. In periods when the ruling elite was trusted, on the other hand, mass compliance with government policies and preferences prevailed. These effects can be phrased in terms of the connections between opinions, attitudes, beliefs, trust, legitimacy, and control.[2]

Opinions are elementary statements or expressions made by individuals about the state of the world.[3] They affect public policy in several ways. Most directly, they reflect individual and collective experiences—not a perfect reflection, to be sure, because such experiences are perceived and interpreted differently throughout the population. Opinions help identify and order current and emerging problems, especially as these become more intense and persistent. And opinions contribute to the language of policies and social discourse by providing the substance of policy preferences.[4]

Whereas opinions are nearly limitless, the attitudes they represent are far fewer in number. Attitudes are summary categories that tend to be relatively stable for individuals and groups over time. Attitudes interpreted according to their supposed meanings are called beliefs. Opinions are identified with statements, attitudes with clusters or factors, and beliefs with the interpretation of these clusters.[5] This pyramiding up from statements to clusters to beliefs forms the basis for the design of the questionnaire used in our investigations.

Beliefs are often treated in oversimplified terms, no attempt being made to get at their opinion or attitudinal bases. C.P. Snow's famous characterization of the "Two Cultures" was a broad-brush attempt to explain differences in attitude, outlook, and policy preference among Britian's wartime elites.[6] It quickly became conventional wisdom. But the two cultures did not discriminate important differences in attitude and opinion. It did not really help much in understanding or formulating policy preferences.

Also not helpful is the common attempt to divide the complexity of attitude into simple categories of optimism and pessimism. Pessimists, in this contrite rendition, view the world as doomed—with science, technology, big government, or other favorite bete noir as the root cause. Optimists see the world as salvageable. Science and technology will lead the way to salvation.[7] This loads the dice in favor of gross policy preferences, such as more science versus less science or more technology versus less technology—a simplistic representation that does little to aid understanding.[8]

Conflict between belief systems is sometimes characterized as "humanistic" versus "analytic." Thus we find Robert Heilbroner, who embraces an ecological point of view, opposing Simon Ramo, who seeks technological solutions. Both share a relatively holistic perception of the world and its problems.[9] There is also a third pattern of belief that brings humanistic and analytic together, as in the work of the Limits to Growth school.[10]

Some have tried to organize beliefs according to disciplinary backgrounds. "The optimist camp is clearly headed by the economists, who, among all other intellectuals, retain a much greater faith than the natural scientists themselves."[11] Representation on the contentious CONAES study of participants from many fields supported the idea of conflicting beliefs rooted in disciplinary differences.

Finally, policy preferences have themselves been used to categorize beliefs. In the characterization of Amory Lovins, preference for a centralized, large-scale, apparently impersonal system of control (the "hard path") was pitted against preference for a small-scale, decentralized, allegedly more humane system (the "soft path"). These distinctions served as symbolic rallying cries for competing groups seeking power and dominance.[12]

INTENSIVE AND EXTENSIVE MEASUREMENT APPROACHES

The distinction between intensive and extensive observation and measurement has long existed in the social sciences.[13] Extensive observation is the relatively brief examination of numerous subjects, as in a public opinion poll, whereas intensive observation investigates small numbers of subjects for lengthy periods of time, as with in-depth interviewing and longitudinal (panel) studies. An intensive approach can help identify issues for later illumination by extensive means. "Extensive methods are eminently suited for counting things and computing proportions which exist in the universe at large ... Intensive methods are best equipped to determine which things are worth counting in the first place."[14] To study the energy elite, we used interviews and an intensive attitude questionnaire patterned after one given earlier to citizen groups.

MASS OPINIONS AND ATTITUDES

During the period from 1975 to 1978, numerous surveys were conducted on popular opinions about energy matters.[15] Opinions about specific energy technologies such as nuclear, coal, and solar—based on relatively little accurate knowledge—did not display much stability. Attitudes about large institutions showed more coherence. National surveys left little doubt that the majority of Americans did not have much confidence or trust in local, state, and federal governments, in electric or gas utilities, and in major oil firms. Trust, consensus, and the power they confer were seriously eroded[16] in the wake of Vietnam, Watergate, and the energy crisis. Close to one-third of the public regarded the energy crisis not as a resource problem but as a manipulation by vested interests.

Popular mistrust of the federal government fixed on suspected waste and individual privilege. People were upset about government's spending tax dollars on questionable programs, inefficient bureaucracy, high taxes, concentration of power in the hands of a self-interested few, huge budget deficits, and the problems all of this meant for them.[17] Public opinion about energy corporations and their officials was even more negative.

Opinions were based in part on the ways individuals sensed and interpreted their own experiences. Energy was a surrogate for other preoccupations of Americans during the 1970s,[18] and it did not hold popular attention nearly so much as economic problems more generally. "The public's concern about energy has been greatly overshadowed by economic concerns, especially the high cost of living (inflation) and, more recently, unemployment."[19] Consideration of energy filtered through the prices paid to heat homes and drive cars. Crime, lawlessness, dissatisfaction with government, among other issues, all took priority over energy in the public consciousness at one time or another.

The low level of citizen attention to energy can be related to the on-again, off-again way the problem was treated by government and the press. The oil embargo and dramatic presidential statements made an impression but were diluted by mixed messages transmitted about the problem's seriousness. "At one moment, media sources, government, and the oil companies stress the urgency of the problem; at another, these same sources tell the American public that they are about to enter a period of an oil glut."[20] Such inconsistency did little to enhance the popular credibility of individuals and institutions relied on for energy expertise.[21]

Surveys of the period gave clear evidence of a general malaise, thus sounding an alarm of sorts, but they failed to provide a clear interpretation of what was cause and what was effect.[22] Policymakers needed better guidance. There was need for a more perceptive look at attitudes to discover which things mattered, and for whom.

Intensive Investigation of Public Attitudes

One of the more illuminating in-depth examinations of public attitudes was conducted by a market-research firm in late 1975 in a series of two-hour focused discussions among heterogeneous groups selected from around the country. Broad issues pursued in these discussions included the following:

- Attitudes on the credibility of various parties involved in energy;
- Sources of information relied on for energy information and their relative influence in shaping attitudes;
- Understanding of economic issues related to energy;
- Receptivity to energy conservation;
- Understanding the tradeoffs that attend increasing energy costs and new technologies; Feelings of personal responsibility about energy use and conservation;
- General level of understanding of the main factors that comprise the energy problem.

At the time, the citizenry was experiencing frustration, anger, and helplessness: frustration about rising energy prices; anger about feeling manipulated by the institutional and individual forces thought to be responsible for the energy crisis; and helplessness in the absence of any consistent information about the true extent of the problem or what individuals could conceivably do to lessen its effects. An "us-versus-them" theme recurred frequently, underscoring the victimization felt by many about rising costs, lost convenience, and other personal manifestations of the energy problem. "Them" referred to big business and government, both of which were regarded as manipulative and exploitive of "us," the average citizen.

The energy crisis was not believed in 1975. It was mainly seen as a means to increase prices. Most respondents felt that energy shortages were the result of mismanagement or unscrupulous profiteering by oil companies and exporting countries. The higher profits reported by oil companies during this period did little to dampen these suspicions. The wish to find a scapegoat dominated serious attention to market forces:

> What kind of angers me is that no one seems to know why or whom to pin it on; whether it's the oil companies or the Middle Eastern countries, or what is causing all this ... Take for example gasoline: when the price of everything goes up, there's plenty of it; when the prices go down, there's a shortage.[23]

It is interesting that the majority of respondents did not blame the Arabs nearly so much as they did the oil companies, public utilities, and government, whom they viewed as self-centered, insensitive, and devoid of concern for consumer needs. They repeatedly accused these institutions of flagrant violations of their powers, of greed, incompetence, and lack of respect for the consumer. The responses showed striking erosion of trust in government—trust being an essential ingredient in consensus-building and policymaking. The public, expecting government to protect citizen interests and preserve the general welfare, felt seriously betrayed and let down.

So far as the public was concerned, prices had less to do with the laws of economics than with manipulation and conspiracy. Bickering among experts about alternative forms of energy were perceived more as jockeying for future dominance of one powerful interest over another, at the expense of the consumer, than as meaningful technical debate. The irresolve of government was perceived more as a means to benefit privileged and powerful special interests, also at the public's expense, than as an honest expression of the difficulties and uncertainties of the energy problem. The problem, in short, was commonly considered more political than technical or economic.

Citizen Questionnaires

In March 1977, the Energy Policy and Planning Staff of the Executive Office of the President invited nineteen citizens to Washington to express their views on energy prior to the completion of the National Energy Plan.[24] Thousands had responded to a questionnaire about energy mailed out by the federal government and printed in the *Federal Register*. Ronald Brunner and Garry Brewer invited each of these nineteen persons to participate in a pilot study of citizen viewpoints designed to provide input to energy policy planning and evaluation. It was an intensive approach. The purpose was not to estimate the proportion who held this or that opinion but rather to clarify how these opinions got organized into coherent viewpoints.[25]

The pilot effort, as well as a parallel one conducted by Brunner in Indiana, and a follow-up study by Brunner and Vivian on an expanded nationwide sample of 101 citizens, all used "Q-sort" techniques[26] rather than survey methods. A Q-sort questionnaire starts with a set of statements on the issues. Sixty statements about energy policy culled from public discourse were typed on individual cards. Respondents sorted the cards according to relative intensity of agreement or disagreement with each statement. They assigned values of from $+5$ (most agree) to -5 (most disagree) to the statements to conform with a standard frequency distribution specified in advance. An algorithm was used to determine associations among responses and to identify clusters of people holding similar viewpoints.[27] These viewpoints were reflected by the statements with which people in a cluster either most agreed or disagreed.

A surprising consensus emerged in the Brunner-Vivian study: one core viewpoint with several peripheral variations.[28] The statements most agreed or disagreed with by holders of this viewpoint are reproduced in Table A-1. The numerals in parentheses are the average agreement and disagreement scores.

Health and safety elicited great concern in the core viewpoint, not surprising in light of the fact that Three Mile Island occurred just a month before the study was undertaken. Distrust was dominant. Credibility had clearly been severely strained. Citizens did not know whom or what to believe, but they yearned for more

Table A–1. Core Viewpoint in the 101-Citizen Study.[a]

Most Disagree

I have no objections to the location of a new nuclear power plant in the area where I live. (−4.1)

We should accept additional risks to public health and safety in order to deal with the energy crisis. (−4.0)

Consumer prices of gasoline, oil, natural gas, and electricity are too low. (−3.8)

The oil and gas companies deserve the trust of people like me. (−3.4)

Over the next twenty years, solar energy just can't make much difference in meeting the nation's energy needs. (−3.2)

I won't cut back on my energy use until others make the same sacrifice. (−2.7)

Most Agree

An increase in gas and electric bills means more hardship for the poor and those on fixed incomes. (+3.7)

Not even the experts know how to safely dispose of radioactive wastes from nuclear plants. (+3.4)

Energy research and development should emphasize renewable energy resources like solar energy. (+3.3)

We should do whatever we can to make our jobs and income less dependent on foreign oil. (+3.3)

Oil and natural gas have been withheld from the market to force consumer price increases. (+3.3)

I want better information about how the energy crisis affects me and my community and what we can do about it. (+2.6)

I don't know whom or what to believe about the energy situation. (+2.5)

---

[a] Number in parenthesis denotes average score in cluster.

and better information about how the crisis affected them and their communities.

There was support for research and development efforts to create alternative energy sources, and there was willingness to conserve even though others might not be doing so. The connection to jobs and incomes was pronounced, as was the sense of personal insecurity engendered by continuing dependence on foreign oil. Price increases were basically rejected as a means to ensure market efficiencies or fair distribution of energy supplies. In follow-up interviews, respondents revealed that they believed higher prices inflicted hardship on the poor and those on fixed incomes and represented a manipulation of supply by industry for its own advantage.

There were three peripheral viewpoints within the core. The first stressed opposition to federal initiatives, and specifically disagreed with the idea that market forces could be mobilized through deregulation, preferential subsidies, or tax benefits to business without doing harm to the public. The second emphasized environmental matters, rejected the expansion of nuclear power, expressed confidence in the feasibility of alternatives, and valued conservation highly. The third reflected heightened anxiety about the world energy situation and its likely impact on jobs and personal security.

The public did not view the issue of pricing as did many economists. Feeling distrust and cynicism, it saw prices as the problem, not the solution. Authoritative national sources of information about energy matters were simply not believed; perceptions of what was happening were mainly derived from close-at-hand experiences, not expert analyses.

For a populace frustrated in its desire for guidance, conservation satisfied the wish to do something positive. Experts, meanwhile, were searching among a variety of alternatives. They were experiencing frustrations too.

ELITE QUESTIONNAIRE

To examine expert attitudes on energy, we adapted the Brunner-Vivian approach to a sample of elite. Our sample consisted of 150 individuals representing a cross section of: (1) persons responsible for commissioning, funding, conducting, or using one or more of the energy studies we were reviewing; (2) persons prominent in energy policymaking, namely, government officials or key members of staffs concerned with energy matters; (3) persons otherwise visibly associated with energy issues; and (4) persons generally knowledgeable in adjunct areas of science and technological policy. There was no attempt to obtain proportionate numbers of people adhering to particular points of view.

To find suitable opinion statements for our questionnaire, we perused those used in the prior citizen questionnaires. Of these, we judged twenty-eight generally pertinent to elite conceptions and views about energy. We retained them in whole or in substantial part, making a few modifications for purposes of sharpness or clarity. Another nine we altered or elaborated more drastically to make them more appropriate for subjects knowledgeable about energy. We devised a number of new statements specifically focused on the relationship of analysis and experts to decisionmaking, and added several more on important issues not already adequately covered. Testing the draft instrument on sympathetic colleagues, we replaced a few of the statements, made some final editorial changes, and were ready to roll with the 60 statements listed in Appendix B.

We administered some questionnaires by mail, some in person, some before personal interviews, some during interviews, some after. Our subjects were busy people. We did not insist on a rigid routine. We were guided by what was feasible and practical.

Analysis of the responses was accomplished by Ronald Brunner on a computer with the requisite software at the University of Michigan. A visual display from this analysis, called a phenogram, is given in Figure A-1. The 150 subjects are arrayed along the horizontal axis. The vertical axis, reading from bottom to top, represents consecutive steps taken by the algorithm to form the clusters shown by the horizontal ties. As the processing moved up the vertical axis, individuals and prior groupings were reordered and combined into successively larger groupings, until all individuals and clusters merged into a single all-encompassing group at the final step—designated step 1 (from the end). In general, the sooner a cluster coalesced in the process, the more homogeneous the views held by its members. It was a matter of judgment to decide when a grouping was a significant cluster. Only clusters deemed significant are shown in Figure A-1.

Clusters were interpreted as viewpoints. A and B were the two main groupings. There were 111 members in A (including A.0) and 39 in B (including B.0). Within A was core cluster A.c (49

Figure A–1.   Formation of Clusters.

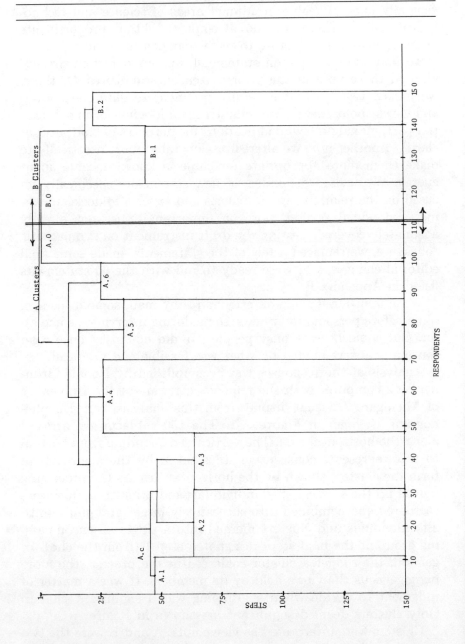

Note: Clusters A.c, A.4, A.5, A.6, B.1, and B.2 correspond to distinctive viewpoints, as described. A.0 and B.0 consist of individuals unassigned to these clusters.

members), and within B, core cluster B.1 (17 members). These clusters constituted the most representative of the A-like and B-like perspectives. A.4 (22 members), A.5 (18 members), and A.6 (11 members) were peripheral clusters within A, while B.2 (10 members) was a peripheral cluster within B. A.1 (11 members), A.2 (20 members), and A.3 (17 members) were subgroups within A.c. A.2 was in effect a core within a core. Finally, A.0 (12 members) and B.0 (12 members) were nonassigned groups of individuals only loosely aligned with the main groupings.

The analysis also provides correlations between the statement scores of cluster averages. Table A-2 presents these inter-cluster correlations.

Table A-2 has several interesting features. First A.c and B.1, the two core clusters, were dissimilar but not predominantly in opposition. They were in fact uncorrelated (0.01). We found that holders of these two core viewpoints appeared to have been focusing on separate issues or attending to different aspects of the problem. Second, A.5 correlated positively with both B.1 and B.2 (.58 and .79), as did B.2 with both A.4 and A.5 (.62 and .79), suggesting possibilities for policy accommodation. Third, no two clusters were uniformly antagonistic across all statements. The only negative correlation, that between A.6 and B.1, was hardly significant (− .08).

Let us look now at the content of the two core clusters. The statements receiving the largest scores by members of these clusters are presented in Tables A-3 and A-4. In the interest of having something more mnemonic than A.c and B.1 to call the corresponding viewpoints, and at the risk of obscuring their complex

Table A–2.   Correlation among Cluster Profiles.

Clusters	A.c	A.4	A.5	A.6	B.1	B.2
A.c	1.00	0.88	0.73	0.75	0.01	0.42
A.4		1.00	0.79	0.66	0.23	0.62
A.5			1.00	0.46	0.58	0.79
A.6				1.00	−0.08	0.24
B.1					1.00	0.81
B.2						1.00

Table A–3.   The A Core Group Viewpoint.[a]

Most Disagree

* With the commercialization of solar energy over the next decade, the end of the energy crisis will be in sight. (−4.3)
* If another oil embargo is imposed on the United States, the government should invoke military force to deal with it. (−4.0)
* Nuclear power cannot be expanded without infringing on the civil liberties of many people. (−3.7)

Major oil companies should be prevented by law from acquiring control over coal, uranium, and the equipment needed to use solar energy. (−3.4)

If we go ahead with the breeder, there is no hope of checking the proliferation of nuclear weapons. (−3.3)

* People should trust the federal government to find a solution to the energy crisis. (−3.2)

The people have confidence in the oil and gas industry based on its contributions to economic growth and prosperity. (−2.7)

Energy industry executives should be prevented from serving as public officials in the Department of Energy. (−2.6)

Most Agree

* We should reduce the vulnerability of our economy to cut-offs of oil supplies from abroad. (+3.7)
* Oil and natural gas prices should be removed from U.S. Government regulation. (+3.7)
* The prices of oil and natural gas produced in the United States are too low. (+3.2)

Increasing dependence of the United States on imported oil since 1973 has put American security in serious jeopardy. (+3.2)

Scare talk about the problem of nuclear waste disposal overlooks several technical options for solving the problem. (+2.5)

The likelihood of accidents involving radioactive materials at U.S. nuclear power plants has been greatly exaggerated. (+2.5)

When natural gas prices are no longer regulated by the federal government, the market will allocate supplies of natural gas efficiently. (+2.3)

---

[a] Number in parenthesis denotes average score in cluster. * Designates statements emphasized in all A clusters

Table A–4.   The B Core Group Viewpoint.[a]

Most Disagree

* Meeting projected demand for electricity requires placing primary reliance on the construction of nuclear generating plants. (−4.2)
* If another oil embargo is imposed on the United States, the government should invoke military force to deal with it. (−4.2)
* Improvements in the American standard of living depend upon production and use of substantially more energy in future years. (−3.8)

The likelihood of accidents involving radioactive materials at U.S. nuclear power plants has been greatly exaggerated. (−3.7)
* The people have confidence in the oil and gas industry based on its contributions to economic growth and prosperity. (−3.1)

We should accept additional risks to public health and safety in our choice of energy policies. (−3.0)

Legal maneuvers of environmental groups have been the principal cause of recent increases in the cost of nuclear power. (−2.9)

The effect on the upper atmosphere of fossil-fuel combustion is too long-range a hazard to warrant serious concern today. (−2.8)

People should trust the federal government to find a solution to the energy crisis. (−2.8)

Environmental and safety regulations are impacting adversely on American productivity and enterprise. (−2.6)

Most Agree

* Energy research and development should emphasize renewable sources such as solor and biomass. (+3.7)

Environmental protection laws and regulations should be vigorously enforced. (+3.2)
* We should nurture a resource-conserving ethic in which human satisfaction is much less dependent on material consumption. (+3.2)
* Decisions about energy made today will have profound and irreversible consequences for future generations. (+2.7)

We should reduce the vulnerability of our economy to cut-offs of oil supplies from abroad. (+2.5)
* People resist sacrifices like those proposed in the National Energy Plan whenever they feel they have been manipulated and exploited. (+2.4)

[a] Number in parenthesis denotes average score in cluster. * designates statements emphasized in all B clusters.

and often subtle components, we refer to them as "traditionalist" and "reformist," respectively. These terms are not meant to signify opposition, only difference of perspective.

Traditionalists strongly disagreed that solar energy would solve the energy problem within the next decade and that another oil embargo would justify the use of military force. Although 80 percent of them took the questionnaire after the incident at Three Mile Island, they nevertheless favored nuclear power and thought the likelihood of accidents at nuclear plants had been exaggerated. They doubted that nuclear proliferation must result from development of the breeder and felt that nuclear waste disposal was amenable to technical solutions. They did not believe that expansion of nuclear power would pose a threat to civil liberties.

The traditionalists were convinced that the marketplace and higher energy prices could help solve the energy problem and thought deregulation of oil and natural gas was the key to efficient allocation. They lacked confidence in the ability of the federal government to deal with the problem and supported freeing the energy industries from governmental restrictions. They acknowledged that the public lacked trust in these industries.

Reformists agreed with traditionalists about lack of public trust and the inadvisability of using military force in the event of another oil embargo. They also agreed that the United States must reduce its vulnerability to cut-offs of oil supplies from abroad; but their ideas on how best to accomplish this were unlike those of the traditionalists. On nuclear accidents, reformists strongly disagreed that the risks had been exaggerated.

Reformists had great sensitivity to environmental concerns. They believed that environmental protection laws should be enforced vigorously, that additional risks to public health in framing energy policies were not acceptable, that legal activities of environmental groups were not resulting in increased nuclear power costs, that upper-air effects of fossil fuel combustion were worth worrying about today, and that environmental and safety regulations were not adversely affecting the American economy.

Reformists favored a resource-conserving ethic and were troubled by the profound implications of today's energy decisions for future generations. They did not regard the American standard of living as dependent upon substantially more energy in the future, they were against primary reliance on nuclear power and were for greater emphasis on renewable sources of energy such as solar and biomass.

Statements scored low by both core clusters suggested, in one possible interpretation, lack of concern for the average citizen. Traditionalists did not give high scores to statements dealing with the longer term consequences of energy decisions, the problem of the public's inability to grasp the overall significance of the energy problem, the desirability of ensuring broad representation of opinion in energy study groups, the impact of increased energy prices on the poor and those with fixed incomes, the depersonalization of energy consumers by energy experts (treating consumers like statistics), and the human risks associated with various energy options. Reformists did not give high scores to statements concerned with whether energy solutions would require sacrifices throughout the country for all classes of people, the short-term stresses the public would feel as a result of changed energy policies, the human risks to coal miners in switching back to coal as a source of energy, the personal distress created by gas lines and cold homes, and the difficulty of communicating the truth about the energy crisis to the public. The general orientation of both core viewpoints was distinctly more national than individualistic.

Peripheral A clusters had much in common with the core traditionalists but there were shades of difference in views and cohesion was looser. Members of the A.4 cluster, for example, disagreed more intensely than did the core traditionalists that large-scale energy systems fail catastrophically. Members of A.5 were not as unreservedly enthusiastic about the development and expansion of nuclear energy. They favored coal. Members of A.6 displayed especially vigorous market and free-enterprise orientations. But there was an underlying similarity of viewpoint among A clusters on a range of economic, institutional, and technical matters. The seven statements that received relatively high scores from all A clusters are starred in Table A-3.

As for the B group, members of peripheral cluster B.2 were less enthusiastic about the prospects for solar energy than were core reformists (B.1), and they were not as ready to disagree that the chance of nuclear accidents at power plants had been exaggerated. B.2 seemed to be a somewhat pessimistic or cynical variant of B.1, yet, as with the A group, differences were outweighed by similarities. The eight statements to which all members of the B group gave relatively high scores are starred in Table A-4.

Searching for Conflicts and Reconciliation

We stated earlier that the views of traditionalists and reformists were not totally in opposition. Yet there was certainly discord between them. We investigated if there might be opportunities for accommodation.

To pinpoint areas of greatest disagreement between traditionalist and reformist viewpoints, we singled out those statements whose core A and B scores were furthest apart. Table A-5 lists such statements. Half deal explicitly with nuclear policy issues (1st, 2nd, 7th, 9th, and 10th), and three more—having to do with renewables and the future consumption of energy (5th, 6th, and 8th)—bear directly on the nuclear question. Nuclear was clearly the premier subject of controversy. The remaining two contentious statements (3rd and 4th) concern the oil companies and price deregulation.

We also looked for statements (having significant scores) with which traditionalists and reformists agreed. There was only one, and it was on invoking military force in the event of another oil embargo (scored −4.0 by traditionalists and −4.2 by reformists). A lower rated statement for which there was agreement (scored 1.4 and 1.3, respectively) criticized energy policy analysts for failing to take adequate account of political infeasibilities and problems of implementation.

The picture changed appreciably when we shifted the comparison from the A and B cores to the two most highly correlated A and B peripherals, A.5 and B.2. Completely missing from the major disagreements between these two peripheral clusters was the divisive nuclear conflict. The dispute centered now primarily on the price deregulation issue. Agreements were more significant than before, as we would expect from the correlation of .79, yet did not yield much specific policy insight or practical guidance. There was no apparent substantive issue on which the two groups could collaborate, only an ethical stand (energy experts should be forthright), an equivocation (the government should support risky research and development, but that is not likely to benefit

Table A–5.   Disagreements between A and B Core Groups[a]

1. The likelihood of accidents involving radioactive materials at U.S. nuclear power plants has been greatly exaggerated.
$|+2.5 - -3.7| = 6.2$

2. Nuclear power cannot be expanded without infringing on the civil liberties of many people.
$|-3.7 - +1.6| = 5.3$

3. Major oil companies should be prevented by law from acquiring control over coal, uranium, and the equipment needed to use solar energy.
$|-3.4 - +1.7| = 5.1$

4. Oil and gas prices should be removed from U.S. Government regulation.
$|+3.7 - -1.4| = 5.1$

5. Energy research and development should emphasize renewable sources such as solar and biomass.
$|-1.3 - +3.7| = 5.0$

6. Improvements in the American standard of living depend on production and use of substantially more energy in future years.
$|+1.1 - -3.8| = 4.9$

7. Scare talk about the problem of nuclear waste disposal overlooks several technical options for solving the problem.
$|+2.5 - -2.4| = 4.9$

8. We should nurture a resource-conserving ethic in which human satisfaction is much less dependent on material consumption.
$|-1.5 - +3.2| = 4.7$

9. Meeting projected demand for electricity requires placing primary reliance on the construction of nuclear generating plants.
$|+0.2 - -4.2| = 4.4$

10. If we go ahead with the breeder, there is no hope of checking the proliferation of nuclear weapons.
$|-3.3 - +1.1| = 4.4$

---

[a] Following each statement is the absolute difference between its average A and B core scores:

$$\text{absolute difference} = |\text{average A.c score} - \text{average B.1 score}|.$$

the citizenry), and two generalities (use no military force, even if provoked; people don't trust the oil and gas companies).

Nor were the agreements between A.4 (the peripheral most like the core traditionalists) and B.2 (whose correlation with A.4 was .62) any more encouraging. There was no obvious ground for productive interaction or consensus building. The most that can be said is that these two clusters were not notably hostile to each other's viewpoint.

Consensus is not the same as absence of conflict, nor is it achieved when a dominant group is successful in squelching opposition. True consensus occurs in response to a redefinition of problems and a clarification of interests, values, and options. The questionnaire results by themselves did not point the way to how such consensus might be achieved.

How about the apparent chasm between the elites and the public? Since the B.2 cluster reflected concerns that mattered to the citizenry, and still correlated markedly with both the A.4 and A.5 groups, we looked to see if the viewpoints of these three peripheral clusters might intermediate between the attitudes of the general public and those of the elite. Such judgments require additional information about the individual elite.

Specimens and Other Selected Individuals

Energy problems are human problems. They assume a life only when human factors are taken into account. We now examine selected members of the elite sample (unnamed) who most nearly corresponded to the viewpoints identified. We also consider the following additional classes of individuals:

- "Hostiles," whose strong attachment to one set of viewpoints and antipathy to another made them seem especially likely to be combative in defending their positions;
- "Compromisers," who by associating themselves with both traditionalist and reformist views suggested an ability to empathize with different conceptions of the energy problem;
- Responsible officials and representatives from industry, whose primary occupation was direct involvement in energy decisionmaking;

- "Spectators," who demonstrated no particular affinity for any of the identified viewpoints; and
- Without veil of anonymity, the five principal collaborators in this study.

Table A–6.   Simplified Characterization of Distinctive Viewpoints.

A.c Nuclear power can be expanded without undue harm. Solar will not solve the energy crisis, at least soon. Deregulation of oil and natural gas is essential. The federal government has been ineffective. It would benefit from the infusion of energy industry executives. American vulnerability to cut-offs of energy supplies must be reduced, but military force ought not be used in reaction to a cut-off.

B.1 Protection of the environment including conservation of natural resources is required and need not diminish economic productivity or prosperity. Nuclear power and the nuclear industry are not to be relied on to meet future needs. Vulnerability must be reduced, but not by using military force. Renewable sources of energy are desirable.

A.5 Oil and gas supplies are limited and their use must be moderated by raising their price and increasing the use of coal. Nuclear power cannot be relied on primarily. Institutions of government have contributed to the energy problem, as have divisive interest groups. Citizens do not trust the energy industry or the government. Industry performance would improve with less government controls.

A.6 The marketplace, if allowed to function properly, will go a long way toward solving the energy problem. There is scant need to enforce conservation. One need only increase prices sufficiently to attain efficient allocations.

B.2 Conservation and the environment matter, but the failure of energy-related institutions (including the government) matters at least as much. Energy problems are real and substantial. They defy easy answers. The public is right in feeling that it has been manipulated and ill-served.

A *specimen* for a cluster, or the viewpoint it represents, was the individual whose statement ratings correlated highest with the averages of that cluster while exhibiting low correlations with the averages of clusters holding dissimilar viewpoints. A completely pure specimen would be an individual perfectly correlated with one cluster and uncorrelated with all others. There were none, though some individuals approximated that ideal. Table A-6 summarizes the viewpoints of each of the five clusters for which specimens with distinctive positions were identified.

Traditionalists.   The specimen for the traditionalists was associated with several of the energy studies treated in this book. Based on these studies and despite initial skepticism about economic analysis, this individual was convinced that price played a major role and that controls had induced excessive consumption. This person also:

- Felt that to the extent government had interfered with prices, it had performed a disservice. "Equity considerations should be handled by other policies than the setting of energy prices."
- Reconsidered—based on involvement in the studies—what had once been ardent support for nuclear power, but still held nuclear to be a valuable and viable energy source. "Further slippage in nuclear power construction will result in increases in oil imports."
- Did not regard proliferation as a problem and believed the breeder option needed to be maintained, especially if demand did not abate in response to increased prices.
- Thought the significance of Three Mile Island had more to do with political resistance than reactor safety.
- Shifted focus from nuclear to growth and opposed those who advocated a policy of no-growth. "Large but gradual increases in future energy prices will result in greatly reduced energy/GNP ratios." Did not believe energy and the economy were in a one-to-one relationship.
- Favored development of synthetic fuels and coal, but doubted that solar and other "radical" technologies would be of much help in the foreseeable future.

- Had a decidedly incremental perspective. "Future world stability does not depend on drastic changes in lifestyle in the industrial countries, nor on reallocation of available energy supplies to developing countries."
- Enthusiastically endorsed the educational value of energy studies in preparing participants for government service.

The specimen for the traditionalists, a paragon of rationality, was the subject of numerous attacks by reformists we interviewed who did not appreciate this individual's depth of understanding and efforts to achieve a working consensus.

Reformists.   Still more vigorous was the attack on the specimen for the reformists, whose intense vision of the future was interpreted as requiring vast alterations in society. The viewpoint of the specimen for the reformists was founded on several general principles:

- Energy use is the appropriate starting point in dealing with the energy problem. "We already have more electricity than we can use to economic advantage."
- Incremental changes in the status quo mainly reinforce the supply-driven system which has brought society to the brink of the energy abyss. "Business as usual does not work." Only a wholesale shift in direction will save society from disaster.
- Reduced vulnerability must come from society's gaining greater ecological resiliency and adaptability in the evolutionary sense.
- Nuclear power epitomizes the technocratic sources of the energy crisis. "If nuclear power stations are turkeys at the margin, you need not argue whether they are safe turkeys. Nuclear power entails the proliferation of nuclear weapons."
- Renewables are not as expensive as nuclear. They offer more options (and therefore greater social resilience), and allow for less centralized economic and political control. "Centralized energy systems tend to increase inequity, technocracy, and autarchy."

Appearances notwithstanding, the specimen for the reformists was more technologically oriented in some respects than many core traditionalists.

A.5 Specimen.   The A.5 specimen found fault with both economists and engineers. "Economists have made a major contribution toward causing an energy problem by using simple models that overestimated supply. Engineers have made a major contribution by making cost estimates of new supplies that were much too low." The A.5 specimen valued environmental laws highly, could imagine no reason for reducing what was presently being done, and had reservations about both nuclear and solar energy. Nuclear was risky and solar uncertain. Conservation was important to reduce drawdowns of supply. It warranted widespread attention and support.

A.6 Specimen.   The specimen for the A.6 cluster was an unabashed proponent of the unfettered marketplace. In this individual's view, oil and gas prices were artificially low because of misguided regulation that obstructed exploration for economically sound energy sources. With the market hobbled by government interference, society was unlikely to secure the much-increased amounts of energy it would need in the future, presenting a grave danger to the United States and other oil- and gas-consuming nations. The environmentalists had it wrong. Conservation and sacrifices were not the answer. Just let the market work. Continual bickering about the energy crisis was pointless. "The large-scale economic models have had the greatest impact on society and decisionmaking." They have lots of problems, are limited and misleading, but are still widely used.

B.2 Specimen.   The B.2 specimen was a well-informed, well-respected government official with lengthy exposure to energy matters. "The United States cannot go on as it has forever, becoming more and more dependent on imported fuels." Decontrol was one means to stem the tide, but the supporting analyses were difficult to understand and unnecessarily complicated. Many solutions needed to be explored: renewables, increased production of domestic oil and gas, and other promising alternatives. No single one could carry the full load. Prolonging the energy crisis hurt, and the public had only limited tolerance. It needed outlets for its anxieties and some way to help in resolving the problem. Clear and constructive guidelines were not yet being offered. Society must cut back on its profligate consumption of energy. Not doing so boded ill for future generations and world stability. Some heavy political costs would have to be paid to change things; the sooner the better.

Hostiles.   About 15 percent of elites holding an A viewpoint displayed negative correlations with the B clusters, while a similar percentage of those holding a B viewpoint showed negative correlations with the A clusters. Of these "hostiles," the most hostile was, for A, a prominent energy industry official with a deep commitment and stake in nuclear power; for B, it was a young staff member who worked for a congressman generally regarded as liberal in energy matters. The industry official was noted for taking public positions against any and all nuclear opponents while the young aide, who was not well known outside of the congressional staff setting, felt that anyone who attempted to argue for a technological standpoint was merely confirming the status quo and disregarding questions of social equity. So far as we know, these two individuals never confronted one another directly. Were they to have done so, the fireworks would have been spectacular.

Compromisers.   Just under 9 percent of the elite demonstrated notable tolerance for both A and B perspectives by having correlations with A.5 and B.2 peripheral clusters of at least +.60. (No other A and B clusters produced the same degree of dual alignment.) These thirteen respondents, apparently conciliatory in bent, were heterogeneous in background, responsibility, and other identifiable characteristics.

Responsible Officials and Representatives of Industry.   More than half of the government officials in the sample held some form of B viewpoint, reflecting the position of the Carter Administration. Among elected officials, the B.2 position was most common. The largest number of civil servants fell in the A.4 cluster, and all of them exhibited high correlations with that cluster, suggesting a certain faith in technical and economic formulations of the energy problem. Civil servants were also scattered among most other clusters, indicating a diversity of viewpoint and relative incohesiveness reminiscent of the A.4 cluster itself. All but one member of industry held the A viewpoint; the majority were core traditionalists.

Spectators. "Spectators" were individuals who failed to correlate highly with any of the distinctive viewpoints. One was a modeling enthusiast. Despite the efforts of the interviewer to broaden the discussion, this person insisted on focusing on the technical and practical details of analysis, avoiding matters of policy.

Collaborators in the Study. Defying the laws of probability, the five initial collaborators in this study fell in five separate Q-sort clusters: Manne in A.2, Russell in A.3, Greenberger in A.5, Hogan in A.6, and Brewer in B.0. This spread, put in terms of correlations with the A and B core clusters, is plotted in Figure A-2, a characterization that we find reasonable. (It reassures us on the meaningfulness of the distinctions drawn by the Q-sort analysis.) The wide range of viewpoints we held was freely expressed in our meetings and subjected our project to some of the same travails we were observing in the studies. Gaining an appreciation for our differences deepened the meaningfulness of the research for us all. We suspect the final product is the better for it.

Characteristics of the Sample

In administering the attitude questionnaire and in interviewing respondents, we collected information on their background, organizational affiliation, age, sex, and other distinguishing characteristics. These attributes shed light on viewpoints and the composition of clusters.

Most members were in their 30s and 40s (3/20s, 50/30s, 54/40s, 27/50s, and 16/60s). They were overwhelmingly male: there were only 8 women among the total of 150. With respect to organizational affiliation, 41 were associated with universities, 30 were civil servants, another 30 worked for research organizations, 22 were elected or appointed government officials, 11 were connected with industry and financial institutions, and the remainder were spread among foundations, consulting companies, law firms, and public interest groups. Economics, the most prevalent specialization, had 40 representatives, followed by engineering and the sciences (excluding nuclear) with 29, operations research and

Figure A–2.   Characterization of the Responses of the Five Collaborators.

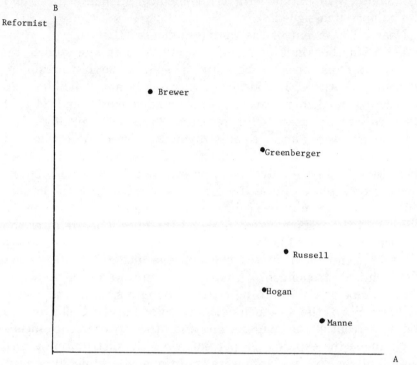

mathematical sciences with 19, management science with 17, nuclear engineering and physics with 14, social sciences (excluding economics) with 13, law with 11, biomedical sciences with 3, and the balance unassigned.

Nearly 80 percent of the sample indicated that they were familiar with the kinds of models encountered in the studies reviewed in this book, suggesting a general sophistication in modeling. Of these, 35 considered themselves model builders (26 of them worked in research, 9 were policymakers), and 84 said they were model users (49 in research, 35 in policymaking). The respondents were almost equally divided between those who had and had not worked in Washington, D.C. (63 had for more than five years, 16 had for five years or less, and 71 had not). Interviewing took place during the period November 1978 to October 1979, much of it occurring during the summer of 1979. Over 70 percent of the respondents completed the questionnaire after the accident at Three Mile Island.

As to membership in the clusters, age made little difference. Nontechnical respondents (many of whom were congressional staff members) tended to be on the younger side, as did model builders. Six of the 8 women belonged to one of the B clusters, as did 40 percent of the civil servants and other government officials (well above the expected 26 percent, were the distribution even). Only 3 of the civil servants were traditionalists, while 30 percent each were located in the A.4 peripheral and B groupings. Over half of the elected and appointed officials (about twice the expected number) were in one of the B clusters. B.2 was composed mainly of government officials (60 percent), fully half of whom were not civil servants. The university component was spread almost evenly throughout the clusters.

The A.1 cluster was composed largely of persons working in research organizations, while A.2, A.3, and A.5 consisted of people from both research organizations and universities. Civil servants and university people dominated A.4. Economists were overwhelmingly located in A clusters (95 percent), whereas social scientists who were not economists were almost as prominently collected in B clusters (77 percent). Of the 43 engineers and scientists, 37 were in an A cluster (86 percent), and only 1 with nuclear associations was in a B cluster (B.2). Operations researchers were also A-oriented (84 percent); none were reformists. Lawyers, many of them elected and appointed officials, were primarily B types.

Within the A core, A.2 seemed to be the special province of economists and management scientists, who together accounted for three-quarters of the membership. Including non-nuclear engineers and scientists raised this to 90 percent. A.3 was decidedly engineering and science-based, with more than half of its members associated with one of these disciplines; two other members (labeled operations researchers) had strong engineering slants. Over half of the peripheral A.4 cluster had worked in positions in Washington D.C. for more than five years. The market-oriented A.6 peripheral cluster was dominated by economists and operations researchers, all but one of whom were modeling specialists. Reformists were very different in disciplinary make-up, leaning toward the humanities and (noneconomic) social sciences. The B.2 peripheral cluster drew from all disciplines except economics. Nearly two-thirds of reformists did not have modeling experience, whereas 80 percent of traditionalists did. Indeed, more than half of those not possessing technical modeling skills were in one of the B groups. While all A groups had heavy concentrations of modeling-literate members, the percentages in the A.2, A.4, and A.6 clusters were especially high, exceeding 90 percent.

Value Analysis

One significant feature of the Q-sort technique is that respondents can attach whatever meaning they wish to opinion statements. The statement, "When large-scale energy systems fail, they fail catastrophically," will mean something different to a New Yorker who has experienced a blackout firsthand than to a farmer in Kansas who benefits from rural electrification but also has a back-up power supply. Likewise, a transmission engineer well acquainted with the intricacies of large-scale energy systems will interpret the statement differently from the average citizen who does not have intimate knowledge of the subject. Two individuals could belong to the same cluster, but still have quite different perceptions and interpretations of the statements that distinguish their alliance.

It is nevertheless possible to assign approximate meanings to statements that will apply across many people. This can be done according to values expressed in the statements.[29] Certain statements reflect concerns about power or wealth. (Don't use military force, even if provoked; U.S. oil and gas prices are too low.) Others reflect issues of information and skill. (Energy experts should reveal policy failures as well as successes; the federal government should inform the public better about its energy policies.) And still others emphasize personal perspectives of respect, well-being, and deference. (Energy decisions should respect future generations; people have confidence in their institutions.)

In certain statements, values interact. Consider the statement: "Were industry to switch to coal, despite government regulation, more fatalities and air pollution would result." Some respondents might key in on economic or power aspects from the statement's reference to government regulation. Others are likely to view the statement more in terms of safety and health. The assignment of primary value is then a matter of analytic judgment.

We made such judgments in arranging the sixty statements according to three possible value emphases: *manipulative*, stressing power and wealth; *contemplative*, concentrating on skill and enlightenment; and *deference* or respect, including personal safety and well-being. An individual who showed relatively strong agreement or disagreement with statements emphasizing power and wealth would be coded in the manipulative category, and so on. Of the sixty statements, twenty were classified as manipulative, thirteen as contemplative, and twenty-seven as deferential. They are shown, in that order, in Appendix B.

Fully 60 percent of the elite, including 80 percent of traditionalists and 70 percent of B.2 peripherals, gave high scores to statements about wealth and power. Contemplative concern was present in all groups save the reformists (B.1) and was especially prominent in A.4. Reformists cited deference values about twice as often as expected by chance. Representatives from industry primarily emphasized manipulative concerns while economists emphasized both manipulative and contemplative aspects. Nonmodelers tended to stress deference values more than modelers whereas modelers were more likely to stress manipulative aspects. Modelers also dominated nonmodelers in the information and skill category.

CONTRASTING THE MASS AND ELITE

Attitudes of experts about what was at the heart of the energy problem generally diverged from those of average citizens. The elite, though certainly not always agreeing with each other on central questions, showed more kinship among themselves (in their attention to pricing policies, technologies, and technical details) than they did with the public at large.

Mistrust of institutions thought to be responsible for the energy crisis was prominent in the public's conception of the energy problem. It was nearly absent from the attitudes of the elite.[30] Popular mistrust of the energy industry is long-standing.

> Ask almost any American today what causes the nation's difficulties, and high on his list will be the oil companies, who, he believes, charge too much for oil and pay too little in taxes. Had the same question been asked seventy years ago, the answer would have been remarkably similar—with this exception: He would have said The Oil Company—Standard Oil. The dramatic nature of this identification tended to obscure more complex issues and conflicts and solidify overly simplified polarities. The repetitive cycle of accusation and mistrust began early on and has formed the narrow arena of debate in which, except for two world wars, policy has been confined ever since.[31]

The public opinion surveys indicated that such suspicions continue to present times, added to by flagging confidence in government. Energy problems, in particular, "have absorbed a disproportionate amount of attention, and have fostered suspicion and mistrust," with the result that "governing institutions based on majoritarian principles have bogged down in an issue that has no clear-cut choices."[32] The public found its leadership unable to supply guidance on positive, constructive, feasible things to do to maintain comforts and make the country less vulnerable to foreign suppliers of oil. Focusing on technical aspects of the problem, the leadership tended to ignore citizen desire for relief from insecurities.

The public viewed experts with a jaundiced eye partly because citizen interests and needs were not central features of energy analysis. The public interpreted bickering among experts as an indication that no one really understood. Citizens did not know whom to believe. Their right to share in decisions affecting their lives and futures had little meaning in the absence of clear, coherent, consistent portrayals of the energy situation. If popular mistrust and anger had a taproot, it was embedded in the soil of denied participation and was fertilized by misinformation.

Energy decisions are complex, but that must not be taken as a reason for keeping the citizenry ill informed. It is rather a challenge for those with special knowledge to make what they know (and do not know) comprehensible to as large a segment of the population as possible. As far as the average citizen was concerned, momentous decisions having real and personal consequences were being made by a small, unknown group of individuals with suspicious motives and little concern for citizen participation. This was no basis for improving credibility. Analysts need to recognize how their work has tended to omit citizen concerns.

Citizens were far less interested during the seventies in the technical details of energy than in the discomforts, injuries, and feelings of helplessness that rising energy prices and deprivations imposed. They felt manipulated and exploited. Most of the energy debate, including the analyses that fed it, concentrated heavily on national dimensions of the energy problem. But local and regional perceptions were what mattered to individuals. One reason for federal ineptness in energy programs was the large variation around the nation in local needs and resources, much of which could not be accounted for in national policies.

By tending to be national in scope, technical, and politicized, analysis can fail to reflect popular needs and potentialities. Rather than clearing the way for compromise or consensus, it may then just contribute to disarticulation and confusion.[33] Society functions best "when citizens voluntarily work together toward a commonly accepted goal."[34] Analysis can help if it is credible to nonspecialists and mindful of what is important to people.

APPENDIX B
THE SIXTY STATEMENTS OF THE
ELITE QUESTIONNAIRE

MANIPULATIVE: Power and Wealth

* If another oil embargo is imposed on us, the government should invoke military force to deal with it.

Increasing dependence of the United States on imported oil since 1973 has put American security in serious jeopardy.

If we go ahead with the breeder, there is no hope of checking the proliferation of nuclear weapons.

* We should reduce the vulnerability of our economy to cut-offs of oil supplies from abroad.

* People resist sacrifices like those proposed in the National Energy Plan whenever they feel they have been manipulated or exploited.

When large-scale energy systems fail, they fail catastrophically.

Government price controls leave producers little incentive to increase domestic oil and gas production.

* The prices of oil and natural gas produced in the United States are too low.

* Legal maneuvers of environmental groups have been the principal cause of recent increases in the cost of nuclear power.

If the price of American crude oil rises to the world price, our dependence on foreign oil will decrease significantly.

---

* designates statements used in both elite and citizen questionnaires.[1]

341

* Unless our growing demand for oil is moderated, there will be explosive price increases and shortages of oil in the 1980's.

Meeting projected demand for electricity requires placing primary reliance on the construction of coal-fired generating plants.

There are abundant supplies of crude oil and natural gas under American soil and offshore.

Energy prices are not nearly as important as attitudes and institutions in affecting residential consumption of energy.

* Oil and natural gas prices should be removed from U.S. Government regulation.

* Major oil companies should be prevented by law from acquiring control over coal, uranium, and the equipment needed to use solar energy.

The U.S. economy depends on the economies of Japan and our European allies, who are more vulnerable to embargoes than we are.

We should do everything possible to reduce the consumption of oil and natural gas in the United States.

* Meeting projected demand for electricity requires placing primary reliance on the construction of nuclear generating plants.

* It would be shortsighted and unfair to future generations to accelerate the domestic production of oil and natural gas in order to curtail imports.

CONTEMPLATIVE: Enlightenment and Skill

* Energy experts have an obligation to communicate clearly and candidly the failures of energy policy as well as the successes.

Energy policy analysts have failed to take account of political infeasibilities and problems of implementation.

You can find a geologist, engineer, lawyer, or economist to back up any organized special interest with a stake in energy policy.

* The federal government should put more effort into selling its energy policy proposals to the public.

Members of an energy study group should be chosen to represent a wide spectrum of opinion.

* The truth about the energy crisis is complex and difficult to communicate to the public.

Government support of energy research and development is likely to produce few benefits for the average consumer.

Our general inability to grasp the energy "big picture" stems from the fact that each of us is exposed to only a small part of it.

* Energy research and development should emphasize renewable sources such as solar and biomass.

No national energy policy can be rational and comprehensive as long as decisions are made one at a time in response to many interest groups.
* With the commercialization of solar energy over the next decade, the end of the energy crisis will be in sight.

The public will experience the burdens of any effective energy policies long before it perceives any benefits.
* The federal government should pay the development costs of promising technologies that are too risky for private investment.

DEFERENCE: Affection, Rectitude, Respect (including Personal Well-being)

Energy experts tend to depersonalize energy consumers and treat them like statistics.
* People should trust the federal government to find a solution to the energy crisis.
* The idea is widely held that the "energy crisis" is another way of ripping off the public.
* The people have confidence in the oil and gas industry based on its contributions to economic growth and prosperity.

Energy industry executives should be prevented from serving as public officials in the Department of Energy.

Nuclear power cannot be expanded without infringing on the civil liberties of many people.

Solutions to our energy problems must ask equal sacrifices from every region, every class of people, and every interest group.

The expansion of U.S. oil imports has had an adverse impact on other oil-importing nations.
* An increase in gas and electric bills means less heat and light for the poor and those on fixed incomes.
* When natural gas prices are no longer regulated by the federal government, the market will allocate supplies of natural gas efficiently.

Energy-economic analyses are typically biased toward the traditional goals of growth and efficiency.
* We should nurture a resource-conserving ethic in which human satisfaction is much less dependent on material consumption.

Decisions made about energy today will have profound and irreversible consequences for future generations.
* Improvements in the American standard of living depend on production and use of substantially more energy in future years.

Only after waiting in line at the gas station or living in cold homes will people accept the need to adapt their life styles.

Americans are unwilling to change their energy-consuming habits, particularly when it comes to the private automobile.

There are very few incentives for the "average man" to reduce his consumption of energy.

Oil and gas should be rationed by the government during periods of shortages.

* The particular energy needs of each community should be met through local resources wherever possible.

* The next time we have oil and gas shortages or an electricity black-out, the public will go looking for a scapegoat.

* We should accept additional risks to public health and safety in our choice of energy policies.

The likelihood of accidents involving radioactive materials at U.S. nuclear power plants has been greatly exaggerated.

If industry switches from oil and gas to coal, no amount of government regulation will prevent more coal-mine fatalities and more air pollution.

* Environmental protection laws and regulations should be vigorously enforced.

* Scare talk about the problem of nuclear waste disposal overlooks several technical options for solving the problem.

The effect on the upper atmosphere of fossil-fuel combustion is too long range a hazard to warrant serious concern today.

Environmental and safety regulations are impacting adversely on American productivity and free enterprise.

NOTES

PREFACE

1. Craufurd D. Goodwin, ed., *Energy Policy in Perspective* (Washington, D.C.: Brookings Institution, 1981), p. 678.
2. Martin Greenberger, "Energy Demand Forecasts" (Opening testimony before the hearings of the Subcommittee on Investigations and Oversight, Committee on Science and Technology, U.S. House of Representatives, June 1, 1981).

CHAPTER 1

1. Another possible paradigm is artificial intelligence. Artificial phenomena, as defined by Herbert Simon, are goal seeking and environment conditioned. Necessities emerge from contingencies by virtue of the limits of the rationality of decisionmakers who cannot be expected to adapt perfectly to situations that are constantly changing. But it is hard to see "how empirical propositions can be made at all about systems that, given different circumstances, might be quite other than they are." Herbert A. Simon, *The Sciences of the Artificial* (Cambridge, Mass.: M.I.T. Press, 1969), p. x.
2. For a discussion of self-referencing systems, see Douglas R. Hofstadter, *Godel, Escher, Bach: An Eternal Golden Braid* (New York: Vintage, 1980).
3. John Weisman, "A New Study of Network Coverage Reveals Disturbing Omissions," *TV Guide* (March 6, 1982): 12. The study reported on is *TV Coverage of the Oil Crises: How Well Was the Public Served?* (Washington, D.C.: Media Institute, 1982). The Media Institute study points out that television correspondents covering the energy problem relied primarily on

government sources for their information. In doing so, they neglected the part played by government price controls in creating energy shortages.

4. See, for example, J. Reston, "Who's to Blame?" *New York Times*, May 13, 1979, p. E21; response by W.D. Burnbaum, *New York Times*, May 23, 1979, p. A26.

5. Economist William Nordhaus cites an econometric study by the Department of Energy to conclude that the net result of government controls, subsidies, and tax provisions affecting energy over the decade was insignificant compared to the effect induced by the price rise. William D. Nordhaus, "Energy Policy: Mostly Sound and Fury," *New York Times*, November 30, 1980.

6. By 1982, Hyman Rickover, father of the nuclear submarine and developer of the first commercially operated light-water reactor, was calling for abandonment of reactor technology because of the radiation hazard. Nuclear sage Alvin M. Weinberg, former director of the Oak Ridge National Laboratory, was acknowledging a de facto nuclear moratorium in the United States and was working on a study designed to develop specifications for doing it right the next time. See Alvin M. Weinberg, *The Second Nuclear Era*, Study funded by the Mellon Foundation (Oak Ridge, Tenn.: Institute for Energy Analysis).

7. Stuart E. Eizenstat, memorandum to Jimmy Carter, *Washington Post*, June 28, 1979.

8. "Mideast Oil Lands Seek Price Stability," *New York Times*, September 24, 1960.

9. The history of the U.S. oil industry is covered in: Alfred D. Chandler, Jr., *The Visible Hand* (Cambridge, Mass: Harvard University Press, 1977); Raymond Vernon, ed., *The Oil Crisis* (New York: Norton, 1976); Anthony Samson, *The Seven Sisters: The Great Oil Companies and the Wealth They Made* (New York: Viking, 1975).

10. For a history of OPEC, see Zuhayr Mikdashi, *The Community of Oil Exporting Countries* (Ithaca, N.Y.: Cornell University Press, 1972).

11. James E. Akins, "The Oil Crisis: This Time the Wolf Is Here," *Foreign Affairs* (April 1973): 462-90. *Foreign Affairs* editor William P. Bundy regards Akins's piece as his "baptism of fire on the energy problem." Bundy says the first draft of the article was "heavily couched in terms of a debate with Professor Adelman of M.I.T." Bundy got Akins "to minimize this debate" and "present his own views in a solid affirmative fashion." Letter from Bundy to Greenberger, September 9, 1981.

12. Arabs prefer the name "Arabian Gulf." By tacit agreement, OPEC members settle for calling this strategic body of water simply "The Gulf."

✳ 13. An interesting set of articles covering various aspects of the oil crisis appeared in *Daedalus*, Journal of the American Academy of Arts and Sciences, Fall 1975.

14. U. S. Senate, *Hearings before the Subcommittee on Multinational Corporations of the Committee on Foreign Relations, Multinational Petroleum Companies*, 93 Cong., 2 sess., 1974 (Washington, D.C.: Government Printing Office, 1975).

15. M.A. Adelman, *The World Petroleum Market* (Baltimore, Md.: Johns Hopkins Press for Resources for the Future, 1972).

16. Personal meeting with Robert Mabro at Oxford, June 19, 1980.

17. Adelman later put forth a plan to use U.S. buying power to counter OPEC's bargaining strength. In a 1976 proposal to President-elect Cart-

er's economic task force and to James Schlesinger, then Carter's energy chief, Adelman suggested that centralizing the procurement of foreign oil would enable the United States to play one country off against another and undermine the cartel. Schlesinger, expecting a tight oil market internationally, did not think the plan would work. Carter resorted instead to personal persuasion and an agreement requested from Saudi Arabia early in his administration to hold oil prices down and keep production up.

18. Personal interview with a former State Department official. Kissinger's recollection is that he did include Akins in the meetings.
19. Interview with Akins, May 1, 1980.
20. Conversation with Ezra Solomon, June 5, 1980.
21. Letter from Joseph W. Twinam to Greenberger, June 27, 1980.
22. Bernard Weinraub, "Shah of Iran Is Seen as a Spur Behind Sharp Advance," *New York Times*, December 25, 1973, p. 31.
23. *London Economist*, August 23, 1975; *Washington Post*, March 14, 1976; *Forbes*, April 15, 1976; as cited in a letter from M. A. Adelman to Mike Wallace of CBS News, May 5, 1980.
24. Interview with Akins.
25. Letter from Yamani to Simon, September 3, 1975.
26. The segment of "Sixty Minutes" dealing with the "Kissinger-Shah Connection" was aired by CBS on May 4, 1980.
27. Phone call from Greenberger to Dan Rather sometime after airing of the program.
28. Henry Kissinger, *White House Years* (Boston: Little, Brown, 1979), pp. 1263-64.
29. U.S. Senate, *Hearings before the Subcommittee on Multinational Corporations of the Committee on Foreign Relations, Multinational Petroleum Companies*.
30. Kissinger, *White House Years*, p. 1263.
31. Thomas J. Bray, Editorial, *Wall Street Journal*, June 6, 1980.
32. Letter from Shultz to William Leonard of CBS News, May, 1980.
33. Letter from M. A. Adelman to Mike Wallace of CBS News, May 5, 1980.
34. U.S. Senate, *Hearings before the Subcommittee on Multinational Corporations of the Committee on Foreign Relations, Multinational Petroleum Companies*.
35. Letter from John Mitchell of British Petroleum to Greenberger, July 8, 1980.
36. Martin Greenberger, "Perspectives on World Oil: Modeling in a Context," in R. Amit and M. Avriel, eds., *Perspectives on Resource Policy Modeling: Energy and Minerals* (Cambridge, Mass.: Ballinger Publishing Company, 1982).
37. William M. Brown and Herman Kahn, "Why OPEC Is Vulnerable," *Fortune*, July 14, 1980, pp. 66-69.
38. Several respondents were conducting modeling studies of world oil. These studies show projected "reference-case" increases of from zero to 80 percent in real oil prices over the next ten years, a range consistent with the views expressed in answers to the questionnaire.
39. Joseph Kraft, "Letter from OPEC," *New Yorker*, January 28, 1980, p. 77.
40. An account of the changing structure of the oil industry was given by J. E. Hartshorn, "From Multi-National to National Oil: The Structural Change," *Middle East Economic Survey* 23, no. 28 (April 1980): 1-10.

41. Oil consultant Walter Levy, "widely quoted in the Middle East," estimated that capital projects built by oil-producing countries cost two or three times as much as competing facilities. "The Changes in OPEC that Are Driving Oil Prices Wild," *Business Week*, October 29, 1979, p. 80.
42. Meeting with Mabro.
43. Suliman S. Olayan, "In Defense of OPEC," *Fortune*, August 13, 1979, pp. 217-220.

CHAPTER 2

1. President's Materials Policy Commission (cited hereafter as Paley), *Resources for Freedom* I-IV (Washington, D.C.: Government Printing Office, June 1952).
2. Energy Resources Committee, *Energy Resources and National Policy*, Report of the Energy Resources Committee to the National Resources Committee (Washington, D.C.: Government Printing Office, 1939), p. 1.
3. Marion Clawson, *New Deal Planning: The Natural Resources Planning Board* (Baltimore, Md.: Johns Hopkins University Press for RFF, 1981), pp. 121-23.
4. Craufurd D. Goodwin, ed., *Energy Policy in Perspective: Today's Problems, Yesterday's Solutions* (Washington, D.C.: Brookings Institution, 1981).
5. "Resources for the Future," *Annual Report for the Year Ending September 30, 1955*, p. 19.
6. Harold Issadore Sharlin, "Energy in Flowing Water and the Public Interest: Public and Private Power at Niagara Falls," December 1977. (Unpublished.)
7. Robert Engler, *The Politics of Oil* (Chicago: University of Chicago Press, Phoenix ed., 1967) contains an exhaustive examination of "the political behavior of the petroleum industry and its impact upon political processes throughout the United States as well as upon foreign policy and public opinion." Goodwin, *Energy Policies in Perspective*, to which we were given access in draft form, was a valuable source of facts and insights on energy policy in the period from the end of World War II to the 1970s. Also useful was the treatment of the history of energy policy and especially oil policy in Douglas R. Bohi and Milton Russell, *Limiting Oil Imports: An Economic History and Analysis* (Baltimore, Md.: Johns Hopkins University Press for RFF, 1978).
8. Robert Engler, ed., *America's Energy: Reports from The Nation* (New York: Pantheon Books, 1980).
9. One 1978 study examining the public's view on energy-related issues reported that from two-thirds to four-fifths of Americans do not have a great deal of confidence in large oil companies. (Jeffrey Milstein, "Soft and Hard Energy Paths: What People Think" [Washington, D.C.: Office of Conservation and Solar Applications, Department of Energy, 1978], p. 22.) The study cited a 1977 Harris poll indicating "three-fifths of the people now feel that bigness in almost anything leads to trouble for individuals who can't stand up to it." (Milstein 1978, p. 33.) Aaron Wildavsky and Ellen Tenenbaum give examples of negative attitudes towards big oil spanning almost a century. In regards to the Teapot Dome Scandal,

Wildavsky writes, "The scandal served to confirm the preservationist, anti-monopolist, and public opinion that a rapacious industry was holding captive a corrupt government." Aaron Wildavsky and Ellen Tenenbaum, *The Politics of Mistrust: Estimating American Oil and Gas Resources* (Beverly Hills, Calif.: Sage Publications, 1981), p. 77.

10. Numerous studies detailing this period have been written. Among those consulted were: Senate Subcommittee on Antitrust and Monopoly, 91st Congress, 1st Session, *Hearings on Government Intervention in the Market Mechanism, The Petroleum Industry* (Washington, D.C.: Government Printing Office, 1969); John M. Blair, *The Control of Oil* (New York: Pantheon, 1976); Staff Report of the Federal Trade Commission, Senate Small Business Committee, 82nd Congress, 2nd Session, *The International Petroleum Cartel* (Washington, D.C.: Government Printing Office, 1952); J.E. Hartshorn, *Politics and World Oil Economics: An Account of the International Oil Industry in its Political Environment,* revised edition (New York: Praeger, 1967); Ralph Hewins, *Mr. Five Percent: The Biography of Calouste Gulbenkian* (London: Hutchinson, 1957); Stephen Hemsley Longrigg, *Oil in the Middle East* (New York: Oxford University Press, 1968); Zuhayr Mikdashi, *A Financial Analysis of Middle Eastern Oil Concessions: 1901-1965* (New York: Praeger, 1966); U.S. Senate, *Hearings before the Subcommittee on Multinational Corporations of the Committee on Foreign Relations, Multinational Petroleum Companies,* 93rd Congress, 2nd Session (Washington D.C.: Government Printing Office, 1974); Anthony Samson, *The Seven Sisters: The Great Oil Companies and the Wealth They Made* (New York: Viking Press, 1975).

11. Gerald D. Nash, *United States Oil Policy 1890-1964* (Pittsburg: University of Pittsburg Press, 1968), p. 202.

12. Douglas R. Bohi and Milton Russell, *Limiting Oil Imports: An Economic History and Analysis* (Baltimore, Md.: Johns Hopkins University Press for RFF, 1978), pp. 16-17.

13. The actual effects of energy measures on the distribution of income by income class, occupation, location, or age group have only recently (and then only incompletely) received systematic attention. See Hans H. Landsberg and Joseph M. Dukert, *High Energy Costs: Uneven, Unfair, Unavoidable* (Baltimore, Md.: Johns Hopkins University Press for RFF, 1981); also, Hans H. Landsberg, ed., *High Energy Costs: Assessing the Burden* (Washington, D.C.: Resources for the Future, 1982); and the works cited therein. The broad distribution of ownership of "bluechip" petroleum and utility stocks (and their popularity with pension funds), the pervasive use of energy products, the substitution possibilities among fuels, and the impact of changes in revenues on factor (including wage) income often make it impossible to anticipate with confidence even the direction of distributional shifts. Yet, beliefs about such effects surely drive decisions. The failure to treat income distribution effects as a serious matter for examination represents a striking indictment of the process that defines the problems to which energy-related analysis has been applied.

14. Tax Reductions Act, Public Law 94-12, 89 STAT 26, 1975, amended by Emergency Compensation and Special Unemployment Assistance Extension Act, Public Law 94-95, 89 STAT 236, 1975, and Energy Tax Act, Public Law 95-618, 92 STAT 3174, 1978.

15. "Adjusting Imports of Petroleum and Petroleum Products," Presidential Proclamation 3279, 24 Fed. Reg. 1781, 1959.
16. Federal Energy Administration Regulation, "Allocation of Old Oil," 39 F.R. 42246, December 4, 1974.
17. David E. Lilienthal, *TVA: Democracy on the March* (New York: Harper and Brothers, 1944).
18. In 1945, Interior Secretary Ickes asked President Truman to convene a world conservation conference in view of the war's heavy toll and the continuing strain on the forests, oil, and coal. The conference would explore the extent of resource depletion, investigate conservation opportunities, and consider development of synthetic fuels. Goodwin, *Energy Policy in Perspective*, pp. 10-11.
19. The twin perspectives of inexorable exhaustion of fossil fuels and unremitting optimism for suitable replacement *if* institutions are appropriate is seen in the work of Erich W. Zimmerman, *World Resources and Industries: A Functional Appraisal of the Availability of Agricultural and Industrial Material*, revised edition (New York: Harper, 1951), pp. 44, 56-58.

> The problem of energy strategy merges with that of good government and the development of the societal arts. Institutions have as much to do with the ultimate efficency of energy use as have engines, machines and logarithm tables ... Energy history probably will appear divided into three great periods or ages: ... animate energy ... inanimate energy derived mainly from fossil fuels, and ... inanimate energy derived from continuous sources such as direct solar radiation, tidal power, carbon dioxide, etc. We are now living in the second age ...[which is] ... rushing on to its *finis* with accelerating speed and demonic force ... What will follow? No one knows, but [here quoting approvingly from a 1948 American Petroleum Institute paper by Eugene E. Ayres] "there is every reason to expect that succeeding eras will provide still greater abundance of energy from our constant sources ... Within a few decades a good start must have been made toward the new systems of energy production and consumption; and while this goes on, our technological rear guard will be engaged in retarding in every possible way the corrosive growth of our energy losses."

20. Goodwin, *Energy Policy in Perspective*, pp. 32-37.
21. Harold J. Barnett, *Energy Uses and Supplies, 1939, 1947, 1965* (U.S. Department of the Interior, Bureau of Mines, Information Circular 7582, October 1950).
22. Harold J. Barnett and Chandler Morse, *Scarcity and Growth: The Economics of Natural Resource Availability* (Baltimore, Md.: Johns Hopkins University Press for RFF, 1963).
23. Goodwin, *Energy Policy in Perspective*, p. 37.
24. Paley, *Resources for Freedom* I (from the President's letter of January 22, 1951 commissioning this study), p. iv.
25. Paley, *Resources for Freedom* I, p. 107.
26. Ibid., p. 129.
27. Ibid., p. 110.
28. Ibid., p. 107.

29. Department of Energy/Energy Information Administration (DOE/EIA), *Annual Report to Congress 1979* II (Washington, D.C.: Government Printing Office, 1979), p. 15.
30. Paley, *Resources for Freedom* I, p. 107.
31. Ibid., p. 107.
32. Harry S. Truman, Inaugural Address, in *Public Papers of the Presidents: Harry S. Truman, 1949*, no. 19, January 20, 1949 (Washington, D.C.: Government Printing Office, 1964); Dwight D. Eisenhower, "Address to the General Assembly of the United Nations on Peaceful Uses of Atomic Energy," in *Public Papers of the Presidents, Dwight D. Eisenhower, 1953*, no. 256, December 8, 1953 (Washington, D.C.: Government Printing Office, 1960).
33. American Petroleum Institute, *Basic Petroleum Data Book* (Washington, D.C.: American Petroleum Institute, 1980).
34. DOE/EIA 1979, Edison Electric Institute, *Statistical Yearbook of the Electric Utility Industry* (Washington, D.C.: Edison Electric Institute, various years), pp. 35, 43, 91; *Economic Report of the President ERP* (Washington, D.C.: Government Printing Office, 1980), p. 206.
35. Sam Schurr and Bruce Netschert, *Energy in the American Economy, 1890-1975: An Economic Study of Its History and Prospects* (Baltimore, Md.: Johns Hopkins University Press for RFF, 1960), p. 5.
36. Warren E. Morrison, *Summary Energy Balances for the United States: Selected Years 1947-62* (Bureau of Mines Information Circular 8242, 1964).
37. Warren E. Morrison and Charles L. Readling, *An Energy Model for the United States Featuring Energy Balances for the Years 1947 to 1965 and Projections and Forecasts to the Years 1980 and 2000* (Bureau of Mines Information Circular 8384, 1968).
38. Some things were done. The Mandatory Oil Import Program and state production controls caused the domestic industry to expand and led to maintenance of standby capacity sufficient to prevent an energy shock in 1967 at the time of the second closure of the Suez canal. It also held oil prices above world levels, thus encouraging development of other fuels. In retrospect, it is uncertain whether the high cost of this program was worthwhile (Bohi and Russell, *Limiting Oil Imports*, pp. 284-300). Low-level federal funding of synthetic fuel research and development was pursued, and the Office of Coal Research maintained a program to encourage new uses of that fuel. Thus some policies even went beyond the recommendations of the original Paley Commission in the direction of preparing for the developments that did occur in the 1970s. For example, the Commission was opposed to any limits on oil imports (Paley, *Resources for Freedom* I, p. 108). And yet the shocks came.
39. M.K. Hubbert, *U.S. Energy Resources: A Review as of 1972*, written for the Committee on Interior and Insular Affairs (Washington, D.C.: Government Printing Office, 1974), pp. 120, 144.
40. Since coal's problems were largely the result of displacement by natural gas and residual fuel oil, and nuclear power threatened to replace coal in the electric utility market, the decline in coal markets and the problems of Appalachia were tied to policies toward other fuels.
41. Irving C. Bupp and Jean-Claude Derian, *Light Water: How the Nuclear Dream Dissolved* (New York: Basic Books, 1978). One alternate technology using heavy water as a cooling and moderating medium can utilize ordinary uranium oxide as a fissionable material, thus avoiding the

enrichment step. The Canadians adopted this design in their Candu reactors. Another technology uses graphite as a moderator and gas as a heat transfer medium.

42. Richard G. Hewlett, "Nuclear Power in the Public Interest: The Atomic Energy Act of 1954" (Presentation at the American Historical Association, Dallas, December 1977); Jack M. Holl, "Eisenhower's Peaceful Atomic Diplomacy: Atoms for Peace in the Public Interest" (Presentation at the American Historical Association, Dallas, December 1977); Goodwin, *Energy Policy in Perspective*, pp. 273-82.

43. The first reactor to produce power in the United States was a breeder reactor, the Experimental Breeder Reactor (EBR-1), whose operation was demonstrated in 1951. Successively larger units have been built in France, the Soviet Union, Germany, Britain, and Japan. As of 1970, the United States held a commanding lead in commercial breeder development, but France became the world leader with its Phenix (operating at 233 megawatt electric) in 1973, and its Super Phenix, designed for start-up in late 1983. American Nuclear Society, "World List of Nuclear Power Plants," *Nuclear News* 24, no. 10 (August 1981).

44. But even including enrichment, the cost of fuel is a small part of the total cost of producing electricity in a nuclear plant. In 1979, fuel costs were about 18 percent of the cost of energy from the newest nuclear plant and 56 percent in a modern coal plant. (*Update*, Nuclear Power Program Information and Data, Office of Nuclear Reactor Programs, Department of Energy, Washington, D.C., July/August 1980.) Nonetheless, for the United States, the breeder's role in lowering prospective nuclear fuel costs is largely what makes it attractive.

45. This discussion draws on a number of sources including Stewart W. Herman and James S. Cannon, *Energy Futures: Industry and the New Technologies* (New York City: INFORM, 1976), pp. 689-99; Irving C. Bupp, "The Nuclear Stalemate," in Robert Stobaugh and Daniel Yergin, eds., *Energy Future: Report of the Energy Project at the Harvard Business School* (New York: Random House, 1979), pp. 115-17; Ford-MITRE Report, Nuclear Energy Policy Study Group of the Ford Foundation, *Nuclear Power: Issues and Choices* (Cambridge, Mass.: Ballinger Publishing Company, 1977), pp. 335-44; Paul MacAvoy, *Economic Strategy for Developing Nuclear Breeder Reactors* (Cambridge, Mass.: M.I.T. Press, 1969); Thomas B. Cochran, *The Liquid Metal Fast Breeder Reactor: An Environmental and Economic Critique* (Baltimore, Md.: Johns Hopkins University Press for RFF, 1974).

46. MacAvoy, *Economic Strategy for Developing Nuclear Breeder Reactors*, p. vii.

47. Atomic Energy Commission, *Cost-Benefit Analysis of the U.S. Breeder Reactor Program*, WASH 1126 (Washington, D.C.: Government Printing Office, 1969), p. iv.

48. Cochran, *The Liquid Metal Fast Breeder Reactor: An Environmental and Economic Critique*, pp. 222, 229.

49. A.M. Weinberg, "Social Institutions and Nuclear Energy," *Science* 177, no. 4043 (July 7, 1972): 34.

50. DOE/EIA, *Annual Report to Congress, 1979*, pp. 93, 59, 113.

51. Phillips Petroleum v. Wisconsin, 347 U.S. 672, 1954; Natural Gas Act of 1938, 15 U.S.C. 717, 52 STAT 821, 1938. The Natural Gas Act placed sales of natural gas for resale in interstate commerce under Federal Power

Commission jurisdiction. The act on its face regulated the sales of interstate pipeline companies and, by extension, the price they were allowed to receive for the natural gas they produced for their own account. As interpreted until the *Phillips* decision, the act did not apply to the sales of independent (nonpipeline company) producers of natural gas. In *Phillips*, the Court held that the Act "gave the Commission jurisdiction over the rates of all wholesales of gas in interstate commerce whether by a pipeline company or not and whether occurring before, during, or after transmission by an interstate pipeline company." (*Phillips* 1954, p. 672.) The issue of whether gas producers should be regulated arose before the *Phillips* decision and continues to absorb congressional attention.

52. Federal Power Commission (FPC), *In the Matter of Phillips Petroleum Company. Proceedings to Determine Whether Phillips Petroleum Company is a "Natural Gas Company" as Defined in the Natural Gas Act*, 10 FPC 246, 1951.
53. Wisconsin v. Phillips Petroleum Company, 205 F. 2a 706 C.A.D.C., 1953.
54. It is intriguing that "independent producer" came to mean two things in the petroleum business. An independent producer of natural gas is a noninterstate natural gas pipeline company. Thus independent producers of gas include such giants of the industry as Exxon, Texaco, and Phillips. An independent producer of oil, on the other hand, is a firm engaged exclusively at the production level, without ownership of pipelines, refineries, or marketing facilities. Hence in the oil business, independent producers are small companies dependent on larger ones for access to markets. Gas producers tried to tap populist sentiments in political battles for decontrol by embracing the "independent" label.
55. Nash, *United States Oil Policy 1890-1964*, p. 226.
56. A useful summary of the state of thinking about gas regulation issues at the beginning of the 1970s is found in Keith C. Brown, ed., *Regulation of the Natural Gas Industry* (Baltimore, Md.: Johns Hopkins University Press for RFF, 1972). The seminar from which this book was drawn was financed by the White House Office of Science and Technology under the direction of S. David Freeman. Freeman, previously assistant to Chairman Joseph Swidler of the FPC, had been at the center of the effort to bring natural gas prices under effective regulation.
57. Among the more important analyses are Leslie Cookenboo, Jr., *Competition in the Field Market for Natural Gas* (Houston: Rice Institute Monograph in Economics, 1958); Franklin M. Fisher, *Supply and Costs in the U.S. Petroleum Industry* (Baltimore, Md.: Johns Hopkins University Press for RFF, 1964); Clark A. Hawkins, *The Field Price Regulation of Natural Gas* (Tallahassee, Fla.: Florida State University Press, 1969); Alfred E. Kahn, "Economic Issues in Regulating the Field Price of Natural Gas," *50 American Economic Review*, no. 2 (1960); Edmund N. Kitch, "Regulation and the Field Market for Natural Gas," *Journal of Law and Economics* II (1968); Paul MacAvoy, *Price Formation in Natural Gas Fields: A Study of Competition, Monopsony and Regulation* (New Haven: Yale University Press, 1962); Edward S. Neuner, *The Natural Gas Industry* (Norman, Okla.: University of Oklahoma Press, 1960).
58. FPC, *Permian Basin Area Rate Proceedings*, 34 FPC 159, 1965; FPC, *Louisiana Area Rate Proceedings*, 40 FPC 530, 1968.
59. FPC, *Opinion and Order Determining Cost of Service and Terminating Proceedings*, p. 542; *Phillips Petroleum Company*, 24 FPC 537, 1960.

60. *Phillips Petroleum Company*, 24 FPC 537, 1960.
61. FPC, *Statement of General Policy no. 61-1 Establishment of Price Standards to be Applied in Determining the Acceptability of Initial Price Proposal and Increased Rate Filings by Independent Producers of Natural Gas*, 24 FPC 818, 1960. For an analysis of this decision see Milton Russell, "An Economic Analysis of the Public-Utility and Area-Price Methods of Federal Regulation of the Independent Producers of Natural Gas" (Ph.D. dissertation, University of Oklahoma, 1963).
62. Among the studies produced during this period were: Edward W. Erickson and Robert M. Spann, "Supply Response in a Regulated Industry: The Case of Natural Gas," *The Bell Journal of Economics and Management Science* 2 (Spring 1971); Edward W. Erickson and Robert M. Spann, "Price Regulation and the Supply of Natural Gas in the United States," in Keith C. Brown, ed., *Regulation of the Natural Gas Producing Industry* (Baltimore, Md.: Johns Hopkins University Press for RFF, 1972); Paul MacAvoy and Robert Pindyck, "Alterative Policies for Dealing with the Natural Gas Shortage," *The Bell Journal of Economics and Management Science* 4, no. 2 (Autumn 1973); Paul W. MacAvoy and Robert S. Pindyck, *Price Controls and the Natural Gas Shortage* (Washington, D.C.: American Enterprise Institute, 1975); Robert Pindyck, "The Regulatory Implications of Three Alternative Econometric Supply Models of Natural Gas," *Bell Journal of Economics and Management Science* 5, no. 2 (Autumn 1974).
63. Milton Russell, "Producer Regulation for the 1970's," in Keith C. Brown, ed., *Regulation of the Natural Gas Producing Industry* (Baltimore, Md.: Johns Hopkins University Press for RFF, 1972); Milton Russell, "Natural Gas Curtailments: Administrative Rationing on Market Allocation," in Harry M. Trebing, ed., *New Dimensions in Public Utility Pricing* (East Lansing, Mich.: Michigan State University, 1976).
64. Edward Mitchell, *Energy Ideology* (Washington, D.C.: American Enterprise Institute for Public Policy Research, 1977).
65. FPC, *National Gas Survey: Curtailment Strategies*, Report to the Federal Power Commission by the Technical Advisory Committee on Curtailment Strategies (Washington, D.C.: Government Printing Office, 1977); American Gas Association (AGA), *Energy Analysis: An Analysis of Oil to Gas Conversion Trends in the Residential Gas Space Heating Market* (Washington, D.C.: American Gas Association, September 18, 1980).
66. Edward Erickson, Testimony before the U.S. Budget Committee, House of Representatives, *Economic and Budget Impact of the President's Energy Proposals* (Washington, D.C.: Government Printing Office, June 29, 1977)
67. The Natural Gas Policy Act of 1978 (P.L. 95-621, 1978) established a partial, phased-in, price deregulation scheme for most previously regulated natural gas but introduced regulation of intrastate natural gas for the first time. The act contains a system of gradual price increases leading to complete decontrol by January 1, 1985 for "new natural gas," "new, onshore production wells" producing below 5,000 feet, and certain intrastate contracts. Certain categories of high-cost gas were deregulated immediately. Old interstate gas, in contrast, would remain under control indefinitely.
68. Edward J. Mitchell, ed., *The Deregulation of Natural Gas* (Washington, D.C.: American Enterprise Institute for Public Policy Research, 1982).
69. Some examples are Breyer, "The Natural Gas Shortage and the Regulation of Natural Gas Producers"; Keith C. Brown, *Federal Power Commis-*

sion Control of Natural Gas Producer Prices, Paper no. 494, Krannert Graduate School of Industrial Administration, Purdue University, Indiana, February, 1975; Erickson, "Price Regulation and the Supply of Natural Gas in the United States"; Helms, *Natural Gas Regulation: An Analysis of FPC Price Controls*; MacAvoy, "Alternative Policies for Dealing with the Natural Gas Shortage"; MacAvoy, *Price Controls and the Natural Gas Shortage*; Russell, "Producer Regulation for the 1970's"; Russell, "Natural Gas Curtailments"; and Patricia E. Starratt, *The Natural Gas Shortage and the Congress* (Washington, D.C.: American Enterprise Institute, 1974).

70. Cabinet Task Force on Oil Import Control, *The Oil Import Question*, prepared by the Cabinet Task Force on Oil Import Control (Washington, D.C.: Government Printing Office, 1970).

71. Tax treatment of the oil industry has always loomed large in policy discussions. However, the issue has most often been drawn as one of "fairness" of the burden, not as one of the effects on output and prices. The competitiveness of the industry has also been the subject of considerable attention.

72. Ohio Co. v. Indiana, 177 U.S. 190, 1900.

73. A standard reference on the economics of petroleum conservation is Stephen L. McDonald, *Petroleum Conservation in the United States: An Economic Analysis* (Baltimore, Md.: Johns Hopkins University Press for RFF, 1971). The following discussion of the conditions under which oil and gas are produced and the economic consequences of these conditions is drawn from Russell, "An Economic Analysis of the Public-Utility and Area-Price Methods of Federal Regulation of the Independent Producers of Natural Gas," pp. 21-57.

74. John M. Blair, *The Control of Oil* (New York: Pantheon, 1976).

75. The success at stabilizing prices was spectacular. Taking the number of month-to-month price changes as an indicator, for twenty years before market-demand prorationing there were changes 29 percent of the months. After this process was established, changes occurred only 5.5 percent of the months over the next twenty-seven years, with the preponderance of their increases in the immediate post-World War II years. Russell, "An Economic Analysis of the Public-Utility and Area-Price Methods of Federal Regulation of the Independent Producers of Natural Gas."

76. Goodwin, *Energy Policy in Perspective*, p. 106.

77. From the Pratt report, "A National Liquids Fuels Policy," as quoted in Goodwin, *Energy Policy in Perspective*, p. 106.

78. Presidential Proclamation, 3279, 1959. For a history and economic analysis of the antecedents, provisions, and effects of MOIP see Bohi and Russell, *Limiting Oil Imports*.

79. Bohi, *Limiting Oil Imports*, pp. 208-30.

80. Voluntary action could not solve the common property problem or prevent periodic gluts and physical waste. Easy entry into the industry led to competitive exploitation of the same reserves. It meant large numbers of producers—too many to get together quietly to control the market. Antitrust laws prevented overt collusion to maintain prices by preventing overproduction, and any such agreement among so many competitors would have been very difficult to enforce. The situation was very different than it was when Rockefeller could control the transportation system as the foundation of the Standard Oil Trust.

81. *Texas Civil Statutes* (Vernon, 1948), 1948.
82. "There is some maximum rate of oil production at each stage of depletion consistent with the fullest exploitation of the expulsive forces present." (McDonald, *Petroleum Conservation in the United States* p. 19.) The MER is thus set as a physical measure unrelated to the economic concept of an efficient rate of production that takes into account the cost of waiting for the revenues from producing the oil.
83. McDonald, *Petroleum Conservation in the United States*, p. 38; National Recovery Administration, *Codes of Fair Competition* 1, nos. 1-47 (Washington, D.C.: Government Printing Office, 1933).
84. Schecter Poultry Corporation v. United States, 295 U.S. 495, 1935.
85. Connally Hot Oil Act, Public Law 74-14, 49 STAT 30, 1935.
86. Interstate Oil Compact to Conserve Oil and Gas, Public Resolution 74-64, 49 STAT 939, 1935.
87. Ernest O. Thompson, "The Oil and Gas Market Demand Law," *The Interstate Oil Compact Commission Committee Bulletin* (June 1960).
88. Ernest O. Thompson as quoted by Ira Butler, "The Texas Market Demand Statute and the Experience in its Application to Practical Conditions," *The Interstate Oil Company Commission Bulletin* (December 1960).
89. McDonald, *Petroleum Conservation in the United States*.
90. Ibid., p. 164.
91. "Torrid Drilling to Surge Past 70,000 Wells," *Oil and Gas Journal* (January 26, 1981): 145.
92. Bohi and Russell, *Limiting Oil Imports*, pp. 99-187.
93. Ibid., inside front jacket.
94. A careful and readable account of all aspects of the special tax treatment of oil and gas is found in Stephen L. McDonald, *Federal Tax Treatment of Income from Oil and Gas* (Washington, D.C.: Brookings Institution, 1963).
95. Harry S. Truman, "Special Message to the Congress on Tax Policy," in *Public Papers of the Presidents: Harry S. Truman, 1950*, no. 18 (Washington, D.C.: Government Printing Office, 1965), p. 123.
96. Harvey O'Conner, "How to Make a Billion (1955)," in Robert Engler, ed., *America's Energy: Reports from The Nation* (New York: Pantheon, 1980), p. 196.
97. Blair, *The Control of Oil*, p. 191. The 27.5 percentage depletion remained for over four decades until 1969 when it was reduced to 22 percent under Title V of the Tax Reform Act of that year. (Tax Reform Act, Public Law 91-172, 83 STAT 487, 1969.) Further modifications came in 1975 when the Tax Reduction Act of 1975 (Public Law 94-12, 1975) changed the form and the scope of the percentage depletion allowance. While its provisions were complex, in essence under the 1975 act the depletion allowance was made available only to independent producers, royalty owners, and for regulated natural gas. It was set initially at the 22 percent level, but the rate phased down to 20 percent in 1981, 18 percent in 1982, 16 percent in 1983, and 15 percent in 1984 and thereafter. Further, percentage depletion was allowed only for the first 2,000 barrels per day in 1975, falling by 200 barrels per day per year to 1,000 in 1980 and remaining at that level thereafter. Other producers were required to base return of capital on actual costs expended, for example, in purchasing producing properties. (Public Law 94-12, 1975.)
98. Bohi and Russell, *Limiting Oil Imports*, p. 188.

99. Schurr and Netschert, *Energy in the American Economy, 1890-1975*, p. 514.

100. National Coal Association (NCA), *Bituminous Coal Facts 1970* (Washington, D.C., NCA, 1971); NCA, *Coal Facts 1978-1979* (Washington, D.C., NCA, 1979).

101. Between 1945 and 1950 consumption of bituminous coal by the railroads declined 51 percent and by 1960 was only 2 percent of the 1945 level. Railroads were 22 percent of the coal market in 1945 but less than 1 percent in 1960. Use of diesel fuel by railroads rose six times in those fifteen years. Association of American Railroads (AAR), *Statistics of Railways of Class I* (Washington, D.C.: AAR, Bureau of Railway Economics, various years). In the years between 1945 and 1960, the proportion of coal loaded mechanically rose from 56 to 86 percent. NCA, *Bituminous Coal Data Book 1962* (Washington, D.C.: NCA, 1963).

102. Harry Caudill, "Appalachia: The Path from Disaster," in Robert Engler, ed., *America's Energy: Reports from The Nation* (New York: Pantheon, 1980), p. 34.

103. In 1950 the unemployment rate for Appalachian Kentucky and West Virginia (a bellwether region for which data are available and in which coal was the dominant industry) was at 4.4 percent, only 0.9 percentage points below the national average of 5.3 percent. In 1960 the unemployment rate for the region stood at 8.4 percent, almost 3 percentage points above the national average. By 1970, after two decades of painful adjustment, this gap had narrowed to 0.5 percent. Population trends tell the adjustment-through-migration story. In this coal region, population declined 10 percent from 1950 to 1960 and a further 7 percent over the next decade. In contrast, the U.S. population rose by 19 and by 13 percent in these same two periods. Statistical data provided by Appalachian Regional Commission, Washington, D.C., 1981; *Economic Report of the President* (Washington, D.C.: Government Printing Office, 1981), p. 269.

104. Appalachian Regional Commission, 1981; *Economic Report of the President*, 1981, p. 262.

105. Goodwin, *Energy Policy in Perspective*, p. 222.

106. Office of Emergency Planning, Memorandum for the President (Relating to the National Security Considerations in the Residual Oil Control Program) (February 13, 1963): 30.

107. Goodwin, *Energy Policy in Perspective*, pp. 138-46. Paley, *Resources for Freedom* I, pp. 115, 117

108. Paley, *Resources for Freedom* I, pp. 115-17.

109. Cabinet Committee on Energy Supplies and Natural Resources Policy, *Report on Natural Gas*, prepared by the Transmission and Distribution consultants of the Oil and Gas Division of the Task Force (Washington, D.C.: Government Printing Office, 1954).

110. Goodwin, *Energy Policy in Perspective*, p. 266.

111. Bohi and Russell, *Limiting Oil Imports*, pp. 151-61.

112. Clean Air Act. Public Law 88-206, 77 STAT 392, 1963.

113. Goodwin, *Energy Policy in Perspective*, p. 225.

114. Ibid., p. 228.

115. Goodwin, *Energy Policy in Perspective*, p. 319, quoting the study.

116. John F. Kennedy, Remarks to a Meeting to Consider the Economic Problems of the Appalachian Region, in *Public Papers of the Presidents: John F. Kennedy, 1963*, no. 127, April 9, 1963 (Washington, D.C.: Govern-

ment Printing Office, 1964), p. 37; Appalachian Regional Development Act of 1965, Public Law 89-4, 79 STAT 6, 1965.
117. Caudill, "Appalachia: The Path from Disaster," p. 36.
118. Ibid., pp. 36-37.
119. Cochrane, "Energy Policy in the Johnson Administration: Logical Order versus Economic Pluralism" in Goodwin, *Energy Policy in Perspective*, pp. 385-87.
120. Appalachian Regional Commission, 1981; Economic Report of the President, 1981, p. 262.

CHAPTER 3

1. Confidential review of the preliminary manuscript for this book, April 2, 1982.

CHAPTER 4

1. Energy Policy Project of the Ford Foundation, Tentative Information Plan, 1972.
2. Cabinet Task Force on Oil Import Control, *The Oil Import Question* (Washington, D.C.: Government Printing Office, February 1970), pp. 20-21.
3. "The Energy Crisis Had Its Prophet," *San Francisco Chronicle*, November 15, 1977.
4. Presidential Proclamation 4210, April 18, 1973.
5. The inconsistencies of interpretation recall the Japanese film "Rashomon" based on a classic Japanese novel about a rape-murder described in several drastically different eye-witness accounts.
6. Energy Policy Project (EPP) of the Ford Foundation, *A Time to Choose: America's Energy Future* (Cambridge, Mass.: Ballinger Publishing Company, 1974).
7. Lindley H. Clark, Jr., "Time and Energy," *Wall Street Journal*, March 31, 1975.
8. Energy Policy Project (EPP) of the Ford Foundation, *Exploring Energy Choices: A Preliminary Report* (Washington, D.C.: EPP, 1974).
9. The outside studies were published by Ballinger Publishing Company. They are: Edward Berlin, Charles J. Cicchetti, and William J. Gillen, *Perspectives on Power, A Study of the Regulation and Pricing of Electric Power*; Donald F. Boesch, Carl H. Hershner, and Jerome H. Milgram, *Oil Spills and the Marine Environment*; Gerard M. Brannon, *Energy Taxes and Subsidies*; Gerard M. Brannon, ed., *Studies in Energy Tax Policy*; Charles J. Cicchetti and John L. Jurewitz, eds., *Studies in Electric Utility Regulation*; Thomas Duchesneau, *Competition in the U.S. Energy Industry*; Foster Associates, *Energy Prices 1960-73*; John E. Gray, *Energy Policy-Industry Perspectives*; Elias P. Gyftopoulos, Lazaros J. Lazaridis, and Thomas Widmer, *Potential Fuel Effectiveness in Industry*; Jerome E. Hass, Edward J. Mitchell, and Bernell K. Stone, *Financing the Energy Industry*; J. Herbert Hollomon and Michel Grenon, *U.S. Energy Research*

and Development Policy; Arjun Makhijani, *Energy and Agriculture in the Third World*; John G. Myers, et al., *Energy Consumption in Manufacturing*; National Academy of Sciences and National Academy of Engineering, *Rehabilitation Potential of Western Coal Lands*; Dorothy K. Newman and Dawn Day Wachtel, *The American Energy Consumer*; Richard Schoen, Alan S. Hirshberg, and Jerome M. Weingart, *New Energy Technologies for Buildings*; Robert Williams, ed., *The Energy Conservation Papers*; Mason Willrich and Theodore B. Taylor, *Nuclear Theft: Risks and Safeguards*; Joseph A. Yager and Eleanor B. Steinberg, *Energy and U.S. Foreign Policy*.

10. Allen V. Kneese, "Review: *A Time to Choose*," *Challenge*, July-August 1975, pp. 57-60.
11. Armen A. Alchian, et al., *No Time to Confuse: A Critique of the Final Report of the Energy Policy Project of the Ford Foundation* (San Francisco: Institute for Contemporary Studies, 1975), p. 1.
12. David Halberstam in *The Best and the Brightest* (New York: Fawcett Crest, 1972), p. 57, calls Bundy "the brightest light in that glittering constellation around the President."
13. Lewis H. Lapham, "The Energy Debacle," *Harpers*, August 1977, pp. 64, 73. Under the Tax Reform Act, foundations are enjoined from taking political stands.
14. David Halberstam, "The Very Expensive Education of McGeorge Bundy," *Harpers*, July 1969, p. 37.
15. S. David Freeman, "Toward a Policy of Energy Conservation," *Bulletin of the Atomic Scientists*, October 1971, pp. 8-9.
16. Ford-RFF Study Group, *Energy: The Next Twenty Years* (Cambridge, Mass.: Ballinger Publishing Company, 1979).
17. Forecasts of future energy use came down markedly during the seventies, and there was a "scramble for low demand estimates." See, Eliot Marshall, "Energy Forecasts: Sinking to New Lows," *Science* 208 (1980): 1356. See also, Martin Greenberger, "Energy Demand Forecasts" (Testimony before the hearings of the Subcommittee on Investigations and Oversight of the Committee on Science and Technology, U.S. House of Representatives, June 1, 1981.
18. EPP, *Exploring Energy Choices*, p. i.
19. Letter from Freeman to Greenberger, April 9, 1981.
20. EPP, *A Time to Choose*, p. 401.
21. Ibid., p. 350. Members of the advisory board were selected to represent a broad cross-section of backgrounds and views. They included: chairman Gilbert White, Institute of Behavioral Sciences, University of Colorado; Dean Abrahamson, Center for the Study of the Physical Environment, University of Minnesota; Lee Botts, Lake Michigan Federation; Harvey Brooks, Engineering and Applied Physics, Harvard; Donald C. Burnham, Westinghouse; J. John Deutsch, Queens University, Canada; Joseph L. Fisher, Resources for the Future; Eli Goldston, Eastern Gas and Fuel Associates; John D. Harper, Aluminum Company of America; Philip S. Hughes, U.S. General Accounting Office; Minor S. Jameson, Independent Petroleum Association of America; Carl Kaysen, Institute for Advanced Studies, Princeton; Michael McCloskey, Sierra Club; Norton Nelson, Institute for Environmental Medicine, New York University; Alex Radin, American Public Power Association; Joseph R. Rensch, Pacific Lighting Corporation; Charles

R. Ross, former member of the Federal Power Commission; Joseph Sax, School of Law, University of Michigan; Julius Stratton, Massachusetts Institute of Technology; William Tavoulareas, Mobil Oil; J. Harris Ward, Commonwealth Edison.

22. William Simon, Transcript of CBS TV special on the energy crisis moderated by Walter Cronkite, New York, August 31, 1977, p. 54.
23. Letter from Harrison to Greenberger, April 2, 1981.
24. Ibid.; letter from Ehrlichman to Greenberger, July 6, 1981.
25. John Noble Wilford, "Nation's Energy Crisis: It Won't Go Away Soon," New York Times, July 6, 1971, p. 1.
26. Interviews with Ames, July 16, 1979 and August 18, 1980.
27. S. David Freeman, Energy: The New Era (New York: Walker, 1974).
28. Harrison, letter to Greenberger.
29. Lapham, "The Energy Debacle," p. 61.
30. Harrison, letter to Greenberger.
31. S. David Freeman, excerpts from January 25, 1973 speech to the Consumer Federation of America, Science, February 23, 1973, p. 15.
32. "Head of Energy Policy Study Criticized by Advisors for Consumer Talk," New York Times, February 23, 1973, p. 15.
33. In fact, judgments" in A Time to Choose are not restricted to the final chapter. They appear throughout the report.
34. "Growth vs. Stagnation: Energy Options for America," Dissent of William P. Tavoulareas, President of Mobil Oil Corporation, to the final report of the Ford Foundation's Energy Policy Project (New York: Mobil Oil Corporation, 1974).
35. William Tavoulareas and Carl Kaysen, A Debate on 'A Time to Choose' (Cambridge, Mass.: Ballinger Publishing Company, 1977), p. xiv.
36. Conversation with Kaysen.
37. Tavoulareas and Kaysen, A Debate on 'A Time to Choose', p. 16.
38. Ibid., p. 74.
39. EPP, A Time to Choose, pp. 361-62. Brooks considers the conservation recommendation well-documented and the most important contribution of the Ford report, but believes it would have been better received had the book not gotten into questions of the economic structure of the oil industry.
40. EPP, A Time to Choose, p. 338.
41. Walter J. Mead, letter to the editor, Harper's, October 1977, p. 12. Mead spent a year as senior economist to the EPP. He was a contributor to the countertract No Time to Confuse.
42. EPP, A Time to Choose, p. 6.
43. Letter from White to Greenberger, April 28, 1981.
44. Harrison, letter to Greenberger.
45. The phrase "destructive engine" was used by Judge J. Skelly Wright in the Calvert Cliffs decision. Calvert Cliffs Coordinating Committee, Inc. v. U.S. Atomic Energy Commission, 449 F. 2d 1109, 1111 (D.C. Cir. 1971).
46. Memorandum to Members of the Advisory Board from S. David Freeman on Project Status and Background for the June 21 Meeting of the EPP, Washington, D.C., June 9, 1972. Also, Monte Canfield, Jr., letter to the editor, Harper's, October 1977, pp. 9, 12.
47. Monte Canfield, Jr., "Normative Analysis of Alternative Futures" (occasional paper, Aspen, Colorado, Aspen Institute for Humanistic Studies, February 1974).

48. Freeman, letter to Greenberger.
49. EPP, *A Time to Choose*, pp. 132, 151.
50. Letter from Jorgenson to Greenberger, June 23, 1981.
51. EPP, *A Time to Choose*, p. 509. Given Freeman's strong public position against lifting price controls, there was an inconsistency in his use of a modeling study that induced energy conservation by artificially raising prices. An EPP staff member says that Freeman never felt comfortable looking at price rises as the mechanism for conservation.
52. An early version of the Hudson and Jorgenson work appears in Dale W. Jorgenson and H. S. Houthakker, "Energy Resources and Economic Growth: Final Report to the Energy Policy Project," in "U.S. Economic Growth, 1973-2000" (Lexington, Mass.: Data Resources Incorporated, 1973), ch. 2. As a commentary on the state of energy/economic research at the time, the report claims its "essential novelty" to be the combination of the "determinants of energy demand and supply within the same framework." Later work is summarized in Dale W. Jorgenson, "Econometric and Process Analysis Models for the Analysis of Energy Policy," in R. Amit and M. Avriel, eds., *Perspectives on Resource Policy Modeling: Energy and Minerals* (Cambridge, Mass.: Ballinger Publishing Company, 1981.
53. George P. Shultz and Kenneth W. Dam, *Economic Policy Beyond the Headlines* (New York: W. W. Norton, 1977), pp. 163-64.
54. Ibid., pp. 5-6.
55. Interviews with Ash, June 28, 1981 and August 5, 1981.
56. William E. Simon, *A Time for Truth* (New York: Readers Digest Press and McGraw-Hill, 1978), p. 47.
57. "Ever since President Nixon axed the President's Science Advisory Committee (PSAC) in 1973 because it insisted on offering advice—sometimes in public—that ran counter to his policies, elders of the scientific community have lamented the lack of a science advisory committee in the White House." Colin Norman, "Pared Down PSAC Proposed," *Science* 214 (December 1981): 1324.
58. Letter from Ehrlichman to Greenberger, April 15, 1981.
59. Neil de Marchi, "Energy Policy under Nixon: Mainly Putting Out Fires," in Craufurd D. Goodwin, ed., *Energy Policy in Perspective* (Washington, D.C.: Brookings Institution, 1981), p. 407.
60. Douglas R. Bohi and Milton Russell, *Limiting Oil Imports: An Economic History and Analysis* (Baltimore, Md.: Johns Hopkins Press for RFF, 1978).
61. Paul McCracken, "U.S. Fuel and Energy Situation," *Conference Board Record*, March 1971.
62. M. A. Adelman, *The World Petroleum Market* (Baltimore, Md.: Johns Hopkins Press for RFF, 1972).
63. The MacAvoy-Pindyck modeling project was initiated before the embargo. Ehrlichman recalls a second model being used in discussions of deregulation at the same time. He says it reached opposite conclusions on the question of elasticity.
64. Ehrlichman, letter to Greenberger; Simon, *A Time for Truth.*
65. The National Fuels and Energy Policy Study was established during the 92nd Congress under Senate Resolution 45. The effort extended over two Congresses and provided a framework for energy legislation following the Arab oil embargo. No final report was ever published.

66. Under Ash's guidance in 1970, the Bureau of the Budget was converted into OMB and the Domestic Council established. See Shultz and Dam, *Economic Policy*, pp. 160-61.
67. Several years later, DiBona was to become chief spokesman for the oil industry as president of the American Petroleum Industry.
68. A commentary on DiBona at the time is found in "Allocation at Long Last," *Time*, October 22, 1973, p. 77.
69. Shultz and Dam, *Economic Policy*, p. 185.
70. Letter from DiBona to Greenberger, January 12, 1981.
71. "Transcript of President's Address on the Energy Situation," *New York Times*, November 8, 1973, p. 32.
72. Edward Cowan, "Retiring U.S. Aide Scores Democrats' Energy Views," *New York Times*, June 13, 1976, p. 26.
73. Conversation with DiBona, June 29, 1981.
74. Wallace Turner, "Love Asserts He Lacked Means to Do Energy Job," *New York Times*, December 9, 1973, p. 31.
75. Leonard Larson, "Love to Quit Washington; Rejects 'Superfluous Job,'" *Denver Post*, December 3, 1973, p. 1.
76. Simon, *A Time for Truth*, p. 53.
77. "Transcript of President's Address."
78. Ernest Holsendorph, "Oil Independence—It Seems Unlikely," *New York Times*, December 2, 1973, section II, pp. 1, 15.
79. Thomas H. Tietenberg, *Energy Planning and Policy* (Lexington, Mass.: Lexington Books, 1976), p. 62.
80. John Herbers, "Ford Is Expected to Shift Cabinet," *New York Times*, October 31, 1974, p. 31.
81. Zarb, letter to Greenberger, April 21, 1981.
82. Federal Energy Administration, *Project Independence: A Summary* (Washington, D.C.: Government Printing Office, November 1974), p. 1.
83. Personal interview, August 15, 1979.
84. Federal Energy Administration, "Project Independence Report: Q's and A's," Washington, D.C.: Government Printing Office, November 1974), p. 39.
85. Edward Cowan, "Sawhill Is Cautious on Synthetic Fuels," *New York Times*, October 12, 1974, p. 39.
86. *New York Times*, October 11, 1975, p. 62. Ronald Reagan would also urge energy independence during his campaign for the presidency five years later. Ironically, Sawhill was at that time interim chairman of a Carter-promoted Synthetic Fuels Corporation established in reaction to the OPEC price hikes following the Iranian revolution.
87. J. Coleman, "Project Independence Evaluation System (B), Wharton School Case Study," University of Pennsylvania, Philadelphia, 1978, p.1.
88. Personal conversation with David Wood.
89. William W. Hogan, "Energy Policy Models for Project Independence," *Computers and Operations Research* 2 (1975): 251-71
90. Policy Study Group of the M.I.T. Energy Laboratory, "Energy Self-Sufficiency: An Economic Evaluation," *Technology Review*, May 1974, pp. 27-32.
91. Federal Energy Administration, *Project Independence Report: A Summary*.
92. Edward Cowan, "Ford Calls Aides to Weigh Energy and the Economy," *New York Times*, December 26, 1974, p. 45.

93. Teitenberg, *Energy Planning and Policy*, p. 87.

94. Committee on Interstate and Foreign Commerce, U.S. House of Representatives, *Analysis of the President's July 17, 1975 Program to Decontrol Domestic Oil*, 94th Congress, 1st Session (Washington, D.C.: Government Printing Office, July 1975), p. ii.

95. Committee on Interstate and Foreign Commerce, U.S. House of Representatives, *A Summary of Economic Analyses of Sudden Decontrol*, 94th Congress, 1st Session (Washington, D.C.: Government Printing Office, August 1975), pp. 17-21.

96. "Energy and Economic Impacts of Alternative Oil Pricing Policies" (Washington, D.C.: Federal Energy Administration, October 22, 1975); also, "Graphics Supplement" (Washington, D.C.: Federal Energy Administration, October 28, 1975). (Private papers.)

97. Simon, *A Time for Truth*, p. 79.

98. Interview with Schroeder, January 16, 1979, and letter to Greenberger, April 14, 1981.

99. Neil de Marchi, "The Ford Administration: Energy as a Political Good," in Craufurd D. Goodwin, ed., *Energy Policy in Perspective* (Washington, D.C.: Brookings Institution, 1981), pp. 492-94.

100. David Nissen, "The Impact of Assessment on the Modeling Process," in Saul Gass, ed., *Validation and Assessment Issues of Energy Models*, Washington, D.C.: Government Printing Office, National Bureau of Standards Special Publication 569, February 1980), p. 268.

101. *Energy Conservation and Production Act*, U.S. Public Law 94-385, enacted August 14, 1976, sections 51a, 142.

102. Professional Audit Review Team (PART), *Activities of the Energy Information Administration*, Washington, D.C.: Government Printing Office, 1979), pp. 25, 37.

103. Barry Commoner, *The Politics of Energy* (New York: Knopf, 1979), pp. 12-17. Commoner's charge is denied by Alvin L. Alm, head of the Schlesinger team. We were just "getting the base case straight," says Alm, and had not yet "decided what to do." Letter from Alm to Greenberger, June 29, 1981.

104. Harvey J. Greenberg, "Modeling the National Energy Plan," July 13, 1978, pp. 2-3. (Private Draft). Also in Fred Roberts, ed., *Proceedings of the Institute of Gas Technology Symposium* (Boulder, Colo.: Institute of Gas Technology, August 1977).

105. Letters from Representative John Dingell to James Schlesinger (June 20, 1978) and Lincoln Moses (July 24, 1978).

106. Letter from Lincoln Moses to William Hogan, July 27, 1978.

107. John P. Weyant, "The Role of Models in the Oil and Gas Pricing Debates" (Palo Alto, Calif.: Stanford University, 1979), p. 17. Chip Schroeder was a critic of the use of PIES by the Ford Administration during the debates on deregulation. For an interesting account, see Michael J. Malbin, *Unelected Representatives: Congressional Staff and the Future of Representative Government* (New York: Basic Books, 1980), pp. 214-19.

108. Executive Office of the President, *The National Energy Plan* (Washington, D.C.: Government Printing Office, April 1977), pp. 53-55.

109. Professional Audit Review Team, *Activities of Energy Information*.

110. Milton Holloway, ed., *Texas National Energy Modeling Project: An Experience in Large-Scale Model Transfer and Evaluation* (New York: Academic Press, 1980).

111. Ibid., pt. III, pp. 2-3.
112. In commenting on the effort to get decisionmakers to consider known data and the "potential implications of various policy options," Zarb says "PIES was only one part of the total epic." Zarb, letter to Greenberger, April 21, 1981.

CHAPTER 5

1. David Burnham, "Hope for Cheap Power From Atom Is Fading," *New York Times*, November 16, 1975, pp. 1, 58.
2. Richard D. Lyons, "A.E.C. Study Finds Hazards of Reactors Very Slight,"*New York Times*, August 21, 1974. The final version of the study appeared as "Reactor Safety Study: An Assessment of Accident Risks in U.S. Commercial Nuclear Power Plants," by Norman C. Rasmussen, WASH-1400, NUREG 75/014, U.S. Nuclear Regulatory Commission, October 1975.
3. U.S. Nuclear Regulatory Commission, Office of Public Affairs, "Nuclear Regulatory Commission Issues Policy Statement on Reactor Safety Study and Review by Lewis Panel," No. 79-19, January 19, 1979.
4. "The China Syndrome" with Jane Fonda and Jack Lemmon.
5. The Atomic Energy Act of 1954 permitted declassification and dissemination of nuclear technology and private ownership of nuclear facilities. See, Charles River Associates, "How U.S. Aid Got the Nuclear Power Industry Started," CRA Research Review (Boston: CRA, January 1980). Eisenhower's "Atoms for Peace" program offered to assist other countries in the development of nuclear energy in return for their agreement to use this assistance only for peaceful purposes.
6. Eliot Marshall, "Problems Continue at Three Mile Island," *Science* 213 (September 1981): 1344-45. General Public Utilities, according to its chairman, was in "desperate shape," with no access to financial markets. The value of its shares dropped from over $17 to $5.
7. The draft report of the second reactor safety study was very controversial. It prompted some 1,800 pages of comments from ninety organizations and individuals. The final version of the report, released fourteen months after the first draft, was not inconsistent with the Brookhaven study. The estimates for a worst-case accident, all revised upward from the draft version, were 3,300 early fatalities, 45,000 cases of early illness, $14 billion in property damage (due to contamination), plus significant long-term health effects. See Philip M. Boffey, "Rasmussen Issues Revised Odds on a Nuclear Catastrophe," *Science* 14 (November 1975): 640; Irvin C. Bupp and Jean-Claude Derian, *Light Water: How the Nuclear Dream Dissolved* (New York: Basic Books, 1978), p. 122.
8. Deborah Shapley, "Reactor Safety: Independence of Rasmussen Study Doubted," *Science* 197 (July 1977): 29. The charges were made by the Union of Concerned Scientists and its president, M.I.T. physics professor Henry Kendall on the basis of 50,000 internal AEC documents released under the Freedom of Information Act, as noted in the text and below.
9. Letter from Benedict to Greenberger, September 11, 1981.
10. Shapley, "Reactor Safety."
11. Ibid., p. 30.

12. In his later role as chairman of TVA, Freeman became a champion of nuclear power, although he never seemed to have much use for the Clinch River Breeder project.

13. David Burnham, "Atom Safety Study by U.S. Is Questioned,"*New York Times,* April 27, 1977, p. B6. This newspaper report was published on the day the AEC documents were made public.

14. Lyons, "A.E.C. Study Finds Hazards of Reactors Very Slight;" and "Reactor Safety Study: An Assessment of Accident Risks in U.S. Commercial Nuclear Power Plants,"WASH-1400, NUREG 75/014, U.S. Nuclear Regulatory Commission, October 1975.

15. Joanne Omang, "Nuclear Experts Fail to Agree On Power Hazards at Hearing," *Washington Post,* February 27, 1979.

16. WASH-1400, "Reactor Safety Study,"Executive Summary, pp. 1, 2, 10.

17. Letter from Frank von Hippel to Representative Morris K. Udall, February 16, 1979; *Reactor Safety Study Review* (Oversight Hearing Before the Subcommittee on Energy and the Environment of the Committee on Interior and Insular Affairs, House of Representatives, Serial No. 96-3, February 26, 1979), pp. 200-201.

18. H. W. Lewis, et al., "Risk Assessment Review Group Report to the U.S. Nuclear Regulatory Commission," NUREG/CR-0400 (Springfield, Va.: National Technical Information Service, 1978), p. 32. Hearings on the Lewis critique were held on February 26, 1979, by Representative Morris Udall's House Interior Subcommittee on Energy and Environment. See "Reactor Safety Study Under Scrutiny In Congress," *The Energy Daily,* February 27, 1979.

19. Interview with Rasmussen, February 26, 1979.

20. "NRC Panel Commends and Criticizes Rasmussen Report," *Physics Today,* December 1978, p. 95. Work on the Reactor Safety Study, started in the summer of 1972 under AEC sponsorship, was completed under the aegis of the NRC, which came into being on January 19, 1975.

21. A ten-year extension to the Price-Anderson amendment was first passed in 1974, but the President vetoed it because Congress had provided for its possible reversal pending revision of the Rasmussen report. Under great pressure, Rasmussen released the final version of his report on October 30, 1975, the day before the Joint Atomic Energy Committee was to reconsider extension of the amendment. See, "JAEC Delays Price-Anderson Markup Pending Review of Complete Rasmussen Safety Study," Bureau of National Affairs, no. 117, November 6, 1975, p. A-3. The year, 1975, incidentally, was both the 30th anniversary of the bombing of Hiroshima and the year of the fire at Brown's Ferry.

22. Joanne Omang, "Nuclear Power Regulator Says Process Is 'Prudent'," *Washington Post,* February 6, 1979. The congressman quoted was Senator Gary Hart of Colorado. He expected that the public view might be "fundamentally altered" by the NRC reevaluation of the Rasmussen report.

23. Comments made by Ahearn to Greenberger, September 1981.

24. Richard Myers, "Reactor Safety Study" (WASH-1400) Enjoys a Renaissance,"*The Energy Daily,* November 21, 1979, p. 4.

25. "Nuclear Regulatory Commission Issues Policy Statement on Reactor Safety Study and Review by Lewis Panel," Press Release, no. 79-19, NRC Office of Public Affairs, Washington, D.C., January 19, 1979; "NRC Official Says Rasmussen Report Did Not Influence Reactor Licensing," *The Energy Daily,* January 30, 1979, p. 5; "NRC Downplays Role of Rasmus-

sen Study in Nuclear Safety," *Current Report,* The Bureau of National Affairs, February 8, 1979, p. 7.

26. In fact, the risks of severe damage to a reactor's core may be significantly higher than WASH-1400 implied, as an analysis based on actual operating experience found several years after TMI. Eliot Marshall, "Using Experience to Calculate Nuclear Risks," *Science* 217 (July 23, 1982): 338-39.

27. "Chief Rasmussen Report Reviewer Believes Risk Lower Than Projected," *Energy Users Report,* February 22, 1979, p. 3; Burt Solomon, "Reactor Safety Study Under Scrutiny in Congress," *The Energy Daily,* February 27, 1979, p. 2.

28. The incorrect reporting is in an article by J. P. Smith with the headline, "'75 Report on Reactor Safety Is Called Unreliable by NRC," *Washington Post,* January 20, 1979, pp. Al, A10. A phone call to the author of the article did not clear up the misunderstanding. A similar error is in Robert W. Deutsch, "Defending the nuclear option," *The (Baltimore) Sun,* March 4, 1979, p. K3.

29. Joanne Omang, "Scientists Call for Closing 16 Atomic Power Plants," *Washington Post,* January 27, 1979; "Scientists Say Government May Have To Shut Down 16 Plants," *Energy Users Report,* February 1, 1979, p. 10.

30. David Burnham, "Federal Actions Renew Questions on Nuclear Power's Future in U.S.," *New York Times,* March 19, 1979, p. A16.

31. Thomas O'Toole, "Estimates of Future A-Plants Revised Downward," *Washington Post,* October 26, 1979.

32. Joanne Omang, "A-Plant Licensing Frozen to Assess Impact of Mishap," *Washington Post,* May 22, 1979, pp. A1, A6.

33. In the early 1970s, the utility industry was projecting an annual average growth in the demand for electricity of over 7 percent. This was subsequently cut back by more than one-half in light of the decline in demand caused by soaring fuel costs and power rates.

34. Michael Brody, "Nuclear Fizzle," *Barron's* August 24, 1981, p. 21.

35. William Ramsay, "Reactor Accidents and Afterthoughts," *Resources,* Resources for the Future, no. 62, Washington, D.C., April-July 1979, pp. 1-6.

36. "Chapter 3: A Swift Rethinking of the 'Unthinkable'," *Washington Post,* April 8, 1979, p. 17.

37. John Kemeny, Chairman, "Report of the President's Commission on the Accident at Three Mile Island" (Washington, D.C.: Government Printing Office, 1979), p. 32 (as excerpted by the staff, Nuclear Safety Analysis Center, Electric Power Research Institute, in Attachment A to a letter from Floyd L. Culler, President, Palo Alto, Calif., November 7, 1979, p. 4).

38. Myers, "Reactor Safety Study (WASH-1400) Enjoys A Renaissance," pp. 3-4.

39. "Randolph Says ERDA Budget Fails to Meet Non-Nuclear Act Requirements," *Energy Users Report,* March 13, 1975, p. A-15.

40. President's Materials Policy Commission, *Resources for Freedom* (Washington, D.C.: Government Printing Office, 1952).

41. Neil de Marchi, "Energy Policy Under Nixon: Mainly Putting Out Fires," in Craufurd D. Goodwin, ed., *Energy Policy in Perspective* (Washington D.C.: Brookings Institution, 1981), p. 459.

42. Dixy Lee Ray, *The Nation's Energy Future,* Wash-1281, Atomic Energy Commission (Washington, D.C.: Government Printing Office, December 1, 1973).

43. Thomas O'Toole, "$11 Billion Asked for Energy Search," *Washington Post,* December 3, 1973, p. A14.

44. Letter from Dixy Lee Ray to Greenberger, April 1, 1981.

45. Ibid.

46. Harold M. Schmeck, "U.S. Urged to Put Brake on Energy," *New York Times,* January 21, 1975, p. 16.

47. Energy Policy Project of the Ford Foundation, *A Time to Choose: America's Energy Future* (Cambridge, Mass.: Ballinger Publishing Company, 1974), p. 343.

48. Interview with Fri, February 6, 1979.

49. Energy Research and Development Administration, *A National Plan for Energy Research, Development and Demonstration, Volume 1: The Plan: Volume 2: Program Implementation* (Washington, D.C.: Government Printing Office, 1975).

50. U.S. Congress, Office of Technology Assessment,*Comparative Analysis of the 1976 ERDA Plan and Program* (Washington, D.C.: Government Printing Office, May 1976).

51. Letter from Paul P. Craig to Kenneth Hoffman with copy to Greenberger, August 9, 1976.

52. Comments by Richard H. Williamson.

53. Interview with LeGassie, February 12, 1979.

54. William D. Metz, "Energy: ERDA Stresses Multiple Sources and Conservation," *Science,* August 1, 1975, p. 369.

55. Letter from Fri to Greenberger, April 1, 1981.

56. The link of the Hudson-Jorgenson model with the Brookhaven system is described in Kenneth C. Hoffman and Dale W. Jorgenson, "Economic and Technological Models for Evaluation of Energy Policy," *Bell Journal of Economics* 8, no. 2 (Autumn 1977): 444-66.

57. Edward G. Cazalet, "SRI-Gulf Energy Model: Overview of Methodology" (Menlo Park, Calif.: Stanford Research Institute, January 1975), p. 1.

58. Stephen Rattien and John Bell "Critical Review of Synfuels Projections—Models and Methodologies" (Washington, D.C.: National Science Foundation, Office of Energy Policy, March 17, 1975).

59. Several energy studies used a computer model of the energy/economy. The Energy Policy Project had the Hudson-Jorgenson model, Project Independence had PIES, ERDA had the Brookhaven model, and the Synfuels Commercialization Study had the SRI/Gulf model. Significantly different from one another, the models were applied and developed for a variety of clients. Updated repeatedly, they played active roles in the energy policy analyses of the seventies. The Hudson-Jorgenson model, for example, was used in some two dozen reports for various government agencies and was integrated with both the Brookhaven and PIES models in work for ERDA and the Department of Energy.

60. John P. Weyant, "Quantitative Models in Energy Policy," *Policy Analysis* 6, no. 2 (Spring 1980): 229.

61. Synfuels Interagency Task Force, Recommendations for a Synthetic Fuels Commercialization Program, vol. II, "Cost/Benefit Analysis of Alternate Production Levels" (Washington, D.C.: Government Printing Office, November 1975).

62. Ibid., vol. I.

63. Brian L. Crissey, "A Framework for the Rational Analysis of Policy Decisions" (Ph.D. dissertation, Johns Hopkins University, Baltimore, Md., 1975).

64. Edward G. Cazalet, et al., "Decision Analysis of the Synthetic Fuels Commercialization Program," draft memo, prepared for the Synfuels Interagency Task Force (Washington, D.C.: Government Printing Office, June 25, 1975), p. 3.

65. "House Kills Loan Guarantees for Synfuels," *Weekly Energy Report,* December 15, 1975, p. 2.

66. "House Kills Synthetic Fuels Program—Or Does It?" *The Energy Daily,* September 27, 1976, pp. 1-2.

67. Ed Edstrom, "General Accounting Office Comes Down Against Commercial Loan Guarantees for Synthetic Fuels," *The Energy Daily,* August 25, 1976, pp. 1-2.

68. Edward Cowan, "Simon Sees Energy Quest Cutting Petroleum's Value," *New York Times,* February 12, 1974, p. 1.

69. David A. Stockman, "The Wrong War? The Case Against a National Energy Policy," *The Public Interest,* Fall 1978, p. 6. An M.I.T. study group recommended against the plans for energy independence even while they were being formulated. See Victor K. McElheny, "U.S. Energy Plan Found Too Costly," *New York Times,* May 11, 1974, p. 39.

70. James L. Cochrane, "Carter Energy Policy and the Ninety-fifth Congress," in Craufurd D. Goodwin, ed., *Energy Policy in Perspective* (Washington, D.C.: Brookings Institution, 1981), p. 548.

71. The CONAES study released a primarily procedural interim report on Inauguration Day, with the final report then thought to be just six months away. "Interim Report of the National Research Council Committee on Nuclear and Alternative Energy Systems" (Washington, D.C.: National Academy of Sciences, January 1977). The fact that the interim report avoided stating conclusions on the issues did not prevent the press from making inferences that alarmed the president of Mobil Oil Corporation and others who had been disturbed by the conservation recommendations of the Ford EPP. Victor K. McElheny, "Specialists Advise Carter to Increase Emphasis on Energy," *New York Times,* January 21, 1977, p. D1. Such news reports caused William Tavoulareas to complain to study cochairman Harvey Brooks, with whom he had served on the EPP's advisory board.

72. David Herst, "The National Energy Plan" (Case for the Kennedy School, Harvard University, Cambridge, Massachusetts, October 15, 1980), pp. 9-10.

73. Edward Cowan, "President Signs Natural Gas Bill, Marking First Legislative Victory," *New York Times,* February 3, 1977, p. 1.

74. Conservation in another context has been defined as the "wise use of resources," prompting President Taft to remark many years ago, "A great many people are in favor of conservation no matter what it means." *Outlook,* May 14, 1910, p. 57.

75. L. R. Klein, Chairman, Economic Task Force, "Final Report on Economic Dimensions of the Energy Problems,"Memorandum to the Carter Administration's Transition Planning Group (Private paper, November 12, 1976), p. 19. Chaired by Lawrence R. Klein of the Wharton School, Carter's economic adviser during the campaign, this task force emphasized the need to get oil prices to approach world levels. Its recommendation against non-

price methods of conservation, such as promulgating standards for auto efficiency and new building insulation (p. 7) went unheeded, as did its endorsement of M.I.T. economist Morris Adelman's proposal that oil import entitlement tickets be auctioned off to producing countries by secret bid to break OPEC's power (pp. 16-17). Klein believes it was Senator Long's political muscle that played the dominant role in energy policy at the time and shaped Carter's approach. Letter from Klein to Greenberger, December 8, 1981.

76. "Transcript of Carter's Address to the Nation About Energy Problems," *New York Times*, April 19, 1977, p. 24.
77. Henry A. Kissinger, *White House Years* (Boston: Little, Brown, 1979), p. 955.
78. Charles Mohr, "Carter Asks Strict Fuel Saving; Urges 'Moral Equivalent of War' to Bar a 'National Catastrophe';" Hedrick Smith, "A Rare Call for Sacrifices: From a Peacetime President to Nation of Consumers, The Message Is Curtailment of the Dream of Plenty," *New York Times*, April 19, 1977, p. 1.
79. Charles Mohr, "Carter Calls for Fight on Energy Waste; Asks Congress to Help in 'Thankless Job'," *New York Times*, April 21, 1977, p. 1.
80. Herst, "The National Energy Plan," pp. 27-28.
81. Two years later, a second round of sharp oil price hikes following upon the Iranian revolution, along with continuing high inflation in the United States, spreading gasoline shortages, and an impending economic recession led to a major crisis of confidence in Carter's leadership. White House domestic affairs adviser Stuart E. Eizenstat cautioned the president on the political perils of the situation in a memorandum leaked to the press after Carter abruptly postponed an energy address, the fifth in his two and a half years in office. Secluding himself at Camp David, Carter conducted secret consultations for a week with 150 prominent Americans on a broad range of economic, moral, and philosphical issues. The initiatives he proposed on his return to Washington in an effort to restore support for his administration were reminiscent of those of his predecessors. He asked for creation of an Energy Mobilization Board empowered to circumvent state and local regulations for large, costly energy projects, and a Synthetic Fuels Corporation to oversee a crash program to establish a synthetic fuels industry. Carter spoke of saving 4 to 5 million barrels of imported oil a day while spending $88 billion to attain a daily production level of 2.5 million barrels of synthetic fuels by 1990. This was considerably larger than Gerald Ford's ten-year goal in 1975 of 1 million barrels, and well beyond the recommendations of the Synfuels Commercialization Study.

The Energy Mobilization Board never got congressional approval but the Synthetic Fuels Corporation did. To lead it, Carter appointed John C. Sawhill as deputy secretary of energy. Five years earlier, Sawhill, as head of the Federal Energy Administration (FEA), had warned Gerald Ford against a crash program to develop synthetic fuels. (Edward Cowan, "Sawhill Is Cautious on Synthetic Fuels," *New York Times*, October 12, 1974, p. 39.) Sawhill returned to the energy bureaucracy as second in command to newly appointed Energy Secretary Charles W. Duncan, Jr., former executive of the Coca Cola company who had been serving as deputy secretary of defense. There was a touch of musical chairs in Sawhill's return. He took the place of John F. O'Leary, who earlier had been his replacement once removed as the head of the FEA (after Zarb). When the FEA was absorbed into Carter's newly created Department of Energy, O'Leary was kept on as deputy secretary and spokesman for the new department. This permitted Schlesinger and other key members of the energy team to concentrate on drafting the energy program.

82. John W. Finney, "James Rodney Schlesinger," *New York Times,* December 24, 1976.

83. James R. Schlesinger, "Uses and Abuses of Analysis" (Memorandum prepared at the request of the subcommittee on national security and International Operations of the Committee on Government Operations, U.S. Senate, 1968), p. 1.

84. For interesting accounts of the history of the National Energy Plan, see Herst, "The National Energy Plan"; and Katie Hope, "The National Energy Plan: Congressional Action," (Case Study, Kennedy School of Government, Harvard University, October 29, 1980).

85. An extensive discussion of the use of analysis in the debate over natural gas pricing is given in Michael J. Malbin, *Unelected Representatives: Congressional Staff and the Future of Representative Government* (New York: Basic Books, 1979), pp. 204-36.

86. Stockman later became head of the congressional advisory panel on energy for presidential candidate Reagan and chaired the group that wrote the energy section of the Republican party's platform. (Eliot Marshall, "Energy Crisis in the Campaign," *Science* 10 [October 1980]: 164.) After Reagan was elected, Stockman was put in charge of OMB, where his indiscretion created problems for him and the president. (William Greider, "The Education of David Stockman," *The Atlantic Monthly,* December 1981, pp. 27-54).

87. Confidential review of the preliminary manuscript for this book, April 2, 1982.

88. "ERDAgate!" *Wall Street Journal,* May 20, 1977. This is the second of five *Wall Street Journal* editorials that used the natural gas projections of MOPPS to score points against the National Energy Plan and gas price regulation. The editorials started on April 27, 1977 with "1,001 Years of Natural Gas" and ran for almost a year. The third one was published as "Jimmy Carter on the Run" in June of 1977, and the fourth one appeared on September 14, 1977 as "20 Million Years of Energy." The last one came out on April 4, 1978. It was entitled, "The Memory Hole."

89. Interview with Fri.

90. Letter from Seamans to Greenberger, November 30, 1981.

91. The White House team told LaGassie that the (PIES) model it was using produced very different results from MOPPS. Skeptics again contended that PIES was being applied for political purposes. It certainly employed different premises. "A difficult time," said LeGassie.

92. Robert W. Fri, "Energy Imperatives and the Environment," in Charles J. Hitch, ed., *Resources for an Uncertain Future* (Baltimore, Md.: Johns Hopkins Press, 1978), p. 50.

93. The Energy Productivity Center of the Mellon Institute continued the MOPPS focus on "energy services" in its analysis aimed at developing a "least-cost energy strategy." See Burt Solomon, "Mellon Institute Preaches Conservation, Free Market," *The Energy Daily* 7, no. 183 (September 24, 1979), pp. 1-3.

94. United States Energy Research and Development Administration, "Market Oriented Program Planning Study (MOPPS)," September 1977. (Unpublished.)

95. Charles J. Hitch, "Review of MOPPS," Letter to Philip C. White, Resources for the Future, March 31, 1978.

96. Fri, "Energy Imperatives," p. 44.

97. J. Frederick Weinhold, who originally led the task force, had been transferred to the White House energy team. Knudsen took his place. When Knudsen left the project, his replacement was Richard H. Williamson, assistant to LeGassie. For an account of the Knudsen incident, see Morton Mintz, "Energy Aide's Controversial Findings Lead to Downfall in Bureaucracy," *Washington Post,* June 4, 1977.

98. Letter from Adams, November 1981.

99. See, for example, James P. Sterba, "Gas Shortage a Fundamental, Long-Term Economic Threat to U.S., Experts Say," *New York Times,* February 22, 1977, pp. 4-6.

100. Aaron Wildavsky and Ellen Tenenbaum, *The Politics of Mistrust: Estimating American Oil and Gas Resources* (Beverly Hills, Calif.: Sage Publications, 1981), p. 240.

101. "1,001 Years of Natural Gas," April 27, 1977.

102. Morton Mintz, "Energy Aide's Controversial Findings Lead to Downfall in Bureaucracy," *Washington Post,* June 4, 1977.

103. The *Wall Street Journal* editorials may have played a part in bringing about the widely publicized hearings on the Knudsen affair by the Senate Committee on Energy and Natural Resources. Committee members Howard Metzenbaum of Ohio and John A. Durkin of New Hampshire "treated Knudsen as a man of integrity who had been wronged by ERDA bureaucrats seeking only to preserve their own interests." Wildavsky and Tenenbaum, 1981, p. 242.

104. "ERDAgate," May 20, 1977.

105. Letter from Fri, November 19, 1981.

106. "The Memory Hole," *Wall Street Journal,* April 4, 1978.

107. Wildavsky and Tenenbaum, *The Politics of Mistrust,* p. 247.

108. Steven Carhart in a letter to Greenberger, January 4, 1982.

109. "The Memory Hole."

CHAPTER 6

1. Nuclear Energy Policy Study (NEPS) Group, *Nuclear Power: Issues and Choices* (Cambridge, Mass.: Ballinger Publishing Company, 1977), pp. xi-xii.

2. In California, the nuclear referendum was known as Proposition 15. For a discussion of the issues, see Martin Greenberger, "The Nuclear Industry Could Go Bust in California," (Baltimore) *Sunday Sun,* May 23, 1976, Perspective Section. See also, Institute for Energy Studies, *The California Nuclear Initiative: Analysis and Discussion of the Issues* (Palo Alto, Calif.: Stanford University Press, 1976).

3. Dick Kirschten, "Nuclear Lobbying—It's Not as Simple as 'Us Against Them'," *National Journal* no. 7, February 18, 1978, p. 264.

4. Letter from Hans Bethe to Greenberger, September 16, 1981.
5. The American Physical Society study drew attention to weaknesses in WASH-1400. It is one of the studies initiated by the Society in its efforts to fill the void left by the demise of PSAC. NSF, in taking over the science advisory apparatus from OST during the Nixon Administration, urged professional societies to promote policy-relevant work. Another study by the American Physical Society, this one titled "Nuclear Fuel Cycles and Waste Management," proceeded concurrently with Ford-MITRE. It was published in *Reviews of Modern Physics* 50, no. 1, pt. II (January 1978).
6. For an account of the formation of the study group, see "Wise Men to Scratch Heads over Nuclear Issues," *Science,* February 6, 1976, p. 449.
7. Bundy, long associated with arms control, attended most monthly meetings and the summer session in Aspen. Another active observer was Larry Ruff, who monitored the project and helped put the economic analysis in readable form. Members of the study group in addition to Keeny were Seymour Abrahamson, University of Wisconsin professor of genetics; Kenneth Arrow, Harvard professor of economics; Harold Brown, president of the California Institute of Technology; Albert Carnesale, Harvard nuclear engineer; Abram Chayes, Harvard professor of law; Hollis B. Chenery, economist at the World Bank; Paul Doty, Harvard professor of science policy and international affairs; Philip Farley, Brookings senior fellow; Richard L. Garwin, IBM Fellow; Marvin Goldberger, Princeton professor of physics; Carl Kaysen, M.I.T. professor of political economy; Hans H. Landsberg, RFF economist; Gordon J. MacDonald, Dartmouth professor of environmental studies; Joseph S. Nye, Jr., Harvard professor of government; Wolfgang K. H. Panofsky, director of the Stanford Linear Accelerator Center; Howard Raiffa, Harvard professor of managerial economics; George Rathjens, M.I.T. professor of political science; John C. Sawhill, president of New York University; Thomas C. Schelling, Harvard professor of political economy; and Arthur Upton, professor of pathology at Stony Brook.
8. Nuclear Energy Policy Study Group, *Nuclear Power,* p. xii.
9. Committee for Economic Development (CED), *Nuclear Energy and National Security* (New York: CED, September 1976), pp. 14-16. The CED study may have influenced the Ford Administration in its decision to defer reprocessing and plutonium recycle. This decision, taken in late October 1976 at a time when Carter was raising nuclear issues in his election campaign, drew on a special task force report by Robert Fri. Fri attended the dinner at which the results of the CED study were presented.
10. See, for example, Albert Wohlstetter, Henry Rowen, et al., *Moving Toward Life in a Nuclear Armed Crowd?* report prepared by Pan Heuristics for U.S. Arms Control and Disarmament Agency, ACDA/PAB-263, PH76-04-389-14, December 4, 1975 (revised, April 22, 1976).
11. Hugh Nash, Review of *Nuclear Power: Issues and Choices,* "Reviews," *Friends of the Earth,* August/September 1977, p. 20.
12. "Theft-Resistant Nuclear System Already Tested, Scientists Say," *New York Times,* March 26, 1977, pp. 25, 27.
13. Nuclear Energy Policy Study Group, *Nuclear Power,* p. 68. Also, see Leonard Silk, "An Encouraging Projection of U.S. Energy and Growth," *New York Times,* April 4, 1977, pp. 45, 47.
14. Institute for Energy Analysis, "Economic and Environmental Implications of a U.S. Nuclear Moratorium" (Oak Ridge, Tenn.: Oak Ridge Associated Universities, 1976).

15. Jerry Ackerman, "Nuclear Power Is Not Essential, Report Concludes," *Boston Globe,* March 22, 1977; "New A's for Nuclear Q's," *New York Times,* March 24, 1977.

16. "We need nuclear energy," *Christian Science Monitor,* March 30, 1977.

17. Edward Cowan, "Schlesinger Against Plutonium Fuel Use in Nuclear Reactors," *New York Times,* March 26, 1977, p. 1. Also, see William D. Metz, "Ford-MITRE Study: Nuclear Power Yes, Plutonium No," *Science,* April 1, 1977, p. 41.

18. Over half of the twenty-one Ford-MITRE participants became associated with the new administration at some point, most in responsibilities directly related to nuclear policy. Brown became secretary of defense. Carnesale was made U.S. representative to the INFCE Technical Coordinating Committee, later to be nominated by Carter as chairman of the Nuclear Regulatory Commission. Chayes, an adviser to the Carter campaign who joined with Carnesale to draft Carter's May 1976 U.N. speech on nuclear policy, served as chairman of the INFCE Technical Coordinating Committee. Doty and Panofsky were named to ACDA's general advisory committee. Farley became deputy to the U.S. Special Representative for Nonproliferation Matters. Keeny took over as deputy head of ACDA. Nye served as deputy to the under secretary of state for security assistance, science and technology, where he chaired the National Security Council Group that formulated Carter's nonproliferation policy. Rathjens replaced Farley under a different title. Sawhill became under secretary of energy, and later president of the Carter-created Synthetic Fuels Corporation. Upton was made director of the National Cancer Institute.

19. The Japanese and West Germans, according to Landsberg, reacted to the Ford-MITRE book (which they translated) as "pure poison." Carter had told Fukuda that the reprocessing plant Japan was building was unnecessary, wasteful, and proliferative. Fukuda disagreed. See Richard Halloran, "Fukuda Presses Case for A-Plant," *New York Times,* March 23, 1977, p. 3.

20. Nuclear Energy Policy Study Group, *Nuclear Pwer,* p. 43. Furthermore, assumptions sometimes depend on desired results. See Alan S. Manne and Richard G. Richels, "Evaluating nuclear fuel cycles: Decision analysis and probability assessments," *Energy Policy,* March 1980, pp. 3-16.

21. James Edwards, former governor of South Carolina, devoted himself to resuscitating the Barnwell reprocessing plant while secretary of energy in the Reagan Administration. (Jack Anderson, *Washington Post,* January 30, 1981.) On October 8, 1981, "with a minimum of fanfare," he and Reagan announced the decision to "reverse the Carter Administration's ban on private reprocessing of used reactor fuel." (Eliot Marshall, Reagan's Plan for Nuclear Power, *Science,* 214 (October 23, 1981): 419.

22. Douglas Colligan, "Plutonium: How Safe?" *Science Digest,* March 1977, p. 63.

23. Mason Willrich and Theodore B. Taylor, *Nuclear Theft: Risks and Safeguards* (Cambridge, Mass.: Ballinger Publishing Company, 1974).

24. Nuclear Energy Policy Study Group, *Nuclear Power,* pp. 22, 271, 299.

25. Kirschten, "Nuclear Lobbying," pp. 261, 263.

26. "Questions about the Ford-MITRE Report, *Nuclear Power: Issues and Choices,*" draft report for comments only, April 11, 1977, p. 12.

27. John H. Barton, Book Review of NEPS, *The American Political Science Review,* 73 (March 1979): 349-50.

28. Martin Greenberger, Matthew A. Crenson, and Brian L. Crissey, *Models in the Policy Process* (New York: Russell Sage Foundation, 1976), pp. 145-46.

29. Rene H. Males and Richard G. Richels, "Economic Value of the Breeder Technology: A Comment on the Ford-MITRE Study" (Palo Alto, Calif.: Electric Power Research Institute, April 12, 1977). Also, see Rene H. Males, "Comments on the Ford-MITRE Study *Nuclear Power Issues and Choices*" (Testimony before the Subcommittee on Nuclear Regulation of the Senate Committee on Environment and Public Works, Washington, D.C., April 28, 1977).

30. Victor K. McElheny, "Lag in Building Breeder Plants Termed Costly," *New York Times,* May 30, 1977, p. 23-24.

31. John P. Holdren, "A Strategy To Buy Time—New Policy Study Provides Some Intellectual Underpinning for Carter's Energy Program," *Bulletin of the Atomic Scientists,* June 1977, p. 58.

32. Report of the Steering Committee of the LMFBR Review, ERDA, Washington, D.C., 1977. The committee included in its membership Chauncey Starr, then president of EPRI; Floyd L. Culler, Jr., who succeeded him at EPRI; Thomas Cochran, antinuclear spokesman for the National Resources Defense Council; and Frank von Hippel, physicist and environmentalist at Princeton. The polarized committee produced two reports, one favoring commercialization of the breeder, the other advising postponement. Coauthor of the second report, von Hippel, felt that by puncturing the assumption of continued electrical growth, the report helped Carter "resist great pressures over the next three years to continue subsidizing the commercialization of the breeder reactor." Frank von Hippel, "The Emperor's New Clothes—1981," *Physics Today,* July 1981, p. 38.

33. Letter from Panofsky to Greenberger, September 8, 1981.

34. Letter from Fri to Greenberger, September 8, 1981.

35. Fri and Nye had met at a December 1976 Council on Foreign Relations seminar, chaired by Nye, at which Fri spoke. Fri was a friend of Nye's college roomate, former FEA head John Sawhill.

36. Joseph S. Nye, "Maintaining a Nonproliferation Regime," *International Organization* 35, no. 1 (Winter 1981): 24.

37. Ibid.

38. Jay James, Jr., "Energy Costs: Nuclear Versus Oil," Letter to the Editor, *Science* 200 (April 28, 1978): 381-82.

39. Eliot Marshall "INFCE: Little Progress in Controlling Nuclear Proliferation," *Science* 207 (March 28, 1980): 1446.

40. INFCE, Summary Volume (Vienna: IAEA, January 1980), p. 1.

41. Amory B. Lovins, L. Hunter Lovins, and Leonard Ross, "Nuclear Power and Nuclear Bombs," *Foreign Affairs,* Summer 1980, p. 1138.

42. Letter from Nye to Greenberger, September 15, 1981.

43. Marshall, "INFCE: Little Progress," pp. 1446-47.

44. Lovins, Lovins, and Ross, "Nuclear Power and Nuclear Bombs,"p. 1138.

45. Interview with Lovins, March 10, 1979.

46. Lovins, Lovins, and Ross, "Nuclear Power and Nuclear Bombs," p. 1143. For a fuller exposition, see Amory B. Lovins and L. Hunter Lovins, *Energy/War: Breaking the Nuclear Link* (New York: Harper Colophon, 1981).

47. Robert Frost, "The Road Not Taken," *The Poetry of Robert Frost,* ed. Edward Connery Lathem (New York: Holt, Rinehart and Winston, 1969).

48. According to figures cited by von Hippel, the utility industry's average annual growth projections for the period 1975 to 2000 dropped from 4.6 percent in late 1977 to 4 percent in 1978 to 3.9 percent in 1979 to 3.5 percent in 1980, representing a reduction of about 40 percent in projected electricity consumption in the year 2000. Frank von Hippel, "The Emperor's New Clothes 1981," *Physics Today,* July 1981, p. 38.

49. Letter from Holdren to Greenberger, September 16, 1981.

50. Holdren, "A Strategy To Buy Time," p. 62.

51. H. A. Bethe and A. B. Lovins, exchange of letters, *Foreign Affairs 55* (April 1977): 636-40.

52. Reply to the Letter to the Editor by Jay James, Jr. (above), "Lovins on Energy Costs," *Science* 201 (September 22, 1978): 1077-78. Also, see J. Michael Gallagher, "Lovin's Data Source," Letter to the Editor, *Science* 202 (December 22, 1978): 1242-44; and Reply by Lovins, "Energy: Bechtel Cost Data," *Science* 204 (April 13, 1979): 124-27.

53. Ellen Frank, "A True Friend of the Earth," *New York Times Magazine,* printed in *The Chronicle,* November 18, 1978.

54. George L. Gleason, "Hard Facts About Soft Energy"(Paper presented at the E. F. Hutton Fixed Income Research Conference on Public and Investor Owned Electric Utilities, New York, March 8, 1979), p. 1.

55. Alvin M. Weinberg and Amory B. Lovins, "Energy Exchange," *The Sciences* (New York: The New York Academy of Sciences, February 1980), p. 13.

56. Alvin M. Weinberg, review of *Soft Energy Paths* by Amory B. Lovins, *Energy Policy,* March 1978, pp. 85-87. Weinberg, in private conversation on February 5, 1982, indicated that the de facto moratorium on nuclear energy in the United States meant that Lovins had "won this part of the battle." Weinberg was busily engaged at the time in research supported by the Mellon Foundation examining "The Second Nuclear Era" fifteen years hence.

57. Allen L. Hammond, " 'Soft Technology' Energy Debate: *Limits to Growth* Revisited?", *Science* 196 (May 27, 1977): 961.

58. Alvin M. Weinberg and Amory B. Lovins, Book Review and Reply, *Non-Nuclear Futures: The Case for an Ethical Energy Strategy* (Cambridge, Mass.: Ballinger Publishing Company, 1975), *Energy Policy,* December 1976, p. 366.

59. Amory B. Lovins, *Soft Energy Paths: Toward a Durable Peace* (New York: Harper Colophon Books, 1979), p. 11-12.

60. By 1981, Lovins was estimating that a total U.S. primary energy demand of around ten to fifteen quads a year could be achieved over the next century with considerable economic growth and cost minimizing investment. See Amory B. Lovins, L. Hunter Lovins, et al., *Least-Cost Energy: Solving the CO_2 Problem* (Andover, Mass.: Brick House, 1981).

61. Lovins, *Soft Energy Paths,* p. 64.

62. Ibid., pp. 65-66.

63. Letter from Carroll L. Wilson to Greenberger, September 14, 1981.

64. Amory B. Lovins, "Energy Strategy: The Road Not Taken?", *Foreign Affairs* 55, no. 1 (October 1976): 65-96.

65. Carroll L. Wilson, "A Plan for Energy Independence," *Foreign Affairs* 51, no. 4 (July 1973): 657-75.

66. A later article in *Foreign Affairs* by Robert Stobaugh and Daniel Yergin ran somewhat longer than Lovins's article and was almost equally anno-

tated. The Lovins, Lovins, and Ross article was longer still, and the most heavily annotated.

67. Amory B. Lovins, "Scale, Centralization and Electrification in Energy Systems," in *Future Strategies for Energy Development: A Question of Scale* (Proceedings of a Conference at Oak Ridge, Tennessee, Oak Ridge Associated Universities, 1977).

68. Lovins, *Soft Energy Paths.*

69. "Soft vs. Hard Energy Paths," *Electric Perspectives,* no. 77/3 (New York: Edison Electric Institute, 1977).

70. "Alternative Long-Range Energy Strategies," *Hearing Record of the Senate Small Business and Interior Committees,* 2 vols. (Washington, D.C.: Government Printing Office, 1978).

71. Hugh Nash, ed, *The Energy Controversy: Soft Path Questions and Answers* (San Francisco: Friends of the Earth, 1979).

72. Lovins was unable to attend this session at the 37th annual meeting of the members and board of trustees of the Institute of Gas Technology, Chicago, November 16, 1978.

73. Amory B. Lovins, "Electric Utility Investments: *Excelsior* or Confetti?"(Invited remarks to the E. F. Hutton Fixed Income Research Conference on Public and Investor Owned Electric Utilities, New York, March 8, 1979). Reprinted in *Journal of Business Administration* (Vancouver) 12, no. 2 (1981).

74. "Springboards to Soft Energy Studies: Amory Lovins's Ingredients for a SEP Do-It-Yourself Kit," *Soft Energy Notes* 2 (San Francisco: International Program for Soft Energy Paths, May 1979).

75. See, for example, Daniel W. Kane, "A Perspective As To The Total Cost of 'Soft' Energy," in Edison Electric Institute, *Electric Perspectives,* p. 12.

76. *The Journals of David E. Lilienthal: The Atomic Energy Years* 2 (New York: Harper and Row, 1964), p. 117. "Carroll" is the name Wilson inherited from his mother's family. Louis was his father's given name.

77. Carroll L. Wilson, "Nuclear Energy: What Went Wrong?", *The Bulletin of Atomic Scientists,* June 1979, p. 13.

78. Letter from Wilson to Greenberger.

79. Jay W. Forrester, *World Dynamics,* (Cambridge, Mass.: Wright-Allen Press, 1971). Also, Donella H. Meadows, et al., *The Limits To Growth* (New York: Universe Books, 1972).

80. For an accounting of Forrester's presentation to the Club of Rome meeting and of how that led to the Limits to Growth study see Greenberger, Crenson, and Crissey, *Models in the Policy Process,* pp. 3-7.

81. Wilson, "A Plan for Energy Independence." Wilson credits Bundy, who read drafts of the speech, with having turned the section on energy independence into an article. Bundy published a closely related piece by James Akins, "This Time The Wolf Is Here," which Wilson's article cites, in the previous issue of *Foreign Affairs.* Chapter 1 provides more information on Akins.

82. Interview with Wilson, September 17, 1979.

83. Ibid.

84. WAES, *Energy Demand Studies: Major Consuming Countries* (Cambridge, Mass.: M.I.T. Press, 1976), Technical Appendix I.

85. Carhart later became assistant director for integrative analysis of the Mellon Institute's Energy Productivity Center. See Roger W. Sant and Steven C. Carhart, "Eight Great Energy Myths: The Least-Cost Energy

Strategy—1978-2000," Energy Productivity Report No. 4, The Energy Productivity Center (Arlington, Va.: Mellon Institute, 1981).

86. "Running Short, No Matter What," *Time,* May 23, 1977, p. 63.

87. Letter from Basile to Greenberger, March 29, 1982.

88. WAES, *Energy: Global Prospects 1985-2000,* Report of the Workshop on Alternative Energy Strategies (New York: McGraw Hill, 1977), pp. 26, 104.

89. Interview with Sternlight, June 13, 1979.

90. WAES, *Energy: Global Prospects,* fig. I-2, p. 7; fig. 1-4, p. 66; and pp. 262-63. The report says that Case D-3 "was based on a highly aggregate analysis, compared to the more detailed analysis for the other 4 year-2000 WAES cases."

91. Ibid., p. xxiv. WAES produced three large technical appendixes to the report: *Energy Demand Studies, Energy Supply to the Year 2000: Global and National Studies* (containing the supply projections), and *Energy Supply-Demand Integrations to the Year 2000: Global and National Studies* (containing the integrations); MIT Press, Cambridge, Massachusetts, 1976-77.

92. Ibid., p. xi.

93. Ibid., p. xxv.

94. Carroll L. Wilson, *Coal—Bridge to the Future,* Report of the World Coal Study, WOCOL (Cambridge, Mass.: Ballinger Publishing Company, 1980).

95. For an overview of the international coal situation at the time, see Edward D. Griffith and Alan W. Clarke, "World Coal Production," *Scientific American* 240, no. 1 (January 1979): 38-47. Griffith and Clarke were both WAES Associates.

96. The fact that the *Wall Street Journal* refused to honor the customary embargo on reviews may have contributed to the overly hasty reporting.

97. Robert Cowen, "Hanging Together: International Energy Strategies," *Technology Review,* July/August 1977, p. 8.

98. Letter from Press to Greenberger, September 10, 1981.

99. Many commentaries compared WAES to the CIA analysis, accusing it of the same "scare tactics."See, for example, *Congressional Record,* Senate, May 20, 1977, for the comments of Congressman McClure; "A Shortage of Intelligence," *National Review,* June 24, 1977. A Senate committee review of the CIA analysis called WAES "perhaps the only study more pessimistic than the reports published by the CIA." See, Senate Select Committee on Intelligence, "The Soviet Oil Situation: An Evaluation of CIA Analyses of Soviet Oil Production," Staff Report, May 1978, Advance News RElease, May 21, 1978, p. 11.

100. House Speaker Thomas P. (Tip) O'Neill, Jr. and Senate Majority Leader Robert C. Byrd were both said to be urging Carter to emphasize the seriousness of the energy situation. See J. P. Smith, "Carter's Energy Policy: A Melding of Pessimism, Optimism,"*Washington Post,* April 21, 1977, p. A24.

101. CIA, "The International Energy Situation: Outlook to 1985," ER 77-10240 (Washington, D.C.: CIA April 1977), pp. 1-2.

102. Thomas O'Toole, "CIA Foresees Global Oil Shortage," *Washington Post,* April 16, 1977, p. A1.

103. Commenting on them the year after they were released, David Stockman referred to the CIA projections as "the central justification for the Carter Administration's National Energy Plan" and called them "almost ludi-

crous." David A. Stockman, "The Wrong War? The Case Against a National Energy Policy," *Public Interest,* Fall 1978, p. 32.

104. WAES, *Energy: Global Prospects,* p. 56.

105. Senate Select Committee News Release, "The Soviet Oil Situation," pp. 3-6.

106. "The International Energy Situation: Outlook to 1985" was made public on April 18, 1977. The companion analysis, "Prospects for Soviet Oil Production," predicting that Soviet oil production would peak in the early 1980s, was made public on April 25, 1977. Follow-up reports include "A Discussion Paper on Soviet Petroleum Production," May 1977; Prospects for Soviet Oil Production, A Supplemental Analysis," ER 77-10425 U, July 1977; and "Soviet Economic Problems and Prospects," prepared for the Subcommittee on Priorities and Economy in Government, Joint Economic Committee, August 8, 1977.

107. Thomas O'Toole and Lee Lescaze, "Soviets Seen Vying for Mideast Oil," *Washington Post,* April 19, 1977, p. A15.

108. Anthony J. Parisi, "Exxon Company Emphasizes Role of Conservation," *New York Times,* April 25, 1978, p. 57.

109. Marshall I. Goldman, "The C.I.A. and Oil," *New York Times,* April 28, 1977, p. 29.

110. "Under any but optimistic scenarios for oil production, and in the absence of a high priority campaign to save oil domestically, the USSR will shift from earning to spending hard currency in its oil trade." CIA, "Soviet Economic Problems and Prospects," August 8, 1977, p. 22.

111. Senate Select Committee News Release, "The Soviet Oil Situation," pp. 16, 26-27.

112. Letter from Ernst to Greenberger, September 24, 1981.

113. Steven Rattner, "Sign Appears That Soviet Supply of Oil Is Tightening," *New York Times,* November 21, 1977, pp. 61, 63.

114. Richard Halloran, "CIA Sees Soviet Importing Oil Soon," *New York Times,* July 30, 1979, pp. D1, Dr.

115. CIA, "The World Oil Market in the Years Ahead," ER 79-10327U, National Foreign Assessment Center, August 1979, pp. 3-4, 40.

116. Marshall I. Goldman, "The CIA Goof on Soviet Oil," *Washington Post,* August 19, 1979, pp. E1-E2.

117. Marshall I. Goldman, "The Soviet-Oil Alarum," *New York Times,* April 17, 1980.

118. "Soviet Oil Projections Up for Grabs," *Science* 213 (September 18, 1981).

119. See, for example, Robert C. Toth, "CIA's Estimate of Soviet Petroleum Production Raised," *Los Angeles Times,* May 20, 1981, pp. 1, 19; "Soviet Oil Decline Put Off 3 Years," *Science* 212 (June 12, 1981): 1252. These accounts were based on informal remarks by a CIA analyst at a May 14, 1981 seminar sponsored by Harvard's Russian Research Center. A reporter from the *Wall Street Journal* who was present picked up the story.

CHAPTER 7

1. National Academy of Sciences (NAS), *Energy in Transition: 1985-2010,* Final Report of the Committee on Nuclear and Alternative Energy Systems

(CONAES) (Washington, D.C.: National Research Council, NAS, 1979), pp. iii, x.

2. "Energy in the 1980s" (Washington, D.C.: American Association for the Advancement of Science, 1980), pp. 4-6, 19-20, panel discussion at the annual meeting, January 1980; John F. O'Leary, "A Future Without an Oil Cushion," *Washington Post,* January 22, 1980, p. A19.

3. Luther J. Carter, "Academy Energy Report Stresses Conservation," *Science* 207 (January 25, 1980): 386.

4. Mike McCormack, Opening Statement, Hearing on National Academy of Sciences Report: *Energy in Transition, 1985-2010,* Subcommittee on Energy Research and Production, January 25, 1980, p. 1.

5. Daniel S. Greenberg, "Million-Dollar Busyness," *The Washington Post,* January 22, 1980.

6. Letter from Kenneth E. Boulding to Martin Greenberger, April 27, 1981.

7. Letters to Martin Greenberger from Robert C. Seamans, Jr. on April 17, 1981, and from Jack Berga on May 4, 1981.

8. Jack M. Hollander, "Future Policies" (Colloquy on Energy and the Environment, Council of Europe, Parliamentary Assembly, Strasbourg, 1977), p. 1.

9. The National Research Council is the joint research arm of the alliance of the National Academy of Sciences, the National Academy of Engineering, and the Institute of Medicine. Seamans served as president of the National Academy of Engineering before taking on the ERDA position. The role of the National Academy of Sciences is the subject of an extensive evaluation by Philip Boffey under the sponsorship of Ralph Nader's Center for Study of Responsive Law. Philip Boffey, *The Brain Bank of America—An Inquiry into the Politics of Science* (New York: McGraw-Hill, 1975).

10. National Academy of Sciences, *Energy in Transition,* p. iii.

11. Ibid., p. xi.

12. Hollander, "Future Policies," p. 1.

13. Berga, letter to Greenberger, 1981.

14. Seamans, letter to Greenberger, 1981.

15. National Academy of Sciences, *Energy in Transition,* p. xi.

16. Ibid., p. iii.

17. For example, the energy study published in 1979 entitled *Energy: The Next Twenty Years,* was originally conceived by the Ford Foundation as an investigation of the coal prospects for the United States at a time when coal was actively being promoted by the Department of Energy. Later it was decided not to treat coal in isolation from other options and the scope was broadened. Hans Landsberg, et al., *Energy: The Next Twenty Years* (Cambridge, Mass.: Ballinger Publishing Company, 1979).

18. Edward L. Ginzton, Cochairman of the CONAES study and chairman of the Board of Varian Associates, said that he believes the problems encountered by CONAES were "preordained" by the overly broad objectives, "volunteer nature of the effort," and "unknown players." When Ginzton runs a project in his company, he picks compatible people he knows "can work together intellectually." He said the members of CONAES were chosen because of their "reputations in their fields, not because of their ability to contribute and work cooperatively." He did not think the Academy should have "tackled such a broad topic." Ginzton felt the Ford-MITRE study, where members wrote chapters themselves, was a much better approach. Interview with Ginzton, November 20, 1978.

19. National Academy of Sciences, *Energy in Transition,* p. iv. CONAES Cochairman Harvey Brooks agreed that three committee members were "strongly pronuclear" at the start of the study, but he regarded another four (including himself) as "mildly pronuclear" rather than neutral. (Brooks, letter to Greenberger, March 22, 1980.) Berga and Hollander thought that eight of the members were pronuclear and eight were antinuclear. (Berga, letter to Greenberger.) Brooks estimated that at the end of the study five members were "strongly pronuclear" and another five were "mildly pronuclear." Committee member John P. Holdren commented on "the remarkable attrition that occurred" among reformists on the committee, probably from "exhaustion and frustration." "Three of the Steering Committee members most in favor of low-energy futures, most concerned about environmental issues, and most critical of nuclear power virtually dropped out (without resigning) in the last six to 12 months of the study" while another earlier had resigned. Thus, of the five reformist committee members at the beginning of the study, only Holdren was still active by the end. "It was pretty lonely." (Letter from Holdren to Greenberger, March 15, 1982.) Holdren considered fellow committee member Boulding "genuinely in the middle." Wrote Boulding, "As I looked onto the matter, I rather changed my mind in regard to the breeder, seeing it as something that is very nice to have as a reserve for the long run." (Boulding, letter to Greenberger, 1981.)

20. Luther J. Carter, "Nader Queries Handler on Status of CONAES Study," *Science* 202 (December 15, 1978): 1171.

21. Carter, "Academy Energy Report Stresses Conservation," p. 386.

22. One other member of the overview committee resigned. Stephen D. Bechtel, Jr. of Bechtel Corporation (a major builder of nuclear plants) left in the study's first year. He was replaced by Ludwig F. Lischer of Commonwealth Edison Company, an electric utility with a strong nuclear bent. Bechtel vice president, W. Kenneth Davis, later became chairman of the CONAES Supply Panel.

23. Duncan writes, "I am glad you did not depict me as an 'anti-nuke,' although I often took that view as a devil's advocate. In the end, I found that a couple of the nuclear engineers were people that I could talk to." Letter from Duncan to Greenberger, April 13, 1981.

24. Dudley Duncan, "Memorandum to CONAES regarding writing assignments: Meditations on Woods Hole," The University of Arizona, September 17, 1976.

25. If the approximately 300 "volunteers" each spent, on the average, about half the time Jones estimates he spent, over a two-year period CONAES effectively consumed about sixty man years of senior-level effort. This was not included in its operating budget. At $50,000 per man year, it would add $3 million to the cost of the CONAES study, putting it in the $7 million bracket.

26. Charles O. Jones, "If I Knew Then ... (A Personal Essay on Committees and Public Policy)," *Policy Analysis,*Fall 1979, p. 5.

27. *The New York Times,* "Don't Count on the Sun Too Soon," editorial, January 16, 1980.

28. Carter, "Academy Energy Report Stresses Conservation," p. 385.

29. Harvey Brooks, conversation with Martin Greenberger, May 4, 1980.

30. Laura Nader, et al., *Supporting Paper 7: Energy Choices in a Democratic Society,* Committee on Nuclear Alternative Energy Systems, Synthesis

Panel, Consumption, Location, and Occupational Patterns Resource Group (Washington, D.C.: National Academy of Sciences, 1977); Laura Nader, "Barriers to Thinking New about Energy," *Physics Today,* February 1981, pp. 99-100; Amory B. Lovins, "Energy Strategy: The Road Not Taken?" *Foreign Affairs* 55, no.1 (October 1976): 65-96.

31. Jones, "If I Knew Then . . .," p. 5.
32. Letter from Jack M. Hollander to Martin Greenberger, May 11, 1981.
33. Lester Lave, et al., "Synthesis Panel Report," CONAES, July 25, 1977. (Unpublished.)
34. Garry D. Brewer and Martin Shubik, *The War Game* (Cambridge, Mass.: Harvard University Press, 1979), pp. 95-98.
35. National Academy of Sciences, *Energy in Transition,* p. 7.
36. The staff member who assisted Hollander was Richard Silberglitt.
37. National Academy of Sciences, *Energy in Transition,* pp. 636-709.
38. Harvey Brooks, Memorandum to Jack Hollander, October 26, 1976.
39. Henry Kissinger, *White House Years* (Boston: Little, Brown, 1979), p. xxi.
40. Carter, "Academy Energy Report Stresses Conservation," p. 386.
41. Ibid.
42. Ginzton had managed the Committee for Motor Vehicle Emissions, mandated by Congress to determine if the Clean Air Act of 1970 could be met. Ginzton said he was reluctant to accept the CONAES assignment because the nuclear field was new to him and he was not very familiar with reactors.
43. National Academy of Sciences, *Energy in Transition,* p. 732.
44. Counting the total lines taken by each of the members of the overview committee in the appendix of the CONAES report, and attributing each statement there to only the first-mentioned author yields the following breakdown: Holdren, 25 percent; Brooks, 19 percent; Boulding, 17 percent; Lischer, 12 percent; Spinrad, 9 percent; Cannon, 6 percent; Gornowski, 6 percent; Kohn, 3 percent; Houthakker, 1 percent; Rose, less than 1 percent; other four, 0 percent. There were fourteen members on the review committee at the conclusion of the study.
45. John P. Holdren, "The Academy of Sciences Energy Study: One Insider's View," guest editorial, *Christian Science Monitor,* January 30, 1980.
46. Stanley Rothman, Mary Huggins Gamble Professor of Government at Smith College, wrote to the editor of the *New York Times* that "it is becoming increasingly difficult for those of us interested in energy issues to read and understand *The New York Times.*" He was complaining about the conflicting reports of CONAES in the *Times* news article and editorial, as described in the text. Stanley Rothman, Letter to the Editor, *New York Times,* January 21, 1980.
47. Anthony J. Parisi, "Science Study Urges Conservation As First Priority for Energy Policy," *New York Times,* January 13, 1980.
48. Anthony J. Parisi, "Creating the Energy-Efficient Society," *New York Times Magazine,* September 23, 1979, pp. 46-49.
49. Harvey Brooks, "The Multiple Focuses of a National Energy Study," Letter to the Editor, *New York Times,* February 10, 1980.
50. In a January 22, 1980 reply to Rothman's letter to the editor of the *New York Times,* Brooks absolved Parisi of "misquotation" with one "exception": the confusion of energy consumption with the energy/GNP ratio. Parisi had reported that "technical efficiency measures alone could cut domestic energy consumption in half." But it was the energy/GNP ratio that

CONAES suggested could be halved. This ratio falls as GNP increases so long as energy consumption does not increase proportionately.

51. National Academy of Sciences, *Energy in Transition,* p. xii.
52. Frederick Seitz and Miro M. Todorovich, Letter to the Business Editor, *New York Times,* January 17, 1980.
53. Thomas O'Toole, "Energy Analysis Stresses Use of Coal, Atom Power," *Washington Post,* January 14, 1980, p. 1.
54. Robert Stobaugh and Daniel Yergin, eds., *Energy Future: Report of the Energy Project at the Harvard Business School* (New York: Random House, 1979); Landsberg, et al., *Energy: The Next Twenty Years;* Sam H. Schurr, et al., *Energy in America's Future—The Choices Before Us,* Resources for the Future, Washington, D.C. (Baltimore, MD.: Johns Hopkins Press, 1979).
55. Interview with Daniel S. Greenberg, January 25, 1980.
56. Parisi, "Science Study Urges Conservation."
57. Luther J. Carter, "Oh Sweet CONAES, Where Art Thou?" *Science* 205 (July 27, 1979): 377.
58. *Energy Daily,* November 30, 1978, p. 4.
59. O'Toole said that the large private law firms in Washington were particularly interested in having copies of the study since general publication of the report was not scheduled for another two months.
60. *The New York Times,* "Don't Count on the Sun Too Soon."
61. Joel Darmstadter, "Interpretations Differ in Press Accounts of CONAES Energy Study," *Resources* 64, Resources for the Future, Washington, D.C. (January-April, 1980): 4-5.
62. *The Washington Post,* "Energy Uncertainty," Editorial, January 18, 1980, p. A16.
63. One article in a popular business magazine, for example, suggested that economic models and modelers were taking over Washington. On the need for better understanding of the capabilities and limitations of models, see Martin Greenberger, "Humanizing Policy Analysis: Confronting the Paradox in Energy Modeling," in Saul I. Gass, ed., *Validation and Assessment of Energy Models,* NBS Special Publication 616, National Bureau of Standards (NBS) (Washington, D.C.: Government Printing Office, 1981) pp. 25-41.
64. Conversation with the writer of the *Post* editorial.
65. *Wall Street Journal,* "Nuclear Power vs. the Sun," Editorial, January 21, 1980.
66. Conversation with O'Toole, January 25, 1980
67. Ibid.
68. Not all of the reporting was one-sided. The *Christian Science Monitor* presented a well-balanced account *(Christian Science Monitor,* January 14, 1980, p. 9), and the *Wall Street Journal* editorial commented on the "blindman-elephant syndrome" common in the interpretation of reports such as CONAES.
69. Harvey Brooks, for example, came to give more emphasis to the effect of price on energy consumption during the course of the study. He also became less pronuclear (although his reservations had more to do with the stiffening political resistance he saw than to technical factors), and he began to attach more importance to proliferation hazards in considering the breeder.
70. Kenneth Boulding reasoned that energy accounted for only a small part (7 percent) of GNP, so that even a quadrupling of the real price of energy would still leave energy at less than 20 percent of GNP if one assumed pro-

ductivity improvements throughout the economy. (National Academy of Sciences, *Energy in Transition,* pp. 734-35.) Tjalling Koopmans convinced himself based on his feeling that the marketplace response to higher energy prices would induce a shift in production to modes that used less energy. Harvey Brooks was persuaded on the basis of the careful sector-by-sector technological analysis conducted by the Demand Panel and by the fact that the demand elasticity implied by the engineering and microeconomic investigations agreed with the macroeconomic assumptions made in the work of the Modeling Resource Group (MRG). Some credited the MRG modeling runs with *showing* that policies constraining energy supplies would have only minimal impact on the economy. But this question should be addressed with a "general equilibrium" model, and the MRG models were all of the "partial equilibrium" type. They were not able to portray the impact on the economy of changes in the energy sector. Koopmans, uncertain about how much error this introduced into the results, guessed it was minimal given the fact that energy accounted for only a small part of GNP. Interest in the coupling of energy and the economy developed only late in the deliberations of the MRG and it remained an open question. It was never really the subject of MRG analysis, despite impressions to the contrary.

71. John H. Gibbons, et al., "U.S. Energy Demand: Some Low Energy Futures," *Science,* April 14, 1978, pp. 142-52; National Research Council, *Alternative Energy Demand Futures to 2010,* Committee on Nuclear and Alternative Energy Systems, Demand and Conservation Panel (Washington, D.C.; National Academy of Sciences, 1979).

72. Harvey Brooks, Testimony before the Subcommittee on Energy Research and Production, Committee on Science and Technology, U.S. House of Representatives, January 25, 1980, p. 6.

73. A characterization attributed to Frank von Hippel by John Holdren in a letter to Greenberger, March 15, 1982.

CHAPTER 8

1. Robert Stobaugh and Daniel Yergin, eds., *Energy Future: Report of the Energy Project at the Harvard Business School* (New York: Random House, 1979), p. 233.

2. Sam H. Schurr, et al., *Energy in America's Future: The Choices Before Us* (Baltimore, Md.: Johns Hopkins Press, 1979), p. xxvi.

3. Hans H. Landsberg, et al., *Energy: The Next Twenty Years* (Cambridge, Mass.: Ballinger Publishing Company, 1979), p. 3.

4. Stobaugh and Yergin, *Energy Future,* p. 226.

5. John Kenneth Galbraith, "Oil: A Solution," *The New York Review of Books,* September 29, 1979, p. 6.

6. Paul L. Joskow, "America's Many Energy Futures—A Review," *Bell Journal of Economics,* Spring 1980, p. 378.

7. William Tucker, "A Hard-Headed Look at the Energy Debate," *The Wall Street Journal,* August 9, 1979, p. 20.

8. Peter Passell, "The Politics of Energy/Energy Future," *The New York Times Book Review,* July 29, 1979, p. 7.

9. Hobart Rowen, "Still No Energy Policy Consensus," *Washington Post,* September 2, 1979, pp. G1-G2.

10. McGeorge Bundy, Foreword, in Landsberg, et al., *Energy: The Next Twenty Years*, p. xvii.
11. Letter from Bundy to Greenberger, June 29, 1982.
12. Landsberg, et al., *Energy: The Next Twenty Years*, p. 2.
13. Publication of the final reports of the Stobaugh-Yergin, RFF-Mellon, Ford-RFF, and CONAES studies occurred between July 1979 and October 1980. Table 1-4 gives the approximate sales figures of these reports and other commercially published studies through May 1982.
14. Landsberg, et al., *Energy: The Next Twenty Years*, pp. 3-6.
15. Schurr, et al., *Energy in America's Future*, p. 46.
16. Landsberg, et al., *Energy: The Next Twenty Years*, p. 4.
17. Stobaugh and Yergin, *Energy Future*, revised and updated edition, 1980, pp. 275-76.
18. Robert S. Pindyck, "The American Energy Debate," *The Public Interest*, Spring 1980, pp. 100-05; Robert Stobaugh and Daniel Yergin, "A Response to 'The American Energy Debate,' and Pindyck's Reply," *The Public Interest*, Fall 1980, pp. 123-26.
19. Joskow, "America's Many Energy Futures," pp. 378-79.
20. Robert Stobaugh and Daniel Yergin, "A Reply to Paul Joskow's Review of an Imaginary *Energy Future*," The Energy Project at the Harvard Business School, October 1981, pp. 4-5. (Private Distribution.)
21. William W. Hogan, et al., *Energy and the Economy*, EPRI EA-620, Energy Modeling Forum, Stanford University (Palo Alto, Calif.: Electric Power Research Institute [EPRI] 1978).
22. Stobaugh and Yergin, *Energy Future*, p. 176.
23. Ibid., p. 273.
24. Paul R. Ehrlich, et al., *Escoscience: Population, Resources, Environment* (San Francisco: W.H. Freeman, 1977), pp. 953-54.
25. Harrison Brown, "Resources and Environment in the Next Quarter-Century," in Charles J. Hitch, ed., *Resources for an Uncertain Future* (Baltimore, Md.: Johns Hopkins Press for RFF, 1978), pp. 28-29.
26. Stobaugh and Yergin, *Energy Future*, revised edition, p. 273.
27. See, for example, George P. Shultz and Kenneth W. Dam, *Economic Policy Beyond the Headlines* (New York: Norton, 1977); Hans H. Landsberg and Joseph M. Dukert, *High Energy Costs: Uneven, Unfair, Unavoidable?* (Baltimore, Md.: Johns Hopkins Press for RFF, 1981).
28. Tom Redburn, "'Energy Future' Goes Beyond Ivory Tower," *Los Angeles Times*, August 19, 1979, Part VI.
29. Aaron Wildavsky and Ellen Tenenbaum, *The Politics of Mistrust: Estimating American Oil and Gas Resources* (Beverly Hills, Calif.: Sage Publications, 1981).
30. Bruce Hannon, "An Energy Standard of Value," *Annals of the American Academy of Political and Social Science* 410 (November 1973): 139-53.
31. Wildavsky and Tenenbaum, *The Politics of Mistrust*, p. 325.
32. Schurr, et al., *Energy in America's Future*, pp. 536-37.

CHAPTER 9

1. For one attempt to compare these alternative econometric approaches, see EMF4 Working Group, Energy Modeling Forum, "Aggregate Elasticity of Energy Demand," *The Energy Journal* 2, no. 2 (November 1981).

2. Price elasticities at the point of end-use (e.g., motor gasoline delivered to an automobile tank) can be more than two times elasticities measured at the point of primary energy (e.g., crude oil at the wellhead). Differences in conversion efficiencies make it misleading to refer to calories or BTUs at the point of end-use. To standardize terminology, our references to elasticities are in terms of primary energy unless specifically stated otherwise.

3. Both the FEA and CIA studies estimated primary energy demands at about 100 quads for 1985.

4. Energy Policy Project the Ford Foundation, *A Time to Choose: America's Energy Future* (Cambridge, Mass.: Ballinger Publishing Company, 1974). See especially Chapter 3 and Appendix A.

5. Energy Policy Project, *A Time to Choose,* p. 49.

6. Data Resources Inc. (DRI), 1980 Oil Embargo Study (Lexington, Mass.: DRI, 1977), p. 2. Another study of energy-economy interactions several years later concluded that "energy prices made a significant contribution to the economic slowdowns of the last decade, but were not the dominant cause." Office of Policy, Planning, and Analysis, "Interrelations of Energy and the Economy," DOE/PE-0030 (Washington, D.C.: Department of Energy, July 1981), pp. 1-2.

7. Energy Policy Project, *A Time to Choose,* p. 511.

8. CONAES MRG, *Energy Modeling for an Uncertain Future: The Report of the Modeling Resource Group* (Washington, D.C.: National Academy of Sciences, 1978), p. 114.

9. CONAES MRG, *Energy Modeling for an Uncertain Future,* pp. 106-15.

10. William W. Hogan, et al., *Energy and the Economy,* Energy Modeling Forum, Stanford University, EPRI EA-620 (Palo Alto, Calif.: Electric Power Research Institution [EPRI], 1978).

11. Energy Research and Development Administration, *A National Plan for Energy Research, Development and Demonstration, Volume 1: The Plan* (Washington, D.C.: Government Printing Office, 1975).

12. C. J. Hitch, ed., *Modeling Energy-Economy Interactions: Five Approaches* (Washington, D.C.: Resources for the Future, September 1977).

13. *The National Energy Plan* (Washington, D.C.: Executive Office of the President, April 29, 1977), p. viii.

14. B. Miller, et al., "Geological Estimates of Undiscovered Recoverable Oil and Gas Resources in the United States," Geological Survey Circular 725, Reston, Va., 1975.

15. Exxon Corporation, *World Energy Outlook* (New York: Exxon, April 1978).

16. B. Miller, et al., Geological Estimates, p. 1.

17. Nuclear Energy Policy Study Group (NEPS) of the Ford Foundation, *Nuclear Power Issues and Choices* (Cambridge, Mass.: Ballinger Publishing Company, 1977), p. 74.

18. CONAES MRG, *Energy Modeling for an Uncertain Future,* ch. 4.

19. A thoughtful review of this dilemma is D. P. Harris, "Informational and Conceptual Issues of Uranium Resources and Potential Supply" (Tucson, Ariz.: University of Arizona, December 1977).

20. L. Hamilton, "Health Effects of Air Pollution" (Informal Report, Brookhaven National Laboratory, Upton, New York, 1975). D. W. North and M. Merkhofer, "Analysis of Alternative Emissions Control Strategies," in *Air Quality and Stationary Source Emission Control,* National Academy of Sciences (Washington, D.C.: Government Printing Office, 1975).

21. Nuclear Energy Policy Study Group *Nuclear Power Issues and Choices,* p. 57.
22. John Weyant, "Quantitative Models in Energy Policy," *Policy Analysis* 6, no. 2 (Spring 1980): 211-234.
23. CONAES MRG, *Energy Modeling for an Uncertain Future,* pp. 26-27.
24. For further implications of the decision analysis viewpoint, see Howard Raiffa, *Decision Analysis* (Reading, Mass.: Addison-Wesley, 1968).
25. For the reader with a specific interest in breeders and other advanced nuclear reactors, see Chapter 4 of the CONAES MRG report and the following subsequent analyses: Alan Manne and Richard Richels, "A Decision Analysis of the U.S. Breeder Reactor Program," *Energy* 3 (1978): 747-67; Alan Manne and Richard Richels, "Probability Assessments and Decision Analysis of Alternative Nuclear Fuel Cycles," presentation to the American Nuclear Society, Williamsburg, Va., April 1979; William Nordhaus, *Efficient Use of Energy Resources* (New Haven: Yale University, 1979).
26. M. Friedman, "The Role of Government: Neighborhood Effects," Reprinted in R. and N. Dorfman, *Economics of the Government* (New York: Norton, 1972), p. 202.
27. Edward S. Mason, "Resources in the Past and for the Future," in C.J. Hitch, ed., *Resources for an Uncertain Future* (Baltimore, Md.: Johns Hopkins Press for RFF, 1978).
28. A. Lovins, "Energy Strategy: The Road Not Taken?" *Foreign Affairs* 55, no. 1 (October 1976): 65-96.
29. CONAES MRG, *Energy Modeling for an Uncertain Future,* p. 43.
30. Nuclear Energy Policy Study Group *Nuclear Power Issues and Choices,* p. 196.

CHAPTER 10

1. Interview, January 17, 1979.
2. Interview, March 22, 1979.
3. For example, by James R. Schlesinger in "Uses and Abuses of Analysis," Memorandum to the Subcommittee on National Security and International Operations of the Senate Committee on Government Operations (Washington, D.C.: Government Printing Office, April 22, 1968), p. 1; Aaron Wildavsky, *Speaking Truth to Power* (Boston: Little, Brown, 1979).
4. Letter from Edgar S. Dunn, Jr. to Emery Castle, March 22, 1982.
5. Kuhn finds major advances in science resulting from dramatic changes in paradigm, not from the process of analysis. Thomas Kuhn, *The Structure of Scientific Revolutions* (Chicago: University of Chicago Pres, 1970).
6. Martin Greenberger, Matthew A. Crenson, and Brian L. Crissey, *Models in the Policy Process* (New York: Russell Sage Foundation, 1976), pp. 47-48.
7. Schlesinger, "Uses and Abuses of Analysis," p. 2.
8. CONAES MRG, *Energy Modeling for an Uncertain Future: The Report of the Modeling Resource Group* (Washington, D.C.: National Academy of Sciences, 1978).
9. A case for model comparison was made in Greenberger, Crenson, and Crissey, *Models in the Policy Process,* pp. 334-41. Also, Martin Greenberger and Richard Richels, "Assessing Energy Policy Models: Current State and Future Directions," *Annual Review of Energy* 4 (1979): 467-500.

10. Information on the Energy Modeling Forum (EMF) is available from EMF offices in Terman Engineering Center, Stanford University, Palo Alto, Calif. 94305. For an overview of the EMF, see J.L. Sweeney and J.P. Weyant, "Energy Modeling Forum: Past, Present, and Future," EMF Planning Paper 6.1, Palo Alto, Calif., February 1979.

11. These ideas are developed further in William W. Hogan, "Energy Models: Building Understanding for Better Use" (Paper presented at the Second Lawrence Symposium on Systems and Decision Sciences, Berkeley, California, October 1978). Also, Greenberger and Richels, "Assessing Energy Policy Models."

12. As one example, Harvey Brooks notes that the CONAES overview committee made little use of the work of the Modeling Resource Group in its conclusions. "None of the committee members," says Brooks, "were professional analysts." Letter from Brooks to Greenberger, April 14, 1982.

13. Graham T. Allison, *Essence of Decision: Explaining the Cuban Missile Crisis* (Boston: Little, Brown, 1971). The quote comes from course notes by Harvey Brooks (NS 134, Harvard University, March 1981). The rational actor is one of several models of organizational interaction and bureaucratic politics offered by Allison to characterize the decision-making process.

14. A major influx of analysts into government followed the appointment of Robert McNamara as secretary of defense in 1961. See Chapter 11.

15. Herbert Goldhamer, *The Adviser* (New York: Elsevier, 1978).

16. Harold Brown, "Speech on the Occasion of the Dedication of the Stennis Center" (Mississippi State University, Starkville, Mississippi, October 21, 1977). Department of Defense Press Release.

17. Harold Lasswell, *The Analysis of Political Behavior* (London: Routledge and Kegan Paul, 1947), p. 130.

18. Interview, May 2, 1979.

19. Martin Rein, *Social Science and Public Policy* (New York: Penguin, 1976).

20. Michel Crozier, et al., *The Crisis of Democracy* (New York: New York University Press, 1975).

21. Harvey Brooks, "Technology: Hope or Catastrophe?" *Technology in Society,* 1, no. 1 (Spring 1979): 15.

22. Geoffrey Vickers, "Commonly Ignored Elements in Policy-making," *Policy Sciences* 3, no. 2 (June 1972): 266.

23. Roy C. Macridis, *Contemporary Political Ideology* (Cambridge, Mass.: Winthrop, 1980).

24. Robert A. Dahl, *A Preface to Democratic Theory* (Chicago: University of Chicago Press, 1956), p. 132.

25. Interview, March 21, 1979.

26. Gibson Winter, "Toward a Comprehensive Science of Policy," *Houghton Lectures* (Cambridge, Mass.: Harvard University, November 1969), p. 7.

27. Kenneth Boulding, *The Image* (Ann Arbor, Mich.: Ann Arbor Paperbacks, 1961), pp. 25-26.

28. S. Watanabe, "Prediction and Retrodiction," *Review of Modern Physics* 27 (June 1955): 179-86; Abraham Edel, *Method in Ethical Theory,* (Indianapolis: Bobbs-Merrill, 1963).

29. C. E. Lindblom, *The Policy-Making Process* (Englewood Cliffs, N.J.: Prentice-Hall, 1968), pp. 26-27.

30. William Ascher, *Forecasting: An Appraisal for Policy-Makers and Planners* (Baltimore, Md.: Johns Hopkins Press, 1978).

31. Geoffrey Vickers, *Value Systems and Social Process* (New York: Basic Books, 1968).
32. Gunnar Myrdal, *Objectivity in Social Research* (New York: Pantheon, 1969), ch. 10.
33. Daniel Yergin, "Two Intelligence Failures," *Washington Post,* January 30, 1979.
34. Robert A. Dahl and C. E. Lindblom, *Politics, Economics and Welfare* (New York: Harper, 1953).
35. Interview with a consultant to Congress, January 18, 1976.
36. Interview, March 1, 1979.
37. Harry Stack Sullivan, *The Fusion of Psychiatry and Social Science* (New York: Norton, 1964), ch. 17.
38. Harold Lasswell, *Psychopathology and Politics* (New York: Viking, 1960), pp. 184-85.
39. Interview with William F. Kieschnick, June 14, 1979.
40. Ibid.
41. Lasswell, "Psychopathology and Politics," pp. 184-85.
42. Abram Shulsky, "Social Science and Public Policy: The Case of Microeconomics." (Unpublished paper, 1979.)
43. Interview, June 2, 1979.
44. Interview, October 12, 1978.
45. Lewis H. Lapham, "The Energy Debacle," *Harper's,* August 1977, p. 74.
46. H.C. Kelman, ed., *International Behavior* (New York: Holt, Rinehart, and Winston, 1965), pp. 339-53.
47. Leon Festinger, *A Theory of Cognitive Dissonance* (Palo Alto, Calif.: Stanford University Press, 1957).
48. John Steinbruner, *A Cybernetic Theory of Decision* (Princeton: Princeton University Press, 1974).
49. Irving Janis and Leon Mann, *Decision-Making: A Psychological Analysis of Conflict, Choice, and Commitment* (New York: The Free Press, 1977).
50. Interview, March 2, 1979.
51. George Orwell, *Collected Essays* (Garden City, N.Y.: Doubleday, 1954), p. 177.
52. Youssef M. Ibrahim and David Ignatius, "As Oil Use Declines, Experts See a Slowing In Price Increases: Western Nations and Japan Find Growth Is Possible Along With Conservation," *Wall Street Journal,* January 27, 1982, pp. 1-24.
53. Interview, September 19, 1978.
54. Greenberger, Crenson, and Crissey, *Models in the Policy Process,* pp. 332-33.
55. Interview, March 1, 1979.
56. Garry D. Brewer, *Politicians, Bureaucrats, and the Consultant* (New York: Basic Books, 1973), p. 84.
57. Meg Greenfield, "Why We Don't Do Anything," *Newsweek,* December 18, 1978, p. 112.
58. Interview, March 21, 1979.
59. Michael J. Malbin, *Unelected Representatives: Congressional Staff and the Future of Representative Government* (New York: Basic Books, 1980), pp. 233-34.
60. Ibid.
61. Interview, January 8, 1979.

62. Francine Rabinovitz, et al., "Guidelines," *Policy Sciences* 7, no. 4 (December 1976): 399-416.

63. Greenberger, Crenson, and Crissey, *Models in the Policy Process,* pp. 331, 333.

64. Robert A. Dahl, *After the Revolution? Authority in a Good Society* (New Haven: Yale University Press, 1970), part I.

65. Ibid., p. 35.

66. Harvey Brooks, "Technology: Hope or Catastrophe?" *Technology in Society* 1, no. 1 (Spring 1979): 3-17; Harvey Brooks, "Environmental Decision Making: Analysis and Values," in Laurence H. Tribe, et al., eds., *When Values Conflict* (Cambridge, Mass.: Ballinger Publishing Company, 1976), pp. 115-35.

67. Robert A. Dahl, *After the Revolution? Authority in a Good Society* (New Haven: Yale University Press, 1970), p. 37.

68. Alvin Weinberg, "Social Institutions and Nuclear Energy," *Science* 177 (1972): 27-34.

69. William Ophuls, *Ecology and the Politics of Scarcity* (San Francisco: W. H. Freeman, 1977), p. 161.

CHAPTER 11

1. Our copy editor wanted a reference. Greenberger is responsible for this fanciful concoction.

2. Isaiah 40:13-14. Quoted in Herbert Goldhamer, *The Adviser* (New York: Elsevier, 1978), p. 7.

3. A.S. Gaye, ed., *The Essays or Counsels Civil and Moral of Francis Bacon* (Oxford: Clarendon Press, 1911), p. 71.

4. Ibid.

5. Goldhamer, *The Advisor,* p. 7.

6. Arnold J. Meltsner, *Policy Analysts in the Bureaucracy* (Berkeley, Calif.: University of California Press, 1976), p. 268.

7. Richard Lyman, former president of Stanford University, quoted in the *Stanford Campus Report,* January 30, 1980, p. 9.

8. Harvey Brooks, "Environmental Decision Making: Analysis and Values," in Laurence H. Tribe, et al., eds., *When Values Conflict* (Cambridge, Mass.: Ballinger Publishing Company, 1976), p. 134.

9. Martin Greenberger, "Energy Demand Forecasts" (Opening Testimony before the Hearings of the Subcommittee on Investigations and Oversight of the Committee on Science and Technology, U.S. House of Representatives, June 1, 1981), p. 23.

10. Martin Greenberger, Matthew A. Crenson, and Brian L. Crissey, *Models in the Policy Process* (New York: Russell Sage Foundation, 1976), chs. 7-9.

11. James R. Schlesinger, "Uses and Abuses of Analysis," Memorandum for the Subcommittee on National Security and International Operations of the Senate Committee on Government Operations (Washington, D.C.: Government Printing Office, 1968), p. 2.

12. In discussing information overload in Congress, Michael Malbin argues for nonpartisan professional staff to be assigned to each congressional committee. The suggestion contains elements in common with the notion of public

analysis. Michael J. Malbin, *Unelected Representatives: Congressional Staff and the Future of Representative Government* (New York: Basic Books, 1980), pp. 242-43.

13. Ibid., p. 236.

14. Schlesinger, "Uses and Abuses of Analysis," p. 3.

15. Herbert A. Simon, *The Sciences of the Artificial* (Cambridge, Mass.: M.I.T. Press, 1969), p. 15.

16. Malbin, *Unelected Representatives,* p. 226.

17. Ibid., p. 223.

18. Phone conversation with Greenberger, February 21, 1982.

19. Alain C. Enthoven, "Systems Analysis in the Pentagon (Speech to the Association for Public Program Analysis, September 26, 1968), p. 9.

20. Because of the "self-centered energy policy that has ignored legitimate needs and interests of the rest of the world," it was charged, "the United States has lost leadership in nuclear energy and much of its ability to influence the nuclear energy policies of others." Philip H. Abelson, "Energy Policies of the United States and U.S.S.R.," *Science,* editorial, April 11, 1980 p. 135.

21. Norman C. Rasmussen, "Methods of Hazard Analysis and Nuclear Safety Engineering," *Annals of the New York Academy of Sciences* 365 (1981): 20-36.

22. Martin Greenberger and Richard Richels, "Assessing Energy Policy Models: Current State and Future Directions," *Annual Review of Energy* 4 (1979): 467-500. Analysis of the work of model builders (first parties) for the benefit of model users (second parties) by model analysts (third parties) has been underway at the Energy Model Assessment Program of the M.I.T. Energy Laboratory for several years. It is exceedingly difficult for third parties to immerse themselves in the assumptions and formulations of models they did not develop without the active cooperation of the model builders.

23. Philip Handler, "Public Doubts about Science," *Science,* editorial, June 6, 1980, p. 1093.

24. T.C. Schelling, "Policy Analysis As Science of Choice" (Unpublished paper, 1982), p. 14.

25. In a more restricted context, the Energy Modeling Forum (EMF) at Stanford University compares modeling studies on a regular basis. Greenberger and Richels, "Assessing Energy Policy Models," discusses the motivation for this work, as does Hillard G. Huntington, John P. Weyant, and James L. Sweeney, "Modeling for Insights, Not Numbers: The Experiences of the Energy Modeling Forum" EMF Occasional Paper 5.1, (Palo Alto, Calif.: EMF, December 1981).

26. Henry Kissinger, Television interview by Dick Cavett, December 8, 1979.

27. Meltsner, *Policy Analysts in the Bureaucracy,* p. 48.

28. Malbin, *Unelected Representatives,* pp. 243-44.

29. Charles J. Hitch, *Decision-Making for Defense* (Berkeley, Calif.: University of California Press, 1965), p. 28.

30. The separate military services today use systems analysis to mediate among competing weapons programs internally, much as it was used to mediate among the services in the sixties.

31. Alice Rivlin, *Systematic Thinking for Social Action* (Washington, D.C.: Brookings Institution, 1971).

32. Greenberger, Crenson, and Crissey, *Models in the Policy Process.*

33. William Greider, "The Education of David Stockman," *The Atlantic Monthly,* December 1981, pp. 27, 38.
34. Congressman Phil Gramm (D-Texas), Luncheon Talk, Seminar on Natural Gas Deregulation at the American Enterprise Institute, Washington, D.C., March 12, 1982.
35. Francisco R. Parra, "The Pricing of Oil in International Markets, *Middle East Economic Survey,* October 17, 1975, p. 1.

APPENDIX A

1. Charles Darwin, quoted in Charles P. Curtis, *A Commonplace Book* (New York: Simon and Schuster, 1957), p. 24. Also in Peter Szanton, *Not Well Advised* (New York: Russell Sage Foundation and the Ford Foundation, 1981), p. 134.
2. V.O. Key, Jr., *Public Opinion and American Democracy* (New York: Knopf, 1961); Walter Lippman, *Public Opinion* (New York: The Free Press, 1965).
3. G.C. Thompson, "The Evaluation of Public Opinion," in Bernard Berelson and Morris Janowitz, eds., *Reader in Public Opinion and Communication* (New York: The Free Press, 1966), pp. 7-12.
4. Steven R. Brown, "Intensive Analysis in Political Research," *Political Methodology* 1, no. 1 (Winter 1974): 1-25.
5. Steven R. Brown, *Political Subjectivity* (New Haven: Yale University Press, 1980).
6. C. P. Snow, *Science and Government* (Cambridge, Mass.: Harvard University Press, 1961).
7. Gerald Holton, "Scientific Opinion and Societal Concerns." *Hastings Center Report* 5 (December 1975): 39-47.
8. Harvey Brooks, "Technology: Hope or Catastrophe?" *Technology in Society* 1, no. 1 (Spring 1979): 3-17.
9. Robert Heilbroner, *An Inquiry into the Human Prospect* (New York: Norton, 1973); Simon Ramo, *Cure for Chaos* (New York: David McKay, 1969).
10. D.H. Meadows, et al., *The Limits to Growth* (New York: Universe Books, 1972).
11. Brooks, "Technology: Hope or Catastrophe?" p. 4. Charles Hitch points out that the characterization of economists as optimists is paradoxical. Economics, in its insistence that there is "no free lunch," has long been known as "the dismal science."
12. Murray Edelman, *The Symbolic Uses of Politics* (Urbana, Ill.: University of Illinois Press, 1967).
13. Harold D. Lasswell, "Person, Personality, Group, Culture," *Psychiatry* 2 (November 1938): 533-61.
14. Brown, "Intensive Analysis in Political Research."
15. This discussion relies heavily on the efforts of Jeffrey Milstein, who sponsored studies of public opinion for the Department of Energy and summarized the findings in Jeffrey Milstein, "Soft and Hard Energy Paths: What People on the Streets Think" (Office of Conservation and Solar Applications, Department of Energy, 1978).

16. Trust in social institutions was generally high prior to 1966, and could be counted on in making policy recommendations. This allowed for a wider range of political activity and choice.

17. Each of these complaints was made by between 40 and 65 percent of those polled. *What Are the Problems of Big Government?* (New York: Skelly, Yankelovich, and White, Inc., 1976).

18. "It can be argued that energy was tailor-made to become the arena for the clash of opinions that, to be sure, are related to energy but for which energy is at best a proxy." Hans H. Landsberg, "Battling on Energy," *New York Times,* October 6, 1980, p. A-23.

19. Eugene Rosa, "The Public and the Energy Problem," *The Bulletin of the Atomic Scientists* 34, no. 4 (April 1978): 5-7.

20. Ibid, p. 7.

21. Aaron Wildavsky and Ellen Tenenbaum, *The Politics of Mistrust: Estimating American Oil and Gas Resources* (Beverly Hills, Calif.: Sage Publications, 1981).

22. George W. Downs and David M. Rocke, "Complexity, Interaction, and Policy Research," *Policy Sciences* 13, no. 3 (June 1981): 281-95.

23. Bee Angell and Associates, *A Qualitative Study of Consumer Attitudes Toward Energy Conservation* (Washington, D.C.: Federal Energy Administration, Office of Energy Conservation and Environment, November 1975).

24. Executive Office of the President, *Report of the White House Mini-Conference on National Energy Policy* (Washington, D.C.: EOP/Energy Policy and Planning, May 1977), p. 114-19.

25. Ronald D. Brunner and Garry D. Brewer, "Citizen Viewpoints on Energy: Nineteen Citizens Sample" (New Haven: Yale School of Organization and Management, August 1978).

26. Brown, *Political Subjectivity.*

27. The clustering algorithm used product-moment correlation between the statement scores of pairs of respondents—or between the average statement scores of groups of respondents—to measure distance between responses or groupings. At each step of the algorithm, the two groups of responses closest to each other in this sense were merged into a single grouping. P.H.A. Sneath and R.R. Sokal, *Numerical Taxonomy: The Principles and Practice of Numerical Classification* (San Francisco: W.H. Freeman, 1973).

28. Ronald D. Brunner and Weston E. Vivian, "Citizen Viewpoints on Energy Policy," *Policy Sciences* 12, no. 2 (August 1980): 147-74.

29. The general scheme of values used here is that proposed in Harold D. Lasswell and Abraham Kaplan, *Power and Society* (New Haven: Yale University Press, 1950). It has been used by Harold D. Lasswell, Nathan Leites, et al., *Language of Politics: Studies in Quantitative Semantics* (Cambridge, Mass.: M.I.T. Press, 1965). Also, Zvi Namenwirth and Harold D. Lasswell, *The Changing Language of American Values* (Beverly Hills, Calif.: Sage Publications, 1970).

30. Political scientist Ronald D. Brunner observes that some energy analysts perceive government as having "blocked or gutted expert solutions. Distrust of popular government among policy analysts and experts is not often voiced, for obvious reasons," says Brunner, "but nevertheless exists." Letter from Brunner to Greenberger, April 23, 1982.

31. Wildavsky and Tenenbaum, *The Politics of Mistrust,* p. 49.

32. Sam H. Schurr, et al., *Energy in America's Future* (Baltimore, Md.: Johns Hopkins Press, 1979), p. 536.

33. Morris Janowitz, "Disarticulation," *New York Times,* April 26, 1981, p. E-23.
34. Executive Office of the President, *The National Energy Plan* (Washington, D.C.: Government Printing Office, April 1977), p. 26.

APPENDIX B

1. Twenty-eight statements used in both elite and citizen questionnaires (with slight word changes on occasion for editorial reasons) are starred. Another nine statements on each questionnaire had the same general intent or theme. They are not starred.

INDEX

Ford Administration, 160–161, 372
Reagan Administration, 7, 373
Nuclear Regulatory Commission (NRC),
125–128, 156, 364–365
Number need, 202, 255–256, 282, 306,
308
Nye, Joseph S., Jr., 160–162, 279, 372–
374

Oak Ridge National Laboratory, 31, 48,
346
Institute for Energy Analysis
projections, 155, 372
symposium on future energy
strategies, 170, 376
Objectivity in analysis, 96, 255, 290, 293,
309
O'Conner, Harvey, 356
Office of Coal Research, 105, 351
Office of Emergency Planning, 62, 357
Office of Emergency Preparedness, 99–
100, 109
Office of Science and Technology. *See*
OST
Office of Scientific Research and
Development, 173
Office of Technology Assessment, 109,
279, 293, 367
Ohio Company, 355
Oil
alternatives. 26
availability, 151
conservation, 53, 57, 72
cutoff, 214
drilling, 58–59
estimates (*see under* Forecasting)
fees, tariffs, and licenses, 16, 84, 104
glut, 24–25
import program, 15, 55, 73, 105, 350
(*see also* MOIP)
production domestically, 17, 42
production limits, 43, 54, 56, 58, 106,
253
prorationing, 56
revenues, 26
shale, 71, 135
shortage, 11, 60, 93, 101, 253
stocks, 25, 41
waste, 53–54
Oil companies, 346
as concessionaries, 10, 26
as scapegoats, 1, 5, 21, 23, 27, 275, 315
independents versus integrated, 33–35,
353
negotiations with, 12, 16
political power of, 32–35, 348
profits of, 11, 16, 23, 315

rights of, 12, 26
selling to third parties, 26
Oil Policy Committee, 100, 103–104
Oil price, U.S.
collusion, 177, 355
pattern pre–1973, 42, 54
policy, 22, 53, 70, 363, 368
stability, 355
Olayan, Suliman S., 27, 348
O'Leary, John F., 108, 369–370, 379
OMB (Office of Management and
Budget), 104, 107, 121, 308, 362, 370
O'Neill, Thomas, P., Jr. (D–Mass.), 377
OPEC (Organization of Petroleum
Exporting Countries), 24, 113, 253, 274
as scapegoat, 1, 5, 8–9, 23, 27, 32
breaking, 24–25, 110, 346–347, 369
founders of (Venezuela, Saudi Arabia,
Iran, Iraq, and Kuwait), 9, 346
members of (Saudi Arabia, Iran, Iraq,
Kuwait, Qatar, United Arab
Emirates, Algeria, Gabon, Libya,
Nigeria, Indonesia, Venezuela, and
Ecuador), 10
Ophuls, William, 389
Opinions, attitudes, and beliefs, 310–312
Optimists versus pessimists, 311–312
Optimization. *See under* Model
Organization of Petroleum Exporting
Countries. *See* OPEC
Orwell, George, 278, 388
OST (Office of Science and Technology),
89, 92, 99–101, 132, 353, 372
O'Toole, Thomas, 198–201, 366–367,
377–378, 382

Paley, William S., 39, 348
Paley Commission, 62
study, 30, 39, 42, 66, 129, 348, 351
Pan Heuristics 372
Panofsky, Wolfgang K. H., 152–153, 160,
372–373
Parisi, Anthony J., 198–200, 378, 381
Parra, Francisco R., 391
Partial equilibrium. *See under*
Equilibrium
Partisan analysis, 292–293, 307–308
strong pull of, 307, 308
Pasternack, Bruce, 109
Permian Basin Area Rate Proceedings,
353
Perry, Harry, 211, 213
Persian Gulf, 346
producers in, 12, 272
protection of, 19, 21
Phenix reactor, 352
Phenogram, 319–320

ABOUT THE AUTHOR

Martin Greenberger is Professor of Public Policy and Analysis at the UCLA Graduate School of Management, where he holds the IBM Chair. While associated with the Electric Power Research Institute during 1976–77, he initiated a number of research projects, which led to the establishment of the Energy Modeling Forum at Stanford University, the Energy Model Assessment Program at M.I.T., and a Utility Modeling Forum for electric companies. Dr. Greenberger is the author of several books on the analysis of public policy and computer application in decision-making. He has taught at Harvard University, Johns Hopkins University, M.I.T., Stanford University, and the Technion–Israel Institute of Technology. He has served industry in management capacities in both the computer and energy fields.

ABOUT THE COLLABORATORS

Gary D. Brewer is Professor at the Yale School of Organization and Management, where he teaches and does research in a number of public policy areas.

William W. Hogan heads the Energy and Environmental Policy Center at Harvard University's John F. Kennedy School of Government.

Milton Russell is a Senior Fellow and Director of the Center for Energy Policy Research at Resources for the Future.